Megacity Mobility

Megacity Mobility
Integrated Urban Transportation Development and Management

Zongzhi Li
Adrian T. Moore
Samuel R. Staley

CRC Press is an imprint of the
Taylor & Francis Group, an **informa** business

First edition published 2022
by CRC Press
6000 Broken Sound Parkway NW, Suite 300, Boca Raton, FL 33487-2742

and by CRC Press
2 Park Square, Milton Park, Abingdon, Oxon, OX14 4RN

CRC Press is an imprint of Taylor & Francis Group, LLC

Library of Congress Cataloging-in-Publication Data
Names: Li, Zongzhi (Professor of transportation engineering), author. |
Moore, Adrian T. (Adrian Thomas), 1962- author. | Staley, Sam, 1961- author.
Title: Megacity mobility : integrated urban transport / Zongzhi Li, Adrian T. Moore, Samuel R. Staley.
Description: First edition. | Boca Raton, FL : CRC Press, 2022. | Includes bibliographical references and index.
Identifiers: LCCN 2021034741 (print) | LCCN 2021034742 (ebook) | ISBN 9780367363581 (hardback) |
ISBN 9781032181899 (paperback) | ISBN 9780429345432 (ebook)
Subjects: LCSH: Urban transportation. | Urban transportation–Planning. |
Urban transportation–Management. | Transportation demand management.
Classification: LCC HE305 .L568 2022 (print) | LCC HE305 (ebook) | DDC 388.4–dc23
LC record available at https://lccn.loc.gov/2021034741
LC ebook record available at https://lccn.loc.gov/2021034742

ISBN: 978-0-367-36358-1 (hbk)
ISBN: 978-1-032-18189-9 (pbk)
ISBN: 978-0-429-34543-2 (ebk)

DOI: 10.1201/9780429345432

Typeset in Sabon
by codeMantra

To Robert W. Galvin,
Without whose leadership this book would not
have been written nor been as bold.

Contents

Foreword

Traffic congestion is a terrible threat to our economy. The analogy is cardiovascular. If our bodies' arteries got blocked, we have two choices: We can die, or we can surgically rebuild them in a new way. Our cities are on course to die as their arteries are becoming congested. If we continue on this path, properties will devalue, people and jobs will leave the cities, and cultural institutions will decay. As cities die, their economic productivity disappears, crippling our national economy.

I have spent more than half of my life identifying and actively solving allegedly unsolvable fundamental issues from the privileged platform of Chief Executive Officer of Motorola, Inc. (I was with Motorola for 50 years, 30 as CEO, mixed with service on major U.S. federal commissions then and in retirement). I conceived and funded the Galvin Project, of which this book is a product, to help lay the groundwork necessary to eliminate traffic congestion. I was motivated by the following primary convictions:

First, the major cities of the United States and most of the developed countries would *die* by 2050 from severely congested arteries—commercial and private vehicular traffic gridlock—unless we do something now. Goods and people are already seriously compromised. Every ordinary citizen sees it. An inevitable civic fatality is apparent.

Second, this threat to city life is bound to the substantial depreciation of its commercial and personal property values due to their early-on inaccessibility.

Third, city-wide property depreciation is avoidable just as human cardiac and orthopedic surgery can sustain the vitality of our bodies. The essential new procedures and solutions will be:

- *Prosthesis.* Prefabricated, ready-to-install road infrastructure such as flyovers and queue jumpers of existing and prospective time-delaying intersections and improvements to existing roads. Where feasible, a series of single pylon supports will elevate lanes adjacent to or in the median of existing roads or tracks.
- *Major new arteries.* Many underground tunnels with convenient parking terminals will network our metropolitan and suburban areas. Many of the new arteries will employ electronic tolling with variable rates to control flow.

Finally, an enlightened private sector—institutions and individuals—will take timely ownership of the problem and its prospects. Their self-interest will be to honorably protect the useful value of their property, and they would deservedly profit from an affordable, uncongested toll road business. Businesses will appealingly open the entire city and region to timely, cost-effective deliveries, un-stressful job accessibility, and unrestricted family activities from shopping to school, soccer, and social ties.

Herein, a substantial new industry will be founded. Little or no public funds will be required in countries such as the United States, Australia, New Zealand, the United Kingdom, and others to rebuild their transportation networks to mitigate congestion. There will be international competition to invest in this emerging business sector, creating many jobs that naturally must be learned and filled locally while this new strategy generates generic wealth country-wide.

Any city that ignores the threat will become an economic wasteland mid-century. Any previously developed country that thinks it will be immune from this threat will blight itself and do a disservice to other countries it should support and serve. We must want most cities and all countries to responsibly succeed.

Accepting these convictions is one thing; achieving the changes necessary to see them addressed is another. Human culture regarding change is well known. Our experience shows that every change is resisted, major change is at first opposed, and almost all changes "for the good" are eventually adopted. Thus, there is light at the end of the tunnel.

The convictions supporting this project are sound and "for the good." The scores and scores of professional thinkers, who have assembled, advised, stimulated, and synthesized their ideas and judgments periodically through the Galvin Project to end congestion at Reason Foundation are almost of one mind. These primary changes must be introduced shortly and must be accomplished by mid-century to avoid significant damage to our economy.

Of course, as our nation grapples with the immense task of financing our transportation infrastructure, many will wonder about the affordability of such an ambitious program. I have two responses.

First, a city cannot afford *not* to mitigate congestion. If it fails, the loss of productivity and the aggregate of the depreciated value of all its properties will be multiples of investment over what will be required to diminish congestion. That investment will not only preserve property value but also nourish the probability of its appreciation.

Second, although new flyovers and tunnels are more expensive than putting in place conventional surface arteries, the initial technology and means of provisioning will affordably launch the private sector founders of this new business. And just like founders of other sectors before them, they will accelerate improvements in technology, processes, costs, business models, and so forth well beyond current imagination.

After all, haven't the "beyond our original imagination" benefits from past convictions to offer air travel, cell phones, computers, medical solutions, etc.—almost every new service—dazzled us? Haven't they grown to a level of need that most of us cannot do without? And most of us can afford them. Congestion-free ground travel may best them all!

Robert W. Galvin
(1922-2011)
Former Chairman and CEO, Motorola Corporation
January 15, 2008

Acknowledgments

To our families, who stimulated and encouraged our passionate work in transportation systems.

List of abbreviations

2D	2-Dimensional
3D	3-Dimensional
AASHTO	American Association of State Highway and Transportation Officials
ABS	Asset-backed security
ACEA (EAMA)	European Automobile Manufacturers' Association
ADA	Americans with Disabilities Act (of 1990)
ADB	Asian Development Bank
ADOT	Arizona Department of Transportation
AHP	Analytic hierarchy process
AHS	Automated highway system
ANP	Analytic network process
APIRG	Arizona PIRG Education Fund
APM	Automated people mover
APRC	Asia Pacific Risk Center
APTA	American Public Transportation Association
ARF	Additional registration fee
ATM	Active traffic management
AV	Automated vehicles
AVI	Automatic vehicle identification
AVL	Automatic vehicle location
BART	Bay Area Rail Transit
BPR	Bureau of Public Roads
BRT	Bus rapid transit
CAVs	Connected and automated/autonomous vehicles
CBD	Central business district
CBS	Columbia Broadcasting System
CCTV	Closed-circuit television
CEO	Chief Executive Officer
CLO	Collateralized loan obligations
CODOT	Colorado Department of Transportation
COE	Certificate of entitlement
CPI	(London) Centre for Public Impact
CTBUH	Council on Tall Buildings and Urban Habitat
CV	Connected Vehicle
CVT	Continuous variable transmission
DEFI	Decentralized finance
DETR	Department of the Environment, Transport and the Regions
DMS	Dynamic message sign

DOT	Department of Transportation
DSRC	Dedicated short-range communications
ECA	European Court of Auditors
EMDEs	Emerging Market and Developing Economies
ESAL	Equivalent single axle load
ETC	Electronic toll collection
ETFs	Exchange-traded funds
EU	European Union
FAA	Federal Aviation Administration
FAIR	Fast and intertwined regular (lanes)
FHWA	Federal Highway Administration
FTA	Federal Transit Administration
FVC	Fair value commuting
GDOT	Georgia Department of Transportation
GDP	Gross domestic product
GIH	Global Infrastructure Hub
GIS	Geographic information system
GLA	Greater London Authority
GPs	General partners
GPS	Global Positioning System
GSC	Global supply chain
HAR	Highway advisory radios
HBO	Home-based other
HBW	Home-based work
HC	Hydrocarbons
HCM	Highway Capacity Manual
HGVs	Heavy goods vehicles
HOT	High occupancy toll
HOV	High occupancy vehicle
HSR	High-speed rail
IHS	Interstate highway system
IHT	Institute of Highway and Transport
IITs	Infrastructure investment trusts
ITF	International Transport Forum
ITR	Indiana Toll Road
ITS	Intelligent transportation systems
IVNS	In-vehicle navigation systems
KCM	King County Metro
KSDOT	Kansas Department of Transportation
LADOTD	Louisiana Department of Transportation and Development
LAMATA	Lagos Metropolitan Area Transport Authority
LOC	Library of Congress
LOS	Level of service
LWR	Lighthill-Whitham-Richards models
MADM	Multi-attribute decision-making
MAUT	Multi-attribute utility theory
MCDM	Multicriteria decision-making
MCDOT	Montgomery County Department of Transportation
MCMDKP	Multi-choice multidimensional Knapsack problem
MDOT	Michigan Department of Transportation

METRO	(Los Angeles County) Metropolitan Transportation Authority
MGPMR	McKinsey Global Private Markets Review
MLPs	Master limited partnerships
MOC	Ministry of Communications
MOO	Multi-objective optimization
MPA	Maryland Port Administration
MSDOT	Mississippi Department of Transportation
MSUM	Multimodal shared use mobility
MTR	Mass Transit Railway
MTRC	Mass Transit Railway Corporation
MPM	Mobility performance metrics
MTRC	(Hong Kong) Mass Transit Railway Corporation
MVDS	Microwave vehicle detection sensors
NCHRP	National Cooperative Highway Research Program
NHB	Non-home-based
NOX	Nitrogen oxides
NRC	National Research Council
NSTIFC	National Surface Transportation Infrastructure Financing Commission
NURTW	National Union of Road Transport Workers
O-D	Origin-destination
ODOT	Ohio Department of Transportation
ODS	Open Data Services
OECD	Organisation for Economic Co-operation and Development
OMV	Open Market Value
ORDOT	Oregon Department of Transportation
ORF	Observer Research Foundation
PCE	Passenger car equivalent
PCMS	Portable changeable message signs
PM	particulate matters
PMT	Person miles of travel
PPI	Private Participation in Infrastructure
PPPs	Public-private partnerships
PRIS	(Spanish acronym for) Intercommunal urban regulation plan
RBC	Reading Borough Council
REITs	Real estate investment trusts
ROCOL	Road Charging Options for London
ROW	Right-of-way
RTEAN	Road Transport Employers Association of Nigeria
RTPI	Real Time Passenger Information
RWIS	Roadway weather information system
SAE	Society of Automotive Engineers
SCAT	Sydney Coordinated Adaptive Traffic System
SCDOT	South Carolina Department of Transportation
SCOOT	Split Cycle Offset Optimisation Technique
SMART	Specific, measurable, achievable, relevant, and timely
SOV	Single occupancy vehicle
SPS	Signal priority system
SSPS	Snowplow signal priority system
TCRP	Transit Cooperative Research Program
TDM	Travel demand management

TfL	Transport for London
THCEA	Tampa-Hillsborough County Expressway Authority
TIMS	Traffic incident management system
TNC	Transportation network companies
TOPSIS	Technique for order preference by similarity to ideal solution
TOL	Truck-only lane
TPIMS	Truck parking information and management system
TPS	Transit priority system
TRANSIMS	TRansportation ANalysis SIMulation System
TRB	Transportation Research Board
TTI	Texas Transportation Institute
UAV	Unmanned aerial vehicle
UDOT	Utah Department of Transportation
UMI	Urban mobility index
UN	United Nations
UNECE	United Nations Economic Commission for Europe
USCB	U.S. Census Bureau
USDOT	U.S. Department of Transportation
UVAR	Urban Vehicle Access Regulations
V2I	Vehicle-to-Infrastructure
V2V	Vehicle-to-Vehicle
V2X	Vehicle-to-Everything
VDOT	Virginia Department of Transportation
VMS	Variable message signs
VMT	Vehicle miles of travel
VTPI	Victoria Transport Policy Institute
WB	World Bank
WCTR	World Conference on Transport Research
WIM	Weigh-in-motion
WSDOT	Washington State Department of Transportation

Authors

Dr. Zongzhi Li received his BE from Chang'an University, China. He obtained his MSCE and PhD in transportation and infrastructure systems engineering, as well as MSIE in operations research from Purdue University. He holds full professor rank with tenure and serves as director of the Sustainable Transportation and Infrastructure Research (STAIR) Center at Illinois Institute of Technology (IIT). His research interests include multimodal transportation demand and system performance modeling, asset management, and network economics.

Dr. Adrian T. Moore holds a Master's and PhD in economics from the University of California, Irvine; a Master's in history from California State University. He is vice president of policy at Reason Foundation, Washington, D.C. Prior to joining Reason, Dr. Moore served 10 years in the U.S. Army on active duty and reserves. He leads Reason's policy implementation efforts and conducts research on privatization, government and regulatory reform, air quality, transportation and urban growth, prisons, and utilities.

Dr. Samuel R. Staley earned a BA in economics and public policy from Colby College, Maine; an MS in applied economics from Wright State University, Ohio; and a PhD in public administration from The Ohio State University. Dr. Staley was Robert W. Galvin Fellow at Reason Foundation, Washington, D.C., and currently serves as the director of the DeVoe L. Moore Center at Florida State University, with research focuses on urban planning, social entrepreneurship, and urban economics.

Chapter 1

Introduction

1.1 URBAN MOBILITY IN A POST-COVID PANDEMIC WORLD

As the world recovers from the far-reaching impacts of COVID-19, the most widespread pandemic in 70 years, nations and their cities will be grappling with an untold number of new realities and adjustments. These changes are likely to fundamentally reshape transportation policy in terms of the technologies used to move from place to place, the ways transportation systems are managed, and evaluations of benefits of long-term investments in transportation infrastructure.

While the full range and depth of these impacts are unknown, some broader parameters are emerging. In high-income nations with freedom of mobility, wealthier households will use their earnings and savings to move to safer locations. Businesses and their managers will take advantage of technology to allow and even encourage more remote work. Schools and education will adopt teaching models to allow more versatility and student-centered learning approaches. Managers in the global economy, and to some extent their leaders and policymakers, will likely adopt more policies that encourage adaptability and use technologies that allow teams to work in more decentralized and physically fragmented locations.

Forrester, a research and business advisory firm, noted in *Forbes* magazine on October 30, 2020, that "most companies will employ what Forrester calls an 'anywhere-plus-the-office hybrid' model" where employees split their time between a remote location, such as a home office, and their company's main office. They estimate that remote work will be at least three times more prevalent than before the pandemic. A survey of 800 executives by McKinsey and Company in 2020 found that 45% of companies had "significantly accelerated" the "digitization" of "employee interaction and collaboration," nearly half had significantly or somewhat accelerated the digitization of customer channels, and more than one-third had accelerated the digitization of their supply chains. Upwork, an online job and freelancing platform, estimates that 41.2% of jobs were fully remote during the height of the pandemic and a quarter of the workforce will be fully remote by 2025.

These forecasts, if accurate, would imply 26.2 million workers in the United States alone will be working fully remote. This number (and percentage) is twice as large as those working remotely prior to the lockdowns and stay-at-home orders issued to manage COVID-19. The virus and subsequent pandemic, McKinsey's analysts note in a separate report, "has broken through cultural and technological barriers that prevented remote work in the past, setting in motion a structural shift in where work takes place, at least for some people."

While few analysts believe that all work will be remote, the technology enabling these shifts to take place has existed for decades. Indeed, doubling the number of fully remote workers was possible only because the technology already existed in many places to accommodate such a dramatic and significant shift in the workplace. Old management practices

DOI: 10.1201/9780429345432-1

and path dependence helped keep them at bay. This is no longer the case in an increasing number of urban areas.

Cities and transportation planners ignore these long-term shifts at their peril. Urban areas, particularly large urban areas, will be expected to accommodate more than just remote work. Their transportation systems and infrastructure will also be asked to adjust to more varied workdays, leisure travel, and recreation. These shifts suggest urban transportation systems will have to be more dynamic, more adaptable, and nimbler than ever before. Fortunately, transportation engineers, planners, and policymakers have a wide range of tools and strategies available for addressing the mobility needs of a post-pandemic 21st century. This book explores how these systems can change to meet these demands in a post-pandemic world where technology provides an unprecedented level of freedom and flexibility to workers and households.

1.1.1 Definition of mobility

Mobility, for the purposes of this book, is defined and measured in a straightforward, common sense way: the ability for goods and people to move from Point A to Point B and to be able to reach more Point B's in a reasonable time. Another overarching theme of this book is that, for all practical purposes, improved mobility is good for society and the economy. The faster goods and people can move to their destinations, the more efficient and productive the economy is and the more choices individuals have for work and personal and social life (Balaker and Staley, 2006; Staley and Moore, 2009). This is particularly true for megacities—cities with populations of about 10 million inhabitants or more, and those likely to reach that population in the next decade or so—on which this book focuses.

While in some circles, particularly in some segments of the professional urban planning community, question the value of improved mobility, the bulk of the empirical evidence suggests that improved mobility enhances community welfare. Criticisms of mobility center around quality-of-life considerations, such as promoting walkable neighborhoods and encouraging social activities. Thus, critics have focused on accessibility—proximity to goods and services—rather than the speed at which these services or goods can be obtained, or the breadth of choices people can enjoy. But accessibility and mobility are not necessarily substitutes for each other. Indeed, mobility improves accessibility.

More problematic is the question of travel mode. Many critics of mobility argue that a policy emphasis on mobility encourages auto travel rather than alternative modes of travel such as walking, bicycles, and transit. Yet, as subsequent chapters will show, the transportation policy question centers on the efficient choice of modes, not the promotion of one to the exclusion of others. The size and growth rates of global megacities, the habits and tastes of urban immigrants, and emerging technologies combine to make various transit modes increasingly important part of good mobility. The key is to ensure travelers and other transportation system users make decisions about travel mode with sufficient information to make individually and socially beneficial choices. Thus, much of the analysis in the following chapters focuses on improving system performance by optimizing the balance of modes and travel choice (Staley, 2012). Moreover, as COVID-19 clearly suggests, future transportation systems need to use technology and planners need to invest in infrastructure with a focus on adaptability and nimbleness unheard of in previous generations.

1.1.2 People versus freight

Ironically, mobility as a transportation system management goal is less controversial with respect to freight. The ability to transport goods and services within and between urban areas is widely recognized as beneficial to the economy and growing communities. Products and goods are produced with the pretext that a market exists in which they can be sold.

Thus, creating a system or network that allows producers to deliver their goods to market is intuitive and obvious. Not surprisingly, the economic benefits are significant. In the United States, for example, the construction of the U.S. Interstate Highway System (IHS), a limited access highway network connecting metropolitan areas, lowered transportation costs and generated a 15% rate of return on the investment (Shirley and Winston, 2004). Faster shipment of goods meant that inventories could be lower—reducing overhead—even as manufacturing increased to meet rising demand. The U.S. Interstate highways were inspired by Germany's autobahn and initially justified as a military improvement. The primary benefits, however, have been economics with regional and national impacts. Moreover, these benefits inspired interurban highway investments elsewhere, including China, as its urban areas grew dramatically as a result of international trade from its major port cities of Shanghai, Shenzhen, Ningbo-Zhoushan, Guangzhou, Hong Kong, Tianjin, Qingdao, and others. In fact, China has surpassed the U.S. limited access highway system in total mileage in half the time.

But the benefits do not just accrue to manufacturers, freight, and logistics companies. Commuters and people also benefit from improved transportation systems. Transportation planners David Hartgen and Gregory Fields (2009) found significant output and income benefits in U.S. metropolitan areas due to greater accessibility to universities, shopping malls, downtowns, outlying suburbs, and airports from reduced congestion (improved mobility). Perhaps not surprisingly, the biggest impacts were for the largest urban areas. Thus, goods movement is not the only beneficiary of faster travel. As business and commerce becomes more and more labor-intensive, the need to move people faster and more efficiently also increases.

Thus, when considering mobility in megacities, freight, commuters (workers), and non-commuting travelers must be considered as part of the same network. Producers need access to markets, but so do consumers. While at times the needs of each traveling segment are specific, the transportation investments, which are necessarily long term, imply impact on the other segments. The megacity of Los Angeles, for example, is a principal transshipment hub for North America, accounting for more than one-third of the imported goods from Asia. Moving goods out of the port onto trucks or on rail requires substantial infrastructure, all of which must be planned and coordinated within an urbanized area of nearly 18 million people and a workforce of over 7 million. However, Los Angeles has unique challenges by North American standards (Feigenbaum, 2015). Without a single economic center, the region is sprawling but also one of the most densely populated metropolitan areas in the United States.

Los Angeles is not unique in this regard. While megacities are often known to the public by their iconic skylines, buildings, or promenades, such as Shanghai's Oriental Pearl Radio and TV Tower along the Bund, large cities are in fact sprawling urban areas with multiple economic and residential centers (Bertaud, 2018). Neighborhoods are not distributed evenly in density or economic function. These subregions need to be serviced by the transportation system. They require a more decentralized and dynamic view of transportation demand and supply.

1.1.3 Economics of mobility

Transportation planners, engineers, specialists, and policymakers often confront difficulties in explaining the importance of mobility in the growth and development of cities. On one level, the case for mobility is intuitive—if we can move from Point A to Point B faster, we can achieve our goals and objectives more efficiently. And more people can do so simultaneously. However, this intuition is not always born out in the empirical literature or supported by theory in other disciplines. For example, the discipline of urban planning has a significant number of professionals who argue that mobility can be detrimental to healthy urban growth. Most often these arguments focus on what they consider to be the side effects of a mobility-focused policy strategy, such as encouraging the use of automobiles, discouraging the use of public transit, or encouraging less physical activity. Instead, urban planners often focus on accessibility rather than mobility (Prud'homme and Lee, 1999).

However, even in this case, accessibility without mobility leads to a degradation of efficiency, productivity, and quality of life. For example, an emphasis on accessibility presumes that people and businesses can access the range of goods and services they need in a physical and temporal way. In other words, people will not have to walk 45 minutes to get to their jobs, or the products businesses sell can be delivered reliably to avoid shortages. Moreover, many urban planners would also agree that policies that lengthen commuting or destabilize delivery timetables—whether at the manufacturing or distribution level—undermine economic growth.

Why, then, does the empirical literature on mobility, and more specifically traffic congestion, fail to consistently show economic and social benefits? While theoretically the benefits of mobility are intuitive, measuring its economic impacts in urban areas is difficult in practice. For one, most measures of traffic and congestion are regional in scope. Benefits at the subregional level are difficult to identify. Moreover, these benefits may be measured on the local or neighborhood level, but they are rarely aggregated or balanced with changes in transportation network performance. For example, the downtown Yuzhong district (about 0.75 million population) in the main city of Chongqing of about 7 million people (and 32 million people within the municipality's jurisdiction) struggles with managing commercial and commuter traffic because of its hilly terrain. One solution was to widen a road leading up from a major entry point from a limited access highway that skirted the district, which is a peninsula at the confluence of the Changjiang and Jialing Rivers, and then allow for reverse flow in the middle lane at key commute times. This change significantly increased traffic flow, improving commute reliability and speeding up the delivery of goods and services, but would not have shown up in the regional traffic data.

In addition, cities are dynamic social systems. Some have analogized them to organisms. As they grow, they shift and change, creating both positive and negative impacts. If the net impact of benefits versus costs is positive, the city grows. If the net impact is negative, the city stagnates or declines. An example of a positive externality is captured in the term "agglomeration economies," which recognizes the economic benefits of companies physically locating near each other (Graham, 2007; Broersma and Van Dijk, 2008). The clustering of businesses and industries has several potential advantages, from drawing from a common labor pool to increasing the likelihood of network impacts from creatives and executives sharing ideas or strategies. Workers also benefit when they are in cities with "thick" labor markets. These labor markets have multiple job opportunities for a wide range of skills, aptitudes, and talents, allowing workers to move from employer to employer in search of better opportunities (and higher pay). These advantages are particularly evident in places such as the tech center of Silicon Valley in the San Francisco Bay Area or financial centers such as Shanghai, London, or New York City.

Similarly, well-functioning transportation networks provide critical benefits that improve economic efficiency and productivity. London's rail transit system, the Tube, is crucial to managing passenger travel in the United Kingdom's central economic hub. Hong Kong's rail transit system allows the city to manage very high residential and commercial densities.

In contrast, urban areas with poorly functioning transportation networks lessen urban economic growth and productivity. Most economists and planners recognize that traffic congestion's impact is negative—it exacerbates pollution, increases travel times, strains network efficiency, and reduces travel reliability. In fact, many economists refer to it as a "tax" on growth and productivity. Yet, traffic congestion seems to be an accepted feature of megacity urban transportation networks. Many planners and policymakers seem to believe that congestion is an inevitable, perhaps even inherent, characteristic of big city life.

1.1.4 Hidden costs of mobility degradation

The U.S. Federal Highway Administration (FHWA) (2017) provides useful insight into the growing importance of nonwork travel. In the U.S. from 1990 to 2017, while household size remained relatively flat (at 2.55 persons), the number of licensed drivers per household increased from 1.75 to 1.89 as the number of workers per household increased from 1.27 to 1.33. Notably, the number of vehicles per household stayed relatively stable, increasing by just 1.4–1.42 in the years before 1990 the rate of change was dramatic. The number of vehicles per household rose dramatically from 1969, growing 62.9% before peaking at 1.89 in 2001. This was a period when incomes rose dramatically with economic growth as more women joined the U.S. workforce and wealth enabled younger drivers to become licensed. The number of workers per household rose most dramatically from 1969 to 1990 (45.8%) before leveling off.

While news headlines captured dramatic shifts in work patterns because of the 2020 pandemic and worldwide lockdowns, patterns in high-income countries were already shifting. Again, trends in the United States are illustrative. According to the FHWA, traditional and work-related business fell as a proportion of personal miles traveled per household between 1983 (26.0%) and 2017 (22.2%). Travel for shopping, family, and personal errands roughly stayed the same (25.8% versus 26.2%). The real shift was in the category of "other" travel that is not connected with work, business, shopping, personal errands, school or church, or social and recreational purposes. Specifically, it increased from 500 average person miles traveled in 1983 to 6,429 in 2017, accounting for 17.7% of the total travel. Of course, these aggregate data fail to capture the true complexity of travel. The length of trip and the purpose vary significantly within these categories. They don't account for how these trips fit into the hierarchy or priority of the traveler's purpose (e.g., urgent, recreational, and targeted purpose). In other words, one of the largest categories of household travel included trips so diverse and unique they could not be lumped into an existing major category. Transportation planners that fail to factor in these dynamics will find their networks performing increasingly poorly as cities become wealthier and households demand more mobility.

Of course, these statistics refer to national averages. Do large cities experience different travel patterns or trends? The data from the United States suggests they do. In fact, the story of the past 30 years may be how travel in large urban areas has mirrored more general shifts. While much in the U.S. media has heralded an increase in public transit use since 1995, annual average person transit trips by household are still well below levels in 1977. For urban areas larger than 3 million persons (the largest available), person trips have increased from 2,459 to 3,246, while transit use is still below 1977 levels. More interesting, walking in these large cities has increased significantly over the same period. In very auto-dependent America, walking accounts for 2.5 times the number of person trips per household than public transit. In smaller metro areas, the ratio of walking to public transit increases dramatically, six times more frequent or more as the metro area declines.

Of course, the relevance of the U.S. case is debated among transportation planners. The United States is unusual, perhaps even an outlier, because of its dispersed geography, low densities, and heavy investment in highway infrastructure. While this may be a generally accurate assessment, the United States includes several metropolitan cases that are relevant to the more conventional patterns of population and employment density. For example, the

New York metropolitan area is dominated by a highly concentrated central employment center (Manhattan) characterized by very low automobile ownership and use (<5%) and high public transit commuting from outlying regions (e.g., Brooklyn, Jersey City, Westchester County). Los Angeles, a metropolitan area of more than 17 million, is a dense urban area characterized by dispersed population and work centers. While transit is well established in New York City, Los Angeles has invested significantly in fixed guideway transit systems to serve the regional population. Moreover, the nation's largest cities have characteristics similar to megacities around the world: increasingly poly-centric commercial and residential patterns, diverse economic bases, global connectivity, and reliance on inter-regional economic and cultural connections. Thus, the U.S. case, with the proper qualifications, remains an instructive example, particularly if it can serve as a frame for thinking about the challenges of transportation policy and strategies.

Moreover, research increasingly shows that a major driver of mobility is household income. As households command more personal resources, they travel more. They invest in mobility, by purchasing new travel technologies. Of course, walking is the most obvious form of automobility. In terms of transportation technologies, the first wave of greater automobility was the adoption of the horse. Subsequent technological waves that improved mobility were ferries, trains, streetcars, and buses (or motor carriages).

The automobile emerged as a dominant form of mobility in the United States in the 1930s as household incomes rose and technological progress brought cars within their financial reach. The automobile became a tool for expanding employment opportunities and quality of life. In fact, research in the United States found that poorer households adopted automobile technology faster than higher income households, and the benefits of the increased mobility from access to cars were significant (Blumenberg and Pierce, 2012).

In addition, while millennials and younger generations tend to travel less, these trends appear to be related to socioeconomic and demographic factors such as delays in household formation, economic conditions, and later homeownership (Polzin et al., 2014). As these households age, they may take on characteristics similar to previous generations although the data is unclear on the magnitude of this effect. The preference for the automobile appears to be robust. Cervero and Kockelman (1997) for example found that while changes in density and urban design could influence personal vehicle trip rates, the effects were modest. The overall pattern of travel did not change.

More importantly, other regions are not immune to these economic forces and demographic pressures. For example, income is a primary driver of travel in Great Britain (Bastian et al., 2016; Stapleton et al., 2017). Research on travel mode choice and travel demand in several Chinese cities found that travelers significantly valued time savings. Older and automobile travelers needed significant time savings before they would give up their automobile for the less flexible bus system (Wang et al., 2013). Ko et al. (2019) studied bus rapid transit (BRT) systems in 111 cities and found that system and network design that improved reliability and the speed of transit was significant in attracting ridership. Even higher-income high-speed rail (HSR) passengers in China were sensitive to travel times and costs associated with shifting trips to intercity bus and automobile (Wang et al., 2014a). These results are consistent with other research that shows the quality of the transit service, usually measured by time savings and reliability, is crucial to performance (Li and Hensher, 2020).

The research in China on HSR is particularly interesting in this connection. HSR is an important part of the nation's overall economic development strategy. A core goal of national policy is to improve accessibility among its primary cities, particularly those in central and western China. Indeed, research suggests that HSR can improve the efficiency of urban areas (Li et al., 2020). Cities that have large service sectors and high densities appear to benefit the most. However, another study of multiple cities in the HSR network found that HSR did not automatically result in economic impact (Jiao et al., 2020). Rather, cities experienced differential impacts based on the degree to which the rail network increased accessibility to other cities and larger markets. Isolated infrastructure investments did little to boost the economic prospects of cities on the network (Ren et al., 2020). In short, large cities face many of the same challenges, regardless of where they are geographically located. The benefits of global trade are likely to continue over the long term. This growth is likely to propel income and greater demand for travel and mobility.

However, megacities in the 21st century face very different problems and challenges than their counterparts in the 20th century. Environmental sustainability is an increasingly important factor. Public pressure and government policy are prioritizing investments that reduce the environmental impact and footprint of transportation investments. The recent addition of electric vehicle manufacturer Tesla to the S&P 500 is simply a sign of the future. Technological progress, such as connected and automated/autonomous vehicles (CAVs), has the potential to fundamentally change travel options for consumers and businesses. These challenges and others create a unique set of barriers and obstacles to the world's largest cities, the megacities, which will be the engines of economic growth, value creation, and income growth in the 21st and 22nd centuries. Also, the ways we increase mobility for people (passengers) and goods (freight) may differ significantly given origins, destinations, preferred times of travel, and safety considerations related to the mix of users.

As more and more megacities grapple with environmental issues, sustainability objectives may come into conflict, requiring planners, engineers, and policymakers into making uncomfortable tradeoffs in designing and managing the system. While electric vehicles may be effective ways to reduce fossil fuel use, the technologies come with potential drawbacks and disadvantages such as fuel cell disposal, externalities associated with alternative fuels such as biofuels, or reliance on other fossil fuel-based energy sources to create the power to fuel the vehicles. Sustainable solutions may be geographically limited or dependent on the local context.

1.2 CHALLENGES OF MEGACITY MOBILITY

The world's megacities will be tasked with investing trillions of U.S. dollars in transportation infrastructure over the next several decades. Where this investment should go, and for what purpose, will be bounded by the transportation needs and challenges these cities face. Among these challenges are certain realities that must factor into transportation planning at regional, sub-regional, and neighborhood levels (Prud'homme, 2000).

Megacities are an incredibly diverse group of urban agglomerations. As shown in Table 1.1, 42 urban areas typically fall into this category. Nearly 60% are in Asia. Eight (19% of the total) are in China alone, but this likely underestimates the growth of cities in this region. The Chinese national government, for example, has identified nine National Central Cities as targets for investment and growth: Beijing, Tianjin, Shanghai, Chongqing, Chengdu, Guangzhou, Wuhan, Xi'an, and Zhengzhou. Wuhan, Xi'an, and Zhengzhou are not on conventional lists of megacities. While Hong Kong's future is somewhat tenuous given domestic politics, its role is likely to remain a critical cog in the wheel of economic growth in the Pearl River Delta as a complement to Guangzhou and Shenzhen. Given planned investments in these designated "growth poles" and general trends toward urbanization, the list of megacities is likely to expand significantly. The same is true for many large and growing cities around the world.

Perhaps more striking is the wide diversity of urban, economic, and cultural types of cities among megacities. The urban dynamics in Lagos, Nigeria (population 13 million) is vastly different from urban areas with similar populations such as Rio de Janeiro (Brazil), Los

Table 1.1 List of megacities by region

Region	*Country*	*Megacity*
Africa (3)	Democratic Republic of Congo	Kinshasa
	Nigeria	Lagos
	South Africa	Johannesburg
Asia (26)	Bangladesh	Dhaka
	China (9)	Beijing, Chengdu, Chongqing, Guangzhou, Shanghai, Shenzhen, Tianjin, Wuhan, Xi'an
	India (6)	Bangalore, Chennai, Delhi, Hyderabad, Kolkata, Mumbai
	Indonesia	Jakarta
	Japan	Nagoya, Osaka, Tokyo
	Pakistan	Karachi, Lahore
	Philippines	Manila
	South Korea	Seoul
	Thailand	Bangkok
	Vietnam	Ho Chi Minh City
Europe (3)	France	Paris
	Russia	Moscow
	United Kingdom	London
Latin America (6)	Argentina	Buenos Aires
	Brazil	Rio De Janeiro, São Paulo
	Columbia	Bogotá
	Mexico	Mexico City
	Peru	Lima
Middle East/West Asia/North Africa (3)	Egypt	Cairo
	Iran	Tehran
	Turkey	Istanbul
North America (3)	USA	Chicago, Los Angeles, New York

Angeles (United States), or Tianjin (China). Similarly, Mumbai, India, with a population exceeding 20 million, functions differently from New York City, Shanghai, or Tokyo. In many ways, when compared directly against each other at a point in time, the only similarity appears to be their population.

Yet, while at different stages of development and income, each of these cities faces a few largely recognized challenges that transcend their culturally specific context. These large cities still have to move people and goods. How they accomplish this, and with what efficiency, will be critical to their ability to support economic growth as well as their commitment to achieve environmental sustainability. Moreover, the struggles of cities that have already advanced through various stages of development, such as New York City, London, or Hong Kong, may have important lessons for cities grappling with rapid growth and dynamic economic systems. Similarly, other growing cities, such as Chongqing or Bangalore, have the advantage of seeing those struggles and potentially becoming adopters of the newest technology and thinking.

1.2.1 Ever-increasing travel demand

Among the most important long-term challenges is the inevitable return of rising travel demand. As this book goes to press, the world is in the grips of a worldwide pandemic triggered by COVID-19. Countries opted to lock down their economies, inducing an unprecedented global depression. Cities suffered the consequences in many different ways. Around the world, cities experienced a stunning collapse of tourism, hospitality, and recreation industry while some sectors experienced much more modest declines. However, the economic decline is a consequence of policy decisions, not fundamental weaknesses in their economies. Thus, most economists expect the economies to recover relatively quickly. Indeed, the World Bank (WB) (2020) expects the European and Central Asia region to post growth of 3.6% in 2021 even as the effects of the pandemic linger well into the year. It expects the East Asian and Pacific region to see growth rebooted to 6%. Any respite from growing travel demand will be short from the perspective of transportation planning and investment.

The return to growth, however, will be uneven. Countries and cities with weak health care systems will struggle longer to bring workers back into the system. Nations with large public debts will also struggle as the financial strain from ramping up government programs and agencies come to bear on lenders. Perhaps most problematic will be how the pandemic impacts supply chains. If some national and city economies remain "offline" for a significant amount of time, manufacturers and producers will shift their sourcing to more nimble and robust economies (and cities). The WB expects economies in the Middle East and North Africa to struggle the most. These oil-dependent economies face a longer recovery in the wake of oil price collapse induced by the global recession. Sub-Saharan Africa, however, will take less time to recover although substantial uncertainty clouds projections due to weak health care systems and global supply chain links.

Nevertheless, the consensus among economists is that the economy will recover from COVID-19 and re-establish a positive growth trajectory. The question is when travel demand recovers, not whether. Those cities positioned to take advantage of the changes in the global supply chain or reboot their economies fastest will be facing similar or greater pressures on their transportation system by mid-decade.

However, the challenges faced by megacities in rebooting their transportation systems cannot be underestimated. As recently noted by Pisarski (2020), a leading American analyst of commuting patterns, travel demand forecasting was already facing unprecedented challenges prior to COVID-19. New vehicle technology, changing work patterns, shifting demographics, and other factors had already taken a toll on conventional forecasting technologies with significant implications for investment. With the present (COVID-19 pandemic) dominated by uncertainty in all aspects of travel demand, any proposals for capacity expansion have zero credibility. One example? If post-pandemic estimates of work-at-home bear fruit, the reduced demand for travel across all modes will dramatically change, making large-scale investments in capacity pointless (at best) and inefficient and wasteful at worst.

Of course, transportation planning cannot come to a stop, particularly in very large cities where the fundamentals for economic growth were sound prior to the pandemic. The question is the pace of recovery, and where, geographically, the recovery will be felt. In addition, the recovery will likely be accompanied by changes in travel preferences and shifting demand across modes. Indeed, certain new modes (e.g., self-driving cars, on-demand point-to-point service, micromobility) might significantly increase compared to more traditional ones (e.g., fixed guideway and single car occupancy commuting). Thus, the critical question facing transportation policymakers centers more on how resilient and robust transportation systems can be developed. A wide range of policy approaches can help policymakers grapple with this changing policy environment.

Market-based mechanisms will be useful in identifying funding priorities. Markets and market pricing are particularly effective at guiding resources in environments characterized by uncertainty, decentralized information and knowledge, and instability (Staley, 2001). Public-private partnerships (PPPs) may be more effective in harnessing the efficiencies and capital of the private sector to address dynamic transportation needs (Poole et al., 2021).

Travel demand management (TDM) will become essential to optimizing the use of the existing and future roadway capacity. Similarly, transit systems will likely have to re-envision their purpose and mission in the context of the relative efficiencies and desirability of different modes and technologies.

1.2.2 Shifting travel patterns

A greater long-term challenge facing megacities is how they adapt to shifting travel patterns due to fundamental changes in the economy. The WB estimates that manufacturing's value added to economic growth has fallen from 17.5% of the global Gross Domestic Product (GDP) in 1997 to 15.4% in 2018. The Organization for Economic Co-operation and Development (OECD) countries found their manufacturing sector fall from 17.0% to 14.1% with smaller national economies experiencing the biggest declines. Some countries such as Vietnam saw their manufacturing sectors grow. But most, including China and India, experienced declines. Sub-Saharan Africa saw its manufacturing value added fall from 16.3% to 11.0%.

These changes have substantial implications for transportation systems. Manufacturing dependent economies rely on trunk roads and rail systems to carry bulk goods and labor to factories, markets, and transshipment points. The traditional hub-and-spoke transportation network can work very efficiently when goods are produced in concentrated locations with good connections to distribution channels.

Factories are typically long-term investments. Tesla's manufacturing plant for the Model 3 electric car required 3 years to acquire the land, build the factory, and begin producing cars in Shanghai, China. At the time, Tesla expected it to take at least 5 years to get up to full production capacity (Korosec, 2018). The company secured the land in 2018, began construction on the factory, and delivered a nominal number of vehicles in December 2019. This is an extraordinarily aggressive timeline, but consistent with national policies designed to encourage the rapid growth of a domestic automobile industry in China (Chen et al., 2020). Even on this timeline, the Tesla plant was not expected to operate at projected full capacity of 500,000 cars until 5 years after the contract was signed. But Tesla has announced it expects to have the capacity to produce this amount in 2021, just 3 years after securing the land (Kane, 2020).

Retooling factories in more established industries in regions such as North America or Western Europe may take years with the expectation the production lines will operate for years and sometimes decades into the future. As manufacturing becomes less concentrated, and economic value is created by less tangible products and services, the need for substantial investments in dedicated infrastructure changes. Indeed, this shift in infrastructure use is already becoming clearer in high-income cities and is likely to accelerate in the post-pandemic decades.

However, in some cases, the challenge may be in repurposing and managing existing infrastructure more flexibly rather than investing in new facilities. For example, as jobs dispersed to the periphery of the Parisian region, workers began to commute to the suburbs, or "reverse commute" (Aguiléra et al., 2009). The unexpected result was rising demand for commuter rail since job creation tended to cluster around transit nodes. The rail system thus became more balanced rather than less. The challenge for the 21st-century transportation network will be in accommodating a more diverse range of trips in a more dynamic travel setting.

1.2.3 The case of China's megacities

China provides a useful case study of both the challenges and solutions megacities face in their transportation systems and networks as they grapple with growth. With the opening of the economy to markets and private capital, China's economy began to grow rapidly. As one of the poorest countries in the world in 1980, the government simply did not have the capital to invest in extensive roadways, ports, air travel, or commercial rail operations. At the same time, policymakers loosened restrictions on internal migration and opened up international trade by establishing special economic zones to facilitate private investment in manufacturing and transshipment of goods and services.

Rapid development was encouraged on the coasts, with the Pearl River Delta taking advantage of Hong Kong as a financial broker with the west and historic point of entry to the mainland. Within decades, the Shenzhen special economic zone in Southern Guangdong Province and on the border with Hong Kong emerged as a primary manufacturing center. This leveraged investment in the city of Guangzhou. In 1980, Guangzhou's largely isolated and rural population stood at 5 million. Now, the Pearl River Delta which includes Hong Kong, Macau, Shenzhen, Guangzhou, Foshan, and Dongguan exceeds 50 million people.

At the same time, the Chinese national government invested in building Shanghai into a powerful growth center to the north. In 1982, the urban area's population was 11.9 million. By 2019, the metropolitan region had grown to 25 million and its international port boasted the world's busiest container port. Shanghai remains the primary access point into China's interior.

Meanwhile, beginning in the 2000s, China shifted national economic development policy to focus on cities in Central and Western China. Specifically, cities such as Chongqing, Wuhan (Hubei Province), Chengdu (Sichuan Province), and Xi'an (Shaanxi Province) were targeted for investment and development. While each of these cities had international connections and became part of the Global Supply Chain (GSC), they became manufacturing, research, and finance centers serving the needs of the expanding Chinese economy.

In these and other regions, growth was facilitated by massive movements of rural populations into urban areas. These migrants, often without formal access to public services such as public education and housing, became the low-wage workforce that was crucial to luring foreign investors into the country to build the economy.

What is often glossed over is the monumental stress this population migration and broad-based growth put on cities. Since public agencies and officials lacked the resources to self-finance transportation infrastructure, China relied on PPPs to build the infrastructure and keep pace with growth and the demand of investors. The nation built the equivalent of the United States' Interstate Highway System (IHS) in less than half the time by funding road investments using private capital from the WB and the Asian Development Bank (ADB) (Cherry, 2005; Wang et al., 2014b). While the central government is responsible for the review and approval of highway projects, highway projects traditionally were not able to receive government subsidies once they were up and running (Wang et al., 2014b). Thus, toll revenue is used to cover the cost of financing the loans, often with a 15-year loan recovery period. Many of these tolling authorities acted as independent companies. More than a dozen were listed on private stock exchanges in Hong Kong, London, and Shanghai.

China's megacities differ dramatically in topography, income, and mode splits. Each of the cities pursued different strategies based on available resources, geography, and the economic needs. While bicycles and e-bike were major components of travel in Beijing, the hills and inclines of Chongqing limited their usefulness as an alternative in no small measure because of safety concerns. The vast relatively flat geography and high incomes associated with Shanghai's growth, enabled a dramatic increase and demand, and use for automobiles. All cities began to invest in their bus systems, experimenting with bus rapid transit (BRT), light rail, and subways. In addition, transportation engineers and public officials began to experiment with various Intelligent Transportation Systems (ITS) technologies to manage traffic through tolls and reversible lanes, traffic signal timing, and parking (Feigenbaum, 2015). In Chongqing, the city made productive use of civil air-defense tunnels to create an underground traffic circle to reduce congestion on surface streets in the Jiefangbei business circle located at its primary downtown area. In China, as elsewhere, growth propelled higher incomes. These higher incomes led to travelers preferring, and using, more transportation alternatives. Timely investments in transportation, however, meant that infrastructure kept up with travel demand.

Because of the scale of these investments, the amount of research conducted on transportation planning and policy, and the diversity of Chinese megacities, this book will draw on their experiences to help create a critical lens for evaluating different strategies and tactics for managing travel demand and ensuring transportation networks are sustainable.

Still, as anyone who has traveled to China's megacities can attest, the transportation networks and systems could be managed more optimally. Traffic congestion is rising. The demand for automobile travel is increasing, putting pressure on the current network and challenging planners. China's cities are experimenting with different tools and strategies, and not all of them are successful. Indeed, most urban highways are not tolled or priced. A study of expressways in four provinces suggests that these highway systems may have been overbuilt by 22%–38% in part because these roads were not priced (Wang et al., 2014b).

Moreover, it's unclear whether many of their current investments will pay off as efficient ways to address mobility and economic productivity. For example, Beijing's experiment with odd-even license plate rationing may reduce automobile traffic volumes and shift mode share to non-automobile alternatives such as subways and buses. This policy, however, runs the risk of reducing mobility, increasing travel times, and reducing economic productivity by limiting travel. Of course, these effects may well be significantly mitigated post pandemic as technology becomes a partial substitute for commuting and in-office work. Similarly, Beijing's investment in subways tends to draw travelers from slower buses and compete with bus rapid traffic, not automobiles. Automobile travelers in China, like other major cities, put a premium on travel time savings, which most often favors automobiles over transit (Wang et al., 2013). Notably, mode splits between transportation options tend to be situational and context specific (Wang et al., 2013). This implies that megacities have more latitude for implementing multiple strategies depending on the travel demand characteristics of corridors and neighborhoods.

For instance, in the worst traffic city in China, Chongqing in 2016, the average daily motorized trips of residents in the main city increased by 250,000 from the previous year, a year-on-year increase of 2.8%. For motorized travel, cars accounted for 33.5% of trips, rail transit 13.3%, bus transit 46.0%, taxi 6.4%, and other forms 0.8%. Chongqing averages 25 minutes of delay during rush hours at the afternoon peak. Cities in southwestern China like Chongqing are built on multiple rivers and mountains, and their road networks encompass bridges and tunnels (Kottasova, 2015). The most congested bridge in the main city of Chongqing can experience congestion as long as 9 hours. While offering a wider range of transportation alternatives to travelers is important, facility planning and design plays an important role. Bridges, for example, that depend on ramps from surface streets designed for local travel are likely to create more congestion than bridges that feed off of higher volume roadways (Figure 1.1).

1.3 FRAMING THE NEXT WAVE OF TRANSPORTATION SOLUTIONS

The following chapters take a more focused look at the principles, framework, and challenges facing transportation planning and infrastructure investment for megacities in the 21st century. They will discuss approaches and technologies designed to address several core

Figure 1.1 The morning peak traffic condition in Chongqing, China.

issues that will be critical to creating a new generation of efficient, effective, and sustainable solutions to overcoming transportation obstacles in large urban transportation networks. These include:

- *Recognizing full economic costs and benefits of urban mobility.* While the narrative in this book presumes that mobility is beneficial, these benefits are net of costs. Moving people and freight entails costs, not just benefits. An important task facing megacities is fully accounting for the benefits and costs so they can make transportation planning and investment decisions based on net benefits on the margin.
- *Re-examining transportation decision-making in very large cities.* Traditional ways of planning regional and local transportation networks may need to be reconsidered based on the scope and scale of these systems. Their transportation systems and their effectiveness are more than the sum of their parts. More holistic approaches are necessary for transportation policy decision making in megacities. Comprehensively considering how this system is managed, at what level, and with which technologies provides an opportunity to rethink transportation planning in a regional and sub-regional context. Megacities are large, complex systems. They continuously change with the needs and desires of its inhabitants. This occurs even when a city has matured, as neighborhoods and sub-regions evolve to adapt to different needs. Megacity transportation planning should be adopting processes, procedures, and decisions that reflect this organic nature of large urban development.
- *Travel demand management.* Megacities, by definition, are growing. While the source of this growth differs by city, region, and economic context, megacities have found simple solutions such as "build more highways" or "build more transit" are not efficient, effective, and equitable ways of managing traffic and travel. Users respond to

incentives, and technology provides planners, engineers, and decision makers with more tools than ever in helping to calibrate a transportation network for optimal performance, manage demand efficiently, and finance the right infrastructure in the right place at the right time. The ability to use technology to manage and monetize transportation demand (through market pricing) also provides opportunities for innovation and entrepreneurial approaches to address problems. Variable rate pricing technologies, for example, allow for real-time monitoring and facility management that also monetizes value. This monetization allows for a broader and more targeted use of PPPs and provides incentives for innovations that capture this value.

- *Multimodal integrated capacity expansion.* By virtue of their scale, megacities require multimodal solutions and capacity. Even advocates of the automobile-dependent transportation system in the United States recognize that transit, including buses and passenger rail, plays a critically important role in cities such as New York where 70% of commuters enter Manhattan via its subway system. Similarly, while Los Angeles's rail system struggles to attract meaningful ridership numbers, the region's bus system is a critical component of its transportation network. Similarly, just because 80% of Mumbai's commuters travel to their jobs and back to their homes via trains does not mean that road infrastructure is irrelevant or unimportant. The task before megacities is how to create a vertically and horizontally integrated multimodal transportation system that balances and adapts to the needs of the city and its subregions.
- *Market-based, traveler-centric, and technology-focused efficient capacity utilization.* One lesson that becomes apparent in all contemporary transportation networks is the paramount importance of incentives and a traveler-centric focus. Building transportation facilities does not in and of itself change travel preferences or behavior. Incentives matter. Thus, as China has learned, travelers optimize mode choices over a wide range of variables, including time, out-of-pocket cost, convenience, accessibility, and reliability, just to name a few. Getting incentives right is critical for balancing the transportation network, and the power of a market-based approach, which emphasizes incentives, has the potential to radically improve both performance and finance.
- *Dynamic transportation asset management for sustainable development.* Perhaps no issue has become more important and salient than planning, programming, designing, building, and managing transportation systems so that they are more sustainable, environmentally and financially. This implies significant changes in the way a transportation system, comprised of multimodal transportation facilities, various types of vehicles, and users/non-users, is conceived, managed, and maintained. These frameworks include taking on new ways for transportation agencies and partners to secure sustained funding, embrace innovative financing, and implement (field and predictive) data-driven performance-based budget allocation using transportation asset management principles.

The transportation challenges of megacities are in a class of their own, with their own dynamic and requirements. The following chapters draw out these unique issues by drawing on examples from megacities from around the world. The next chapter dives more fully into transportation planning and how an integrated multimodal system works in a 21st-century framework in megacities. Chapter 3 addresses travel management with particular emphasis on destination and arrival time management, mode choice, demand leveling, travel path, and facility choice management. Chapters 4 and 5 examine how transportation facilities can be integrated into a three-dimension (3D) concept of transportation system management. Chapter 6 looks more directly at mobility-centered, performance-based transportation systems management, with an emphasis on ITS and market pricing to regulate the use of facilities. Chapter 7 elaborates on the history of transportation funding and revenue, discussing how traditional sources of revenue are inadequate to keep up with the need for transportation facilities. The varied ways in which market pricing of transportation facilities has helped shore up funding while improving efficiency are explored in depth. Chapter 8 scopes out a new way of thinking about performance-based budget allocation using transportation asset management principles, exploring data needs and budget implications for managing transportation systems on a regional scale. Chapter 9 concludes this book with a summary of lessons learned about transportation planning, infrastructure investment, and system management for megacities in the 21st century.

REFERENCES

Aguiléra, A., Wenglenski, S., and Proulhac, L. 2009. Employment suburbanisation, reverse commuting and travel behaviour by residents of the central city in the Paris metropolitan area. *Transportation Research Part A: Policy and Practice* 43(7), 685–691.

Balaker, T., and Staley, S.R. 2006. *The Road More Travelled: Why the Congestion Crisis Matters More Than You Think, and What We Can Do About It.* Rowman & Littlefield, New Kingstown, PA, USA.

Bastian, A., Börjesson, M., and Eliasson, J. 2016. Explaining “peak car” with economic variables. *Transportation Research Part A: Policy and Practice* 88, 236–250.

Bertaud, A. 2018. *Order Without Design: How Markets Shape Cities.* MIT Press, Cambridge, MA, USA.

Blumenberg, E., and Pierce, G. 2012. Automobile ownership and travel by the poor: Evidence from the 2009 National Household Travel Survey. *TRB Journal of Transportation Research Record* 2320(1), 28–36.

Broersma, L., and Van Dijk, J. 2008. The effect of congestion and agglomeration on multifactor productivity growth in Dutch regions. *Journal of Economic Geography* 8(2), 181–209.

Cervero, R., and Kockelman, K. 1997. Travel demand and the 3Ds: Density, diversity, and design. *Transportation Research Part D: Transport and Environment* 2(3), 199–219.

Chen, Y., Lawell, C.-Y.C.L., and Wang, Y. 2020. The Chinese automobile industry and government policy. *Research in Transportation Economics* 84, doi.org/10.1016/j.retrec.2020.100849.

Cherry, C. 2005. China's Urban Transportation System: Issues and Policies Facing Cities. Working Paper. The University of California- Berkeley, Berkeley, CA, USA. [Online]. Available: https://escholarship.org/content/qt1fx5m1ph/qt1fx5m1ph.pdf [Accessed on: March 21, 2021]

Feigenbaum, B. 2015. Increasing mobility in Southern California: A new approach. Policy Study No. 447. Reason Foundation, Los Angeles, CA [Online]. Available: https://reason.org/wp-content/uploads/2015/11/southern_california_mobility_plan.pdf [Accessed on January 10, 2021].

FHWA. 2017. 2017 National Household Travel Survey. Federal Highway Administration, U.S. Department of Transportation, Washington, DC [Online]. Available: https://nhts.ornl.gov/ [Accessed on January 10, 2021].

Graham, D.J. 2007. Variable returns to agglomeration and the effect of road traffic congestion. *Journal of Urban Economics* 62(1), 103–120.

Hartgen, D.T., and Fields, G.M. 2009. Gridlock and growth: The effect of traffic congestion on regional economic performance. Policy Study No. 371. Reason Foundation, Washington, DC [Online]. Available: https://reason.org/policy-study/gridlock-and-growth-the-effect/ [Accessed on January 10, 2021].

Jiao, J., Wang, J., Zhang, F., Jin, F., and Liu, W. 2020. Roles of accessibility, connectivity and spatial interdependence in realizing the economic impact of high-speed rail: Evidence from China. *Transport Policy* 91, 1–15.

Kane, M. 2020. Rumor: Tesla Model Y Production in China to Start by End of 2020. Auto News. Microsoft Network, Microsoft Corporation, Redmond, WA, USA. [Online]. Available: https://www.msn.com/en-us/autos/news/rumor-tesla-model-y-production-in-china-to-start-by-end-of-2020/ar-BB19Tp66 [Accessed on January 10, 2021].

Ko, J., Kim, D., and Etezady, A. 2019. Determinants of bus rapid transit ridership: System-level analysis. *Journal of Urban Planning and Development* 145(2), 04019004.

Korosec, K. 2018. Tesla Ends Partnership with Home Depot. [Online]. Available: https://fortune.com/2018/06/12/tesla-home-depot-stores-partnership/ [Accessed on January 10, 2021]

Kottasova, I. 2015. World's worst cities for rush hour traffic. CNN Business, Atlanta, GA [Online]. Available: https://money.cnn.com/gallery/news/economy/2015/03/30/worst-traffic-cities/5.html [Accessed on January 10, 2021].

Li, Z., and Hensher, D.A. 2020. Performance contributors of bus rapid transit systems: An ordered choice approach. *Economic Analysis and Policy* 67, 154–161.

Li, Y., Che, Z., and Wang, P. 2020. Impact of high-speed rail on urban economic efficiency in China. *Transport Policy* 97, 220–231.

Pisarski, A.E. 2020. Commuting in America: The Third National Report on Commuting Patterns and Trends. NCHRP Report 550/ TCRP Report 110. Transportation Research Board, National Academies Press, Washington, DC, USA.

Polzin, S.E., Chu, X., and Godfrey, J. 2014. The impact of millennials' travel behavior on future personal vehicle travel. *Energy Strategy Reviews* 5, 59–65.

Poole, R.W., Feigenbaum, B., O'toole, R., Frost, M., Cox, W., and Scribner, M. 2021. Transportation and COVID-19: A state guide to policy and priorities. Reason Foundation, Washington, DC [Online]. Available: https://www.washingtonpolicy.org/library/doclib/Transportation-and-COVID-19-A-State-Guide-to-Policy-and-Priorities-2-FINAL.pdf [Accessed on January 10, 2021].

Prud'homme, R. 2000. Introductory report, Infrastructure-induced Mobility (Round Table 119). European Conference of Ministers of Transport, in Report of the Hundred and Fifth Round Table on Transport Economics, Paris, 7th–8th November, 1996. Economic Research Centre, 81–98.

Prud'homme, R., and Lee, C.W. 1999. Size, sprawl, speed and the efficiency of cities. *Urban Studies* 36(11), 1849–1858.

Ren, X., Chen, Z., Wang, F., Dan, T., Wang, W., Guo, X., and Liu, C. 2020. Impact of high-speed rail on social equity in China: Evidence from a mode choice survey. *Transportation Research Part A: Policy and Practice* 138, 422–441.

Shirley, C., and Winston, C. 2004. Firm inventory behavior and the returns from highway infrastructure investments. *Journal of Urban Economics* 55(2), 398–415.

Staley, S.R. 2001. Markets, Smart Growth, and the Limits of Policy. In *Smarter Growth: Market-Based Strategies for Land-Use Planning in the 21st Century*. Editors: Holcombe, R.G., and Staley, S.R. Greenwood Press, Westport, CT, USA. 201–218.

Staley, S.R. 2012. Practical strategies for reducing congestion and increasing mobility for Chicago. Policy Study No. 404. Reason Foundation, Washington, DC.

Staley, S.R., and Moore, A.T. 2009. *Mobility First: A New Vision for Transportation in a Globally Competitive Twenty-First Century*. Rowman & Littlefield, New York.

Stapleton, L., Sorrell, S., and Schwanen, T. 2017. Peak car and increasing rebound: A closer look at car travel trends in Great Britain. *Transportation Research Part D: Transport and Environment* 53, 217–233.

Wang, Y, Wang, Z, Li, Z, Staley, S.R., Moore, A.T., and Gao, Y. 2013. Study of modal shifts to bus rapid transit in Chinese cities. *ASCE Journal of Transportation Engineering* 139(5), 515–523.

Wang, Y., Li, L., Wang, L., Moore, A.T., Staley, S.R., and Li, Z. 2014a. Modeling traveler mode choice behavior of a new high-speed rail corridor in China. *Transportation Planning and Technology* 37(5), 466–483.

Wang, J., Mao, X., Li, Z., Moore, A.T., and Staley, S.R. 2014b. Determining the reasonable scale of a toll highway network in China. *Journal of Transportation Engineering* 140(10). doi: 10.1061/(ASCE)TE.1943-5436.0000671.

World Bank. 2020. Special focus: Regional macroeconomic implications COVID-19. In *Global Economic Prospects (July 2020)*. World Bank, Washington, DC, 109–130.

Chapter 2

New perspectives of urban transportation decision-making

Megacities have unique challenges. Their size and complexity require more integrated and comprehensive solutions to ensuring transportation network efficiency compared to smaller or less dynamic cities. They are also a diverse group, making generalizations and specific recommendations problematic outside their specific context. Thus, megacities require a different approach to the development and management of transportation facilities and networks because they require an integrated approach to meet the needs of their users. People and goods must be able to move efficiently and safely, and a dynamic and proactive approach to transportation system management will be crucial for achieving these objectives through design and implementation.

2.1 FRAMING THE MEGACITY MOBILITY CHALLENGE

Megacities have several common elements that allow planners, engineers, and policymakers to think about their transportation systems cohesively and rigorously as a group. First, these very large cities are typically among the most concentrated centers of economic activities in their respective national economies. Even though many are not political capitals, their sizes and economic productivities give them signature status as well as influence over resources and access to national and regional assets. For example, as India's largest commercial center and principal gateway to international markets, Mumbai's ports account for more than two-thirds of India's water-borne trade and the nation's financial capital transactions. It also is home to the highest number of millionaires and billionaires. The Brookings Institution in Washington, D.C. estimates the city's region generated about US$370 billion in economic activity in 2014 (after adjusting for purchasing power). The size of Mumbai's economy would rank it as one of the World Bank's top 50 most prosperous countries. Tokyo and New York City metropolitan economies generate about US$1.5 trillion each, easily placing them among the World Bank's top 20 economies based on the value of goods and services produced, on par with the national economies of Australia, Saudi Arabia, Thailand, and Egypt.

Second, a global megacity's economy is complex, in scale and scope. At the micro level, neighborhood businesses provide services to people in blocks. These goods need to be moved within the city to local retail centers that occupy streets and corners. But most goods and services are brought in from outside the city from peripheral areas or wholesale districts distant from the neighborhood. These goods need to be shipped in volume, implying trunk road or rail connections intended for transshipment points that will then be loaded on vehicles for local delivery. Then, at yet another level, interregional or international goods are shipped to regional distribution centers for intra-metropolitan distribution.

DOI: 10.1201/9780429345432-2

For example, Chongqing's Yuzhong district, the urban area's principal downtown, includes nearly 1 million residents. One of the district's challenges has been reorganizing the delivery and distribution of goods within the district. The district reorganized its local freight delivery hubs to create a more centralized delivery system for food and other products from outside the district and serve local vendors and local carriers who deliver their goods and wares to local stores in neighborhoods. Larger trucks more suitable to bulk deliveries are diverted from neighborhood streets which improves traffic flow and travel times. The growth of the city has also led the district to encourage the creation of new commercial centers and facilitate the decentralization of population west of the traditional city center commonly known as the Jiefangbei business circle. This polycentric freight hub system more effectively meets the needs of a growing, dynamic high-density urban environment with more complex needs.

Third, megacities are continuously adapting to the historical legacy of their built environment and topography. Chongqing again provides a useful example. The city served as the China's provisional capital during World War II and benefited from its hilly and challenging terrain. Yet, these advantages are less useful for a growing economy. The efficient and quick construction of housing and commercial centers is important for matching the economic needs of residents and businesses. Yet building into hillsides, or creating flat terrain for high rises, is expensive. Roads originally carved out as paths for a largely immobile society with few demands need to be widened or added to meet the mobility and transportation needs of a modern economy. For older sections of cities, the challenging terrain is complicated by the replacement of older, obsolete structures. Thus, even in cities, such as Beijing, Cairo, or Paris, which are largely flat and can build out concentrically, older residential, commercial, and industrial spaces influence the relative cost and expense of tearing structures down and rebuilding on these sites.

Hong Kong may be one of the best examples of large cities that streamlined its construction and building process to enable near real-time construction to meet the rising demand (Staley, 1994). Indeed, the Hong Kong approach of using property development to underwrite the cost of major infrastructure, also seen in numerous other Asian countries (e.g., Japan), has created a nexus between the investments that appears to facilitate urban growth and development. Still, investments in transportation infrastructure and facilities create their own unique challenges.

A fourth challenge, and perhaps the most obvious, is the sheer complexity and scale of an urban economy struggling to accommodate the needs and aspirations of 10, 15, or 20 million people concentrated in one geographic area. The belief that transportation needs can be met with a simple solution evaporates quickly when faced with the realities of a complex economy and population with diverse goals and expectations. To use an often-critiqued example levied at the U.S. transportation system, the idea of meeting the needs of many of these cities with an automobile-centric transportation network is simplistic. Automobiles are highly efficient in smaller cities with populations wealthy enough for all but a few to afford personal vehicles. But as these cities grow to include millions of residents and workers, more complex transportation systems that incorporate multiple travel modes become essential. And preferences for mode choices will change based on income levels, accessibility, residential and commercial densities, the locations of employment centers, and the efficiency and quality of the transportation services themselves.

Few people argue that Manhattan borough of New York City, one of the highest density urban places in the United States, could adequately be served without an effective high-capacity trunk travel mode such as an underground or elevated rail. Manhattan accounts for more than half of the city's employment. More than two-thirds of workers commute into the city from outside Manhattan. Transporting hundreds of thousands of workers into Manhattan, which is an island at the confluence of the Hudson River on the West and the East River and bounded by the Harlem River to the North, is simply impractical if the city relied largely on a mode more suitable to low-density land uses such as the automobile. Moreover, access from outside is limited to 21 bridges and 15 tunnels. These limits, combined with high residential and commercial densities, mean fewer than 25% of Manhattan residents own a car, and these personally owned vehicles account for less than 5% of travel.

Of course, the flip side of this observation is that Manhattan needs much more than a passenger rail system as well. Manhattan has an extensive system of ferries, buses, heliports, taxis, rideshares, microtransit, and other modes that supplement rail and personal automobiles. While the splits between modes become more balanced as residents and employment centers disperse outward from Manhattan, New York City and the metropolitan area more generally, requires a complex, layered, and sophistical transportation network to support its economic, political, and cultural role in the U.S. urban hierarchy.

Thus, a fifth challenge is the size and complexity of the transportation systems. Figure 2.1 provides a stylized matrix of travel modes when bounded by size (scale) and density in a U.S. context. Notably, the U.S. transportation network has historically relied on various forms of auto- or self-directed mobility, whether walking, riding horses (or carriages), buses, or automobiles. Its large geographic size and income-fueled growth into its peripheries, enabled by substantial investments in roadways, allowed for fast development of automobiles as a primary mode of transportation in the 20th century. Indeed, by 1930, three quarters of Americans owned a personal vehicle. (A large share of these vehicles was on farms.) The growth of automobile travel as the dominant mode of travel for the vast majority of urban households coincided with post-World War II economic growth, a federal commitment to highway development, and a housing construction boom for single-family detached units facilitated by federal subsidies, changes in housing finance, and dramatic improvements in construction technologies that reduce building costs. Thus, walking and bicycling as principal modes of travel declined significantly.

Nevertheless, American cities are not immune to more generalized travel preferences and the consequent mode choices. As densities increase, travel patterns and choices are more supportive and conducive to fixed guideway transit options. Due to cost and flexibility, bus is preferred to rail at lower and mid-level densities in the United States. But as densities increase significantly, typically 15,000 people per square mile or higher, more intensive fixed guideway options such as light rail and heavy rail, as well as bus rapid transit (BRT) become more feasible. As in most cities regardless of income levels, walking has become a preferred mode of travel for a wide range of income levels in very high densities. Indeed, automobile travel remains a very small part of overall travel for residents who work in very high-density locations. In the United States, these densities are typically residential densities of 25,000 people per square mile before these transit options gain significant market shares.

The contrasts among megacities in the United States and other countries is striking, although not unusual for high-income nations such as Canada, Australia, the United Kingdom, or even France. As shown by Bruegmann (2001, 2005), Parisian development has suburbanized dramatically with most employment growth occurring in the suburbs. While still higher density than its American or Australian counterparts, decentralization has led to greater demand for a more flexible transportation network which fits the decentralized, dynamic travel patterns of residents with higher incomes living in lower-density urban regions (Aguiléra et al., 2009). These trends favor transportation management technologies and planning strategies that accommodate an increasingly diverse and nuanced set of alternatives for residents and businesses.

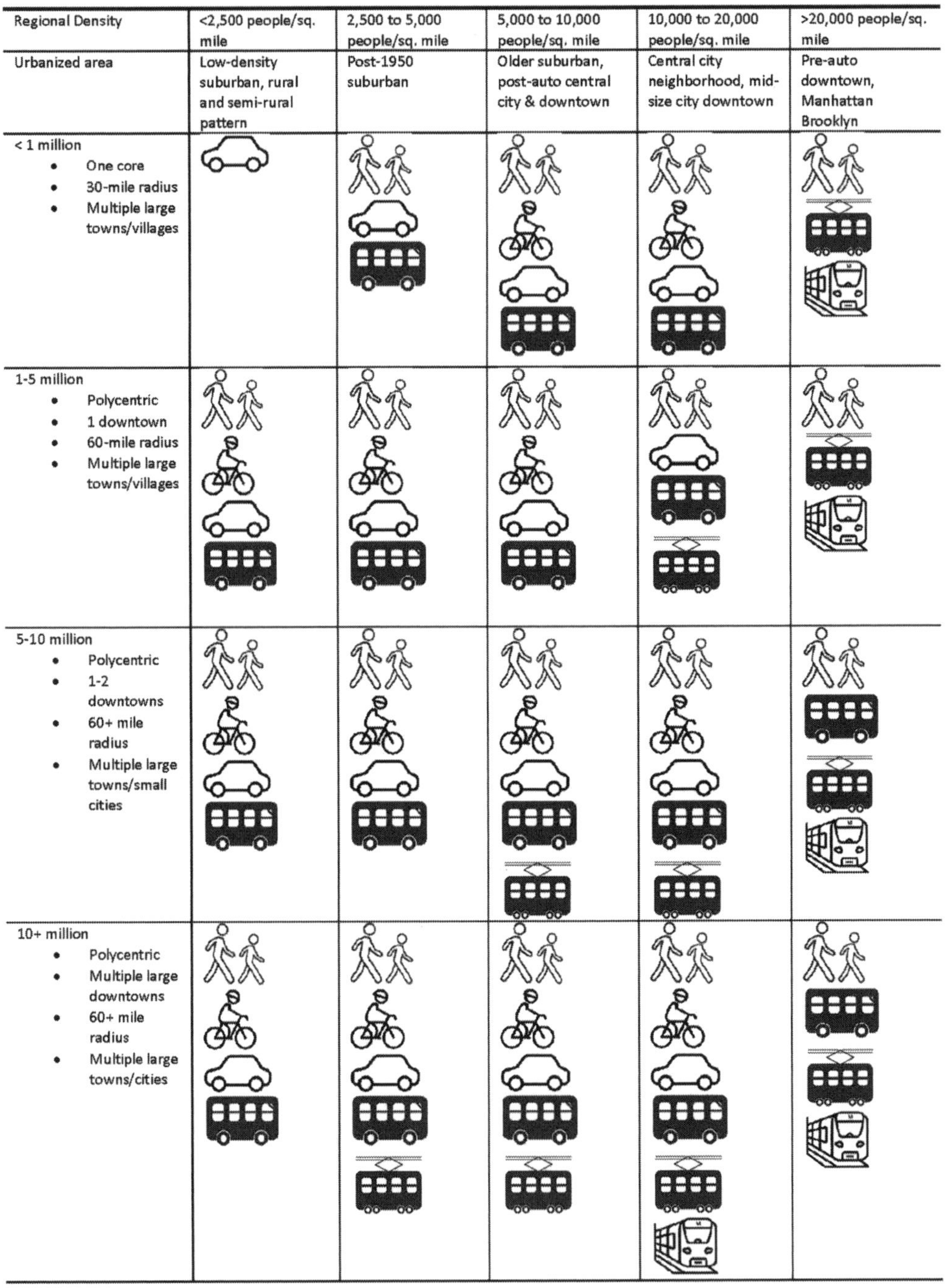

Regional Density	<2,500 people/sq. mile	2,500 to 5,000 people/sq. mile	5,000 to 10,000 people/sq. mile	10,000 to 20,000 people/sq. mile	>20,000 people/sq. mile
Urbanized area	Low-density suburban, rural and semi-rural pattern	Post-1950 suburban	Older suburban, post-auto central city & downtown	Central city neighborhood, mid-size city downtown	Pre-auto downtown, Manhattan Brooklyn
< 1 million • One core • 30-mile radius • Multiple large towns/villages					
1-5 million • Polycentric • 1 downtown • 60-mile radius • Multiple large towns/villages					
5-10 million • Polycentric • 1-2 downtowns • 60+ mile radius • Multiple large towns/small cities					
10+ million • Polycentric • Multiple large downtowns • 60+ mile radius • Multiple large towns/cities					

Figure 2.1 Travel mode choice and urban form in the United States.

2.2 THE NEED FOR COMPLEX AND SOPHISTICATED TRANSPORTATION SYSTEMS

As urban agglomerations become larger and more complex, more sophisticated transportation systems are necessary to provide mobility. The challenges faced by cities rarely center around one or a few more corridors. London, for example, is the United Kingdom's most congested urban area according to big data research firm INRIX (2019), logging one-third more hours lost to congestion annually than its nearest competitor (Belfast) and 41% more than the nation's third most congested city (Bristol). Its 2019 Global Traffic Scorecard found London's 3 most congested corridors alone result in annual hours of delay greater than that of any whole city in the U.K. outside of London. Indeed, London's five most congested corridors account for five out of the top 12 most congested corridors in the U.K.

Of course, big cities are not the only regions faced with inadequate transportation facilities or poor performance. Take Bournemouth, a coastal resort city southwest of London as an example. Now part of the Unitary Authority of Bournemouth, Christchurch, and Poole (BCP), the population's region ranks 17th overall in the United Kingdom. According to INRIX's UK traffic scorecard, the A338 corridor between Hurn Road and St. Paul's Road east of Bournemouth's city center is the fourth most congested corridor in the United Kingdom. But BCP does not rank among the UK's top ten most congested cities.

Fixed route transit tends to be most effective along corridors with high concentrations of work and residential development at the terminus points and at major stops in between. Using Paris as an example again, higher demand for fixed-guideway modes such as commuter rail between high-density living in the central city radiating out toward high-density employment centers on the periphery can be consistent with a growing urban region. Indeed, even for a region with low overall traffic congestion and high mobility, certain corridors can be highly congested. In countries without land use policies to direct development into transit-friendly high-density nodes and with largely flat or unimpeded geographic limits on development, transit supportive densities are unlikely to emerge until after decades of growth and redevelopment.

Thus, one of the primary takeaways from understanding urban mobility patterns is the rising importance of complex, layered, and adaptable transportation systems. Data from travelers in China is insightful in this respect. BRT travelers in five Chinese cities—Beijing, Changzhou, Guangzhou, Jinan, and Kunming—were asked which factors were most important in their decision to shift to BRT from their other modes (Wang et al., 2013). For conventional bus travelers, convenience and low price were consistently the primary motivating factors. Local buses have the benefit of multiple pick up and drop off points as well as being very cheap for most riders. Automobile travelers, on the other hand, indicated that time savings were the dominant factor. Non-motorized travel—bike and walking—was preferred for the flexibility. Income also played a role—wealthier travelers had access to cars—but the mode choices were nonetheless telling. Price, convenience, and flexibility triangulate into these individual decisions. Each of these modes, however, requires different approaches for planning, design, construction, and management. Further research examining travelers' demand and mode choices along high-speed rail corridors found similar results, although, not surprisingly, time savings played a large role in choosing alternative modes (Table 2.1).

2.3 SETTING A NEW STANDARD FOR MEGACITY MOBILITY

In many policy circles, degraded mobility is often considered an inevitable consequence of the size and scale of cities, particularly megacities. Indeed, during much of the 1990s and

Table 2.1 Factors influencing travel mode choices in Chinese cities

	City					
Travel mode	*Beijing BRT1*	*Beijing BRT3*	*Jinan BRT1*	*Changzhou BRT1*	*Guangzhou BRT*	*Kunming BRT1*
Bus	Convenience	Convenience	–	Reliability	Low price	Convenience
	Low price	Low price	–	Convenience	Saving time/ convenience	Low price
Auto	Saving time	Saving time	Saving time	Saving time	Saving time	Saving time
Non-motorized	Flexibility	Flexibility	Flexibility	Flexibility	Flexibility	–

Source: Wang et al. (2013).

early 2000s, many urban planners considered traffic congestion a sign of a healthy city. Robust growth, they reasoned, led to higher population and employment densities as people moved to the city for work and businesses started up and expanded to take advantage of the city's productivity benefits (agglomeration economies). But degraded mobility, as discussed in Chapter 1, carries important costs to cities (Staley and Moore, 2009). Congestion and lower mobility are factors that work against economic productivity because it limits access to labor and undermines the quality of life for residents (Balaker and Staley, 2006). In principle, a city's transportation network should be designed to facilitate the transfer of people and goods from Point A to Point B as efficiently as possible. Mobility, in this sense, is a cornerstone benchmark for evaluating an effective transportation system.

Cities, however, are dynamic social systems. They evolve, respond, adjust, and change to meet the demands, aspirations, and needs of their residents and businesses. Designing a transportation network and implementing a system that meets the evolving requirements of a city is in practice an impossible task. Transportation infrastructure, given current and foreseeable technologies, involves investments that are expected to last decades.

Yet preferred modes of travel can change quickly, depending on the economic, cultural, and political context of the urban areas. Cities in Europe, for example, have cultures that support fixed route and fixed guideway transit despite the high incomes that enable the rapid adoption and wide use of automobility modes such as the car or personal vehicle in other countries. Importantly, however, transit continues to lose overall market share although it remains competitive in key corridors (Balaker and Staley, 2006). In the United States, in contrast, rising incomes and the electrification of street cars enabled the development of "streetcar suburbs." These suburbs easily adapted to automobiles as technology drove costs down for consumers and cities built infrastructure to accommodate them. While never a dominant transit mode in the United States, street cars declined precipitously with the widespread use of the automobile improvements in traffic management, investments in roadways, and labor strikes which significantly disrupted service and raised concerns about public safety.

2.4 IS THE HUB-AND-SPOKE TRANSPORTATION NETWORK DESIGN OBSOLETE?

Despite mode shifts and changing transportation preferences, the legacy impacts of infrastructure investments set broad parameters for transportation planning and engineering that may last much longer than their efficient application may warrant. For example, much transportation planning, particularly in higher-income countries, is built around a "hub and spoke" system with a central urban area at its core. This urban core is typically "the" downtown. In standard urban economic theory, transportation costs lead to the outward migration of industry and residents, reserving the downtown at the commercial and regional services hub. As residents move out, the desire to reduce travel time and transportation costs lead them to cluster along corridors that give them easy access to jobs. Planners and policymakers invest in these corridors to create trunk roads and transit lines, with the access points becoming high-density commercial and residential nodes along the corridor and in the network. As urbanized areas grow into sprawling regions, these corridors are connected by beltways or "ring roads." Employment and residential agglomerations evolve from "sub-urban" centers that serve a cohesive downtown to urban centers. This pattern is remarkably resilient across metropolitan areas (Farahani et al., 2013; Bertaud, 2018).

Figure 2.2 illustrates the complexity of a modern city. The "C" represents central places and urbanized areas that provide regional services within or outside urbanized areas. Quite often, particularly in global megacities, international services are provided from business headquarters outside the central city. In this urban hierarchy, town centers (T) serve subregions. Even more local, but still representing clustered economic and community activity, would be villages (V). Notably, as automobiles have become increasingly ubiquitous in transportation networks, the overall pattern reflects a "spiderweb" rather than a hub-and-spoke wheel. Significant transportation corridors exist within this framework, but the dynamic, organic nature of travel increasingly divorced from traditional 8-hour per day, 5-day workweek commuting means that the traditional road network is used for different trip purposes and destination choices, at different times of the day, and at different intensities than in the 20th-century planning model. The aftereffects of the Pandemic of 2020 will reinforce these shifts to more remote work and e-commerce.

The period when a city's growth accelerates also seems to influence land use and development patterns. Los Angeles, for example, is famous for its diverse and vibrant neighborhoods. It's also well recognized as an "auto-centric" urbanized area, reflecting its growth in

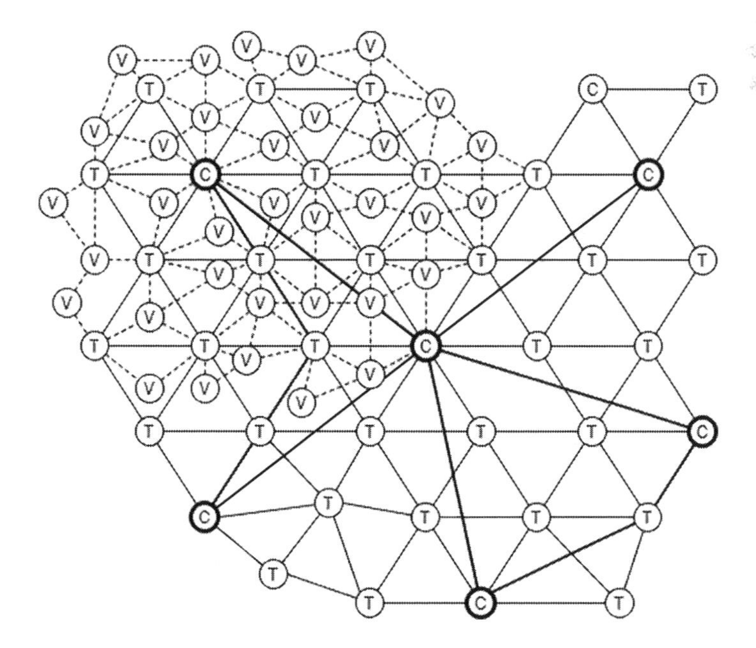

Figure 2.2 Illustration of the complexity of a modern city.

the post-World War II era. The automobile combined with its largely flat geography enabled urban areas growth in corridors and locations outside the traditional downtown (e.g., West Los Angeles, Long Beach, the San Fernando Valley north of the Santa Monica Mountains). Thus, Los Angeles is more aptly described as a "city of villages" (Kotkin, 2014). The region has a downtown, but it's role in the regional economy is remarkably small. Instead, the region's economy is driven by "growth poles" where industry and residents tend to cluster. In the U.S. urban context, Kotkin argued that these "new suburban" developments with higher density, and mixed-used subcenters within larger regions are becoming the dominant urban form (Kotkin, 2005). As a "multipolar" city, its transportation network functions very differently than a city like New York, London, or Paris with a more conventional center city serving as a downtown core (Zhang and Levinson, 2007).

Los Angeles also provides an example of how travel patterns and connectivity might differ from a pattern based on the hub-and-spoke system. Since Los Angeles does not have a traditional central business district (CBD) as its core, travel patterns are much more dispersed than many might believe despite having the highest average metropolitan-wide population densities in the United States. Having transit (and roads) center on the CBD would hinder travel more than improve mobility. Thus, the region's transportation system reflects a grid rather than a hub-and-spoke system. A recent proposal from Reason Foundation identified numerous corridors and opportunities for improving mobility and easing traffic congestion through an expanded network of BRT and managed lanes (Feigenbaum, 2015). As Figure 2.3 illustrates, these proposed improvements reinforce a grid pattern rather than a centralized CBD model of traffic management.

Urbanists believe that China's cities have the potential to forge a more modally balanced approach to urban development and transportation. Despite their rapid growth, China's cities have about one-third of their pavement devoted to roadways compared to North American, Australian, and European cities (Cherry, 2005; Wu, 2007). Moreover, while growing dramatically, automobile travel remains relatively small for most cities compared to other modes such as bus, walking, and bicycle. Another challenge in China, as in many Asian countries, is highly concentrated spikes in demand for specific modes at certain times of the year. In China, hundreds of millions of people board trains and buses for long-distance

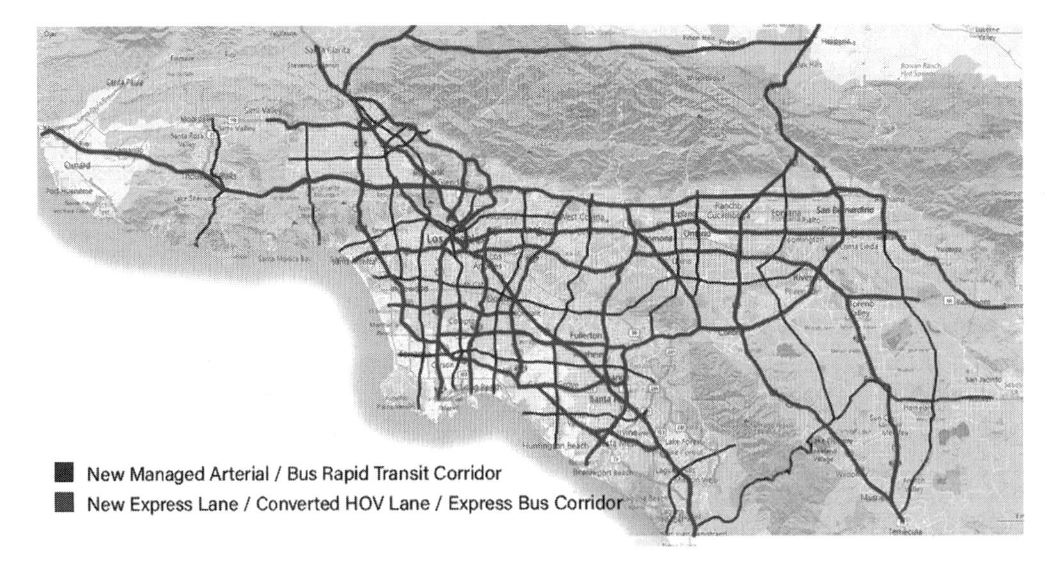

Figure 2.3 Grid network of proposed managed lanes and bus rapid transit lines in Los Angeles.

travel during holidays such as National Labor Day, National Independence Day, New Year's Day, and Spring Festival. Walking, public transit, and non-motorized travel made up more than 90% of travel in major cities such as Beijing, Shanghai, and Guangzhou prior to 2000 (Knapp and Zhao, 2009; Song and Ding, 2009). While automobile travel has increased in many cities, urban planners hope policies that favor mixed uses, higher densities, and transit corridors will moderate the growth of automobile travel by encouraging walking and shifting travelers from bicycles to public transit.

Notably, China provides an example of a country which often has policies working at cross purposes. As some cities, such as Beijing, attempt to limit automobile use, the central government has adopted policies supporting the automobile industry as a major source of economic growth. Indeed, the Tesla automobile factory epitomizes the paradoxes created by competing policy priorities. On the one hand, the growth of the automobile industry is an important source of jobs and income growth. It also helps China position itself as a leader in industry technology and sustainability. On the other hand, these policies spur demand for domestic automobile ownership. Owning an automobile serves a travel purpose by providing individualized, point-to-point travel flexibility while also showing national pride as buyers support official policy objectives. Ironically, these policies, combined with greater urbanization, are reinforcing the need to rethink megacity transportation planning and policy.

The weakening of the traditional hub–and-spoke model is not unique to high-income countries in the West. On the contrary, megacities seem to evolve into complex urban agglomerations with multiple growth centers and nodes over time. Sometimes, these newly formed urban subcenters are intentional. For example, in Chongqing, policymakers and planners have consciously pursued policies to create "new" downtowns. To the west of the traditional downtown (Jiefangbei business circle), which is on the tip of the peninsula that comprises the Yuzhong district, the city is developing the Daping business circle. Daping is expected to be a high-density mixed-use subregional residential and commercial hub. The goal is to reduce residential densities in the eastern tip and create a more balanced urban area. Across from the Jialing River (and north of Jiefangbei) is the Jiangbei business circle.

This decentralization and diffusion are not unique to China. It's also not historically unique. Many established large urban centers emerged out of decades and sometimes centuries of development that incrementally encompassed nearby villages or towns. New York City is in fact the result of consolidating five distinct local governments in 1898: the counties of Richmond (Staten Island), portions of Queens, New York (Manhattan), Bronx, and Kings (Brooklyn). London, England, is a power sharing arrangement among 32 boroughs and the city of London. Each borough functions independently providing local services such as housing, libraries, schools, development control, solid waste collection, and revenue collection. They are part of the Greater London Authority, which administers police and fire, but shares authority over certain types of housing, planning, transportation, and roads. Other large cities, such as Los Angeles and Houston, have simply annexed outlying areas into their city boundaries. The concept of newly formed cities, sometimes called New Towns, such as Shenzhen, Riyadh, or Brasilia, is a relatively new planning phenomenon. As cities evolve into megacities, this decentralization is likely to continue (see also Von Criekingen et al., 2007). Indeed, the Council on Tall Buildings and Urban Habitat (CTBUH) Journal dedicated an entire issue to "polycentric cities" in the Middle East (Safarik et al., 2018).

In other cases, the development is more spontaneous and organic. Houston, Texas, for example, has at least six urban centers. One center, as expected, is the traditional downtown. However, other centers have developed around the international airport, the Texas Medical Center, and neighborhoods such as Westwood or Montrose. While transit use is lower than most global cities, including those in the United States, Houston's land use density and growth suggests future investments in transit may pay off as "thick" corridors with higher densities and mixed uses emerge that favor fixed route travel modes. Indeed, Houston and more conventionally recognized global megacities, such as Chicago and Los Angeles, have used BRT investments to move substantial numbers of passengers along critical corridors in their regions. In fact, BRT systems are often more efficient than heavy and light rail alternatives, particularly in more decentralized, fragmented, and dynamic cities (Staley and Moore, 2009).

These shifting patterns of travel, combined with the size and complexity of megacities as regional economies, create a unique set of challenges. Transportation planning for an agricultural village is relatively simple and stable: create local infrastructure to support walking, trucks, and automobiles while providing connections to commercial destinations for the food or agricultural products its "exports" through commercial trunk roads or rail lines. Megacities are large, complex, and dynamic. These cities include diverse industries and businesses serving intraurban, interurban, and international populations and markets. Bringing these functions into balance is a challenge under the best of circumstances. Thus, it's not surprising that when the Global Positioning System (GPS) firm TomTom (2020) ranked traffic congestion in 34 megacities (out of 416 cities overall), 18 ranked in the top 10%, 25 ranked in the top quartile, and only two ranking below the 50th percentile. Chicago was in the 66th percentile and Wuhan was ranked in the 74th percentile.

However, the TomTom data provides an unexpected insight into the national clustering of megacities and their infrastructure. Among the megacities that TomTom can track, about one-third (12) rank outside the top 20th percentile in its annual traffic congestion index. All three U.S. megacities—New York, Los Angeles, and Chicago—are outside this percentile. Some megacities in this TomTom group are Chinese—Beijing, Guangzhou, Shenzhen, Shanghai, Tianjin, and Wuhan. Wuhan ranks in the 74th percentile! The others are Buenos Aires and Johannesburg. Thus, North America and China cities perform the best on one measure of mobility—vehicular, rubber-tire traffic congestion—compared to other megacities.

TomTom's data, of course, is narrow and limited. It does not take into account public transit, non-motorized travel such as bicycles, or walking. Excluding these alternative modes, however, makes these rankings more interesting, particularly for China where travel modes tend to be more balanced than those in the United States.

American transportation planning is justifiably well known for its investment in highways as its principal transportation investment strategy. New York City, where one half of the population uses public transit to commute to work, is the exception not the rule. Indeed, commuters in the New York metropolitan area make up one-third of the entire nation's public transit users. Among the 50 most populous cities in the United States, fewer than a third of the resident population uses transit in 45 cities according to the 2015 American Community Survey conducted by the U.S. Census Bureau. Fewer than 20% commute via public transit in 34 cities. When looking regionally, the numbers are stark. According to the Center for Neighborhood Technology's *AllTransit™ Performance Score* (https://alltransit.cnt.org/rankings/), 34% of commuters in the New York urbanized area used public transit. Just 5.8% of Los Angeles residents and workers use public transit. Chicago, with the fifth highest transit use as a percentage of travel, is slightly higher with 14%. The San Francisco urbanized area scores the second highest transit use with 20.8%, followed by Washington, D.C. (17.1%), Boston (15.2%). Thus, the TomTom data may be a reasonably reliable indicator of traffic congestion.

The situation in China is different. Megacities such as Beijing and Shanghai register relatively low levels of automobile commuting. About one-third of commuters in Beijing and one quarter in Shanghai commute to work by automobile. Substantial portions of commuting are via walking, bicycles (motorized and non-motorized), and public transit. Notably, fixed guideway transit such as trolleys, subways, and commuter rail are relatively new travel mode choices. As recently as 2010, global megacities such as Shanghai, Beijing, and Guangzhou had modest fixed guideway systems according to the Institute for Regional and Transportation Economy at Chang'an University in Xi'an, Shaanxi, China (Moore et al., 2012). Shenzhen had just 64 km of fixed guideway, Chongqing 17 km, Tianjin 80 km, and Wuhan 46 km. Yet, by 2020, fixed guideway systems were expected to expand by three times the levels in Beijing to reach 1,050 km, nearly double in Shanghai to 877 km, and triple the network in Guangzhou to 677 km. Chongqing grew its fixed guideway transit network to 513 km in just one decade. Tianjin expanded its guideway network by nearly ten times to reach 827 km by 2020. These transit investments were supplemented by 91,000 km of intercity rail track by 2010, 10% of which was dedicated to high-speed rail. China expects its HSR network to exceed 17,000 km as primary connector to major population centers.

At the same time, China invested heavily in its local and intercity roadway system, building the equivalent of the U.S. Interstate Highway System in less than 15 years. China invested in 32 major corridors building more than 161,000 km of expressways by 2020. The state invests about US$20 billion per year. All major expressways are tolled. Most of the roadways were built through toll road financing, first using private capital from outside China and more recently replaced through state financing and state-owned or state-controlled tollway authorities. Notably, 20 Chinese toll agencies were listed on stock exchanges in Shanghai, Hong Kong, and London in 2010 (Staley, 2012a; Wang et al., 2014a, 2014b).

Thus, China invested heavily in multiple modes of transportation to accommodate the burgeoning urban populations. This clearly has had an impact on keeping traffic congestion in check and preventing much more significant degradations in mobility.

China, however, has done more than simply build new transportation infrastructure and scale in a traditional transportation framework. On the contrary, it may be the most ambitious nation in changing the way global megacities integrate their travel modes and systems.

2.5 BUILDING RESILIENCE USING 3D TRANSPORTATION PLANNING

China's megacities are useful examples of new ways to use and opportunities to think about mobility and transportation infrastructure. While the hub-and-spoke framework, and its conceptual prodigy of ring roads, served low-mobility cities with traditional agriculture- and manufacture-based economies well, the dynamic complexity of modern cities requires new ways of thinking about infrastructure. Fortunately, as subsequent chapters will show, many technologies exist for applying a more holistic approach, at least in conceptual form and often in practice. The key is to integrate these technologies and applications comprehensively to ensure they are fully integrated to optimize the performance of the transportation network while maximizing mobility. A foundational step in this process is to move beyond a two-dimensional (2D) transportation planning paradigm into a 3D one (Staley and Moore 2009; Moore et al., 2010).

The 2D framework focuses primarily on two factors—distance and space. This conforms to traditional thinking about urban economics and the geography of space. Conceptually, it organizes the ideas of transportation planning around the geography the facilities are expected to serve. Thus, trunk roads (or rails) connect central cities to outlying towns. Local roads connect the town centers to villages or nearby neighborhoods. Examples include Paris, London, Moscow, Jakarta, Bangkok, Chengdu, and Delhi. Even cities bounded by large bodies of water, such as Buenos Aires or Chicago, reveal a transportation network built around the outward expansion of infrastructure is reasonably concentric rings. In most cases, these ring roads center around a traditional downtown.

Of course, exceptions exist. But most of these exceptions reflect geographic barriers. The economic center and the financial district of Mumbai, for example, are located on a peninsula. Access points are limited because of the wide expanse of water at the confluence of the Arabian Sea and mouth of the Ulhas River, which serves the city's shipping and container port. Mumbai consists of seven separate islands, and much of the city has been reclaimed from the sea. A particularly notable exception is Los Angeles, where major expressways conform more to a grid network. This grid network, it turns out, is a remarkably efficient way to move traffic in dispersed and fragmented metropolitan areas that lack a single dominant core or are characterized by policy centricity (Staley and Moore, 2009).

A 3D transportation network builds on the 2D framework by adding elevation. Rather than seeing tunnels or elevated infrastructure as an extension of the hub and spoke or ring road surface network, the 3D framework integrates new facilities above and below ground in ways that improve the network's efficiency and productivity. Of course, tunnels and elevated infrastructure have been around as long as humans have been engineering roads for traffic. Too often, engineering tradition has seen these facilities as limited enhancements or instrumental ways to overcome an obstacle. The tunnels beneath Versailles, France, near Paris, for example, were developed to maintain the integrity of historical significance of location. Underground interchanges in Sydney, Australia, and Boston were designed to direct through traffic below downtowns. The limited use of these transportation facilities is understandable given their costs often exceed surface roads and infrastructure by three or five times (Staley and Moore, 2009). However, given their size and wealth, megacities have the ability to justify these expenses on scales previously thought unworkable and infeasible.

The Pearl River Delta in southeast China is a particularly useful illustration of how transportation planning and engineering are adapting to megacity travel and mobility needs. This largely continuous urbanized region, a "megalopolis," consists of the megacities of Guangzhou, Foshan, and Shenzhens as well as Dongguan and Hong Kong. Combined, the region includes more than 60 million people. Most of the region's growth emerged from Hong Kong's central role in directing foreign investment into mainland China during the 1980s and 1990s and the decentralization of manufacturing into Guangdong province. The Port of Hong Kong is one of the busiest in the world, a crucial transshipment link between China's manufacturing sector and markets in the West and other parts of Asia. It's also one of the wealthiest regions in China.

Keeping this region connected requires an extensive network of transportation links.

In November 2019, China completed the construction of a 55 km mega seaborne corridor that contains two bridges and underwater tunnels linking Hong Kong, Zhuhai, and Macau. At one point, the bridge submerges for about 3 miles to accommodate ship passage on the surface. After opening to traffic, the bridge was able to cut a 4-hour trip to about a half hour. Thus, the bridge creates an important link in the "spiderweb" of the region's transportation network. It even provides an element of 3D transportation engineering architecture.

However, more emblematic of the transportation complexities for megacities is the engineering and planning going into new expressways. These roads are integrated into a sophisticated multiple-level urban transportation network, as seen in Figure 2.4. On the surface, roads and streets are protected for local traffic and pedestrian access. One level below the street level may be retail and other commercial activities similar to shopping and eating opportunities already available in subterranean rail and bus stations. As Figure 2.4 shows, the second level below the surface is reserved for local traffic. Higher speed through-traffic is directed to the third level with access provided through the second level. The design explicitly recognizes the varied nature of trip purposes and the volume of traffic likely to

Figure 2.4 Illustration of a 3D spiderweb transportation network.

emerge with rising wealth, thickening labor markets, and higher concentrations of population. Rather than trying to accommodate all these needs within one surface-level (2D) transportation networks, the wealth and densities encourage thinking about subterranean options.

Further north, policymakers in Beijing are equally if not more ambitious in pursuing a 3D approach to traffic and transportation management. Beijing currently has developed six ring roads emanating away from the center city. These ring roads connect major precincts and sections of the city but have proven inadequate for managing the volume of east-west and north-south crosstown traffic as incomes have increased and automobile travel has proliferated. Compounding the capital city's transportation management issues is the level of automobile use by public officials, which is subsidized by the government.

Government officials have proposed constructing a series of underground expressways to channel the regional traffic (Zhang et al., 2009). Most recent plans call for four north-south and two east west tunnels to connect outer traffic to points inside the central city. Recent plans suggest these regional expressways may not intersect with each other (Zhang et al., 2009). Rather the expressways will have entrances and exits that provide access to surface streets and districts at key points along the route. The precise number and type of entrances and exits will depend on the network design's ability to meet the primary policy objective. The goal is to minimize traffic congestion, so the exact numbers would be determined by the extent they impact traffic speeds (rather than volume). Of course, the authors note that increasing the number of entrances and exits also increases the number of alternative routes, thus choices and opportunities for travelers to adjust their routes based on their intended destinations (Figure 2.5).

Beijing's efforts compliment a number of other initiatives within China to aggressively address traffic congestion in its megacities. Chongqing, for example, converted a series of civil air-defense tunnels in its Yuzhong district, the traditional downtown, into a 3-km long underground circular road. The circle diverts traffic from crowded surface streets which are

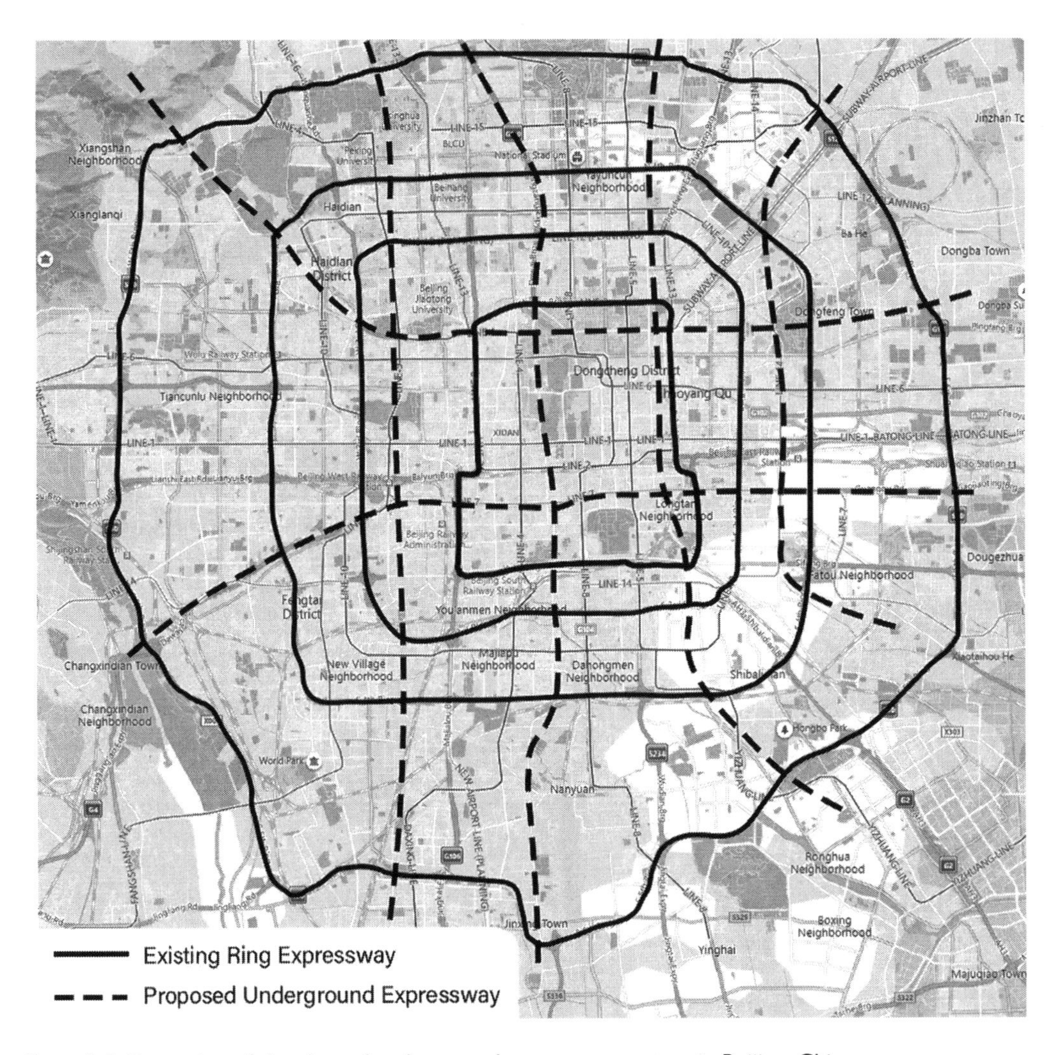

Figure 2.5 Illustration of the planned underground expressway system in Beijing, China.

narrow and difficult to manage given the Yuzhong district's steep and hilly terrain on a peninsula. Major bridges that provide access to the district connect to the circle and a network of underground parking garages.

How might an initiative similar to Beijing's look in a high-income Western megacity? The Reason Foundation conducted a series of studies between 2012 and 2015 using the spiderweb and 3D planning architecture to evaluate its effects on traffic and congestion and mobility. These studies included the global megacities of Chicago (Staley, 2012a) and Los Angeles (Feigenbaum, 2015). Both plans call for significant investments in infrastructure to improve freight and passenger mobility, strengthen the effectiveness of alternative modes, and use the latest Intelligent Transportation Systems (ITS) to actively manage flow on facilities, assess and collect user fees, and manage travel demand. For example, the Southern California plan calls for US$714.1 billion in new investment, including US$102.4 billion in transit operations and maintenance, US$97.2 billion in new expressways with tunnels, and US$105.0 billion in express toll lanes, among others. A key element of all proposed transit and roadway additions is filling in missing links in a spiderweb network using 3D approaches.

The use of off-the-shelf tolling technologies and innovative arterial management strategies allows the new roads to be fully funded by users through sustainable revenue sources and cut in half the tax-funded requirements needed for the current plan using more conventional approaches. For example, cities can elevate over or tunnel under highly congested arterial intersections, funding the infrastructure through all electronic tolling such as license plate recognition or transponders (Poole and Swensen, 2012; Poole, 2018).

Chicago, however, is probably the clearest illustration of how a comprehensive approach to regional transportation planning within this framework might look like. Chicago's economy is the tenth largest among global megacities based on purchasing power, ranking higher than Moscow and Beijing but just under Shanghai and Istanbul according to the Brookings Institution. This puts the city's economy on the same level as Austria, Ukraine, or Sweden. Given its size and geographically central location in the U.S. economy, Chicago is also a critical transshipment point for goods moving from the nation's west to east coasts. Thus, commercial transportation—road, rail, shipping, and air—are central components of the economy.

Like many older global megacities, Chicago has a well-established downtown core. Colloquially known as The Loop, the core is a major commercial, education, and government center. Chicago's congestion levels, however, increase economic costs and lower productivity. Moreover, Chicago represents a major logistics bottleneck as goods sometimes can take several days to move across town on inadequate road and rail links. This inefficiency has national repercussions for economic competitiveness. City and state policymakers are wrestling with how to reinvent a regional transportation system when its essential components were established more than a century earlier. The region's size and complexity requires a comprehensive and integrated approach to planning, designing, implementing, and managing the region's transportation network.

Reason Foundation proposed a significant reworking of the transportation network by adding new capacity in road and transit to complement an existing regional commitment to improving freight capacity and efficiency (Staley, 2012a). The region's transportation network is constrained because the city's eastern edge is bordered by Lake Michigan, one of the five "Great Lakes" and the largest lake by area contained within one country. (Nearby Lake Superior, the world's second largest after the Caspian Sea by surface area, borders Canada). Thus, all land-based surface transportation is forced to go north, west, or south.

A review of the regional transportation network found that Chicago's transportation network is built on a half of a hub–and-spoke system, with corridors emanating from the hub to the north, west, and south. The Illinois Tollway created a much-needed beltway and north-south outer link, but growth was moving rapidly outward. The proposed transportation plan identified a need for more north-south routes, a new road capacity in the northern areas of the region, and great access to the fast-urbanizing western section of the metropolitan area.

The regional proposal is shaped by investments that layer in new capacity to help create a highly networked system (spiderweb) that also uses above- and below-ground (3D) improvements optimized using ITS technologies such as electronic tolling, Figure 2.6. New capacity consisted of three tunnels and a new beltway to accommodate growth on the fringe of the urbanized area. An 11-mile (17.8 km) Crosstown Tunnel would create a critical north-south link to the existing freeway system. A 9.8-mile (15.8 km) tunnel would parallel the Kennedy highway connecting the Downtown to the northwest regions. A 7.3-mile (11.8 km) east-west tunnel below the Eisenhower freeway would serve the urbanized western area. The entire freeway network would be optimized using a 275-lane-mile (442.8 km) high-occupancy toll (HOT) lane project built off existing freeway capacity. The HOT lane network also allows the system to build sufficient capacity to add an extensive, financially sustainable BRT network to the existing transit system. Arterial congestion is addressed by adding 54 queue

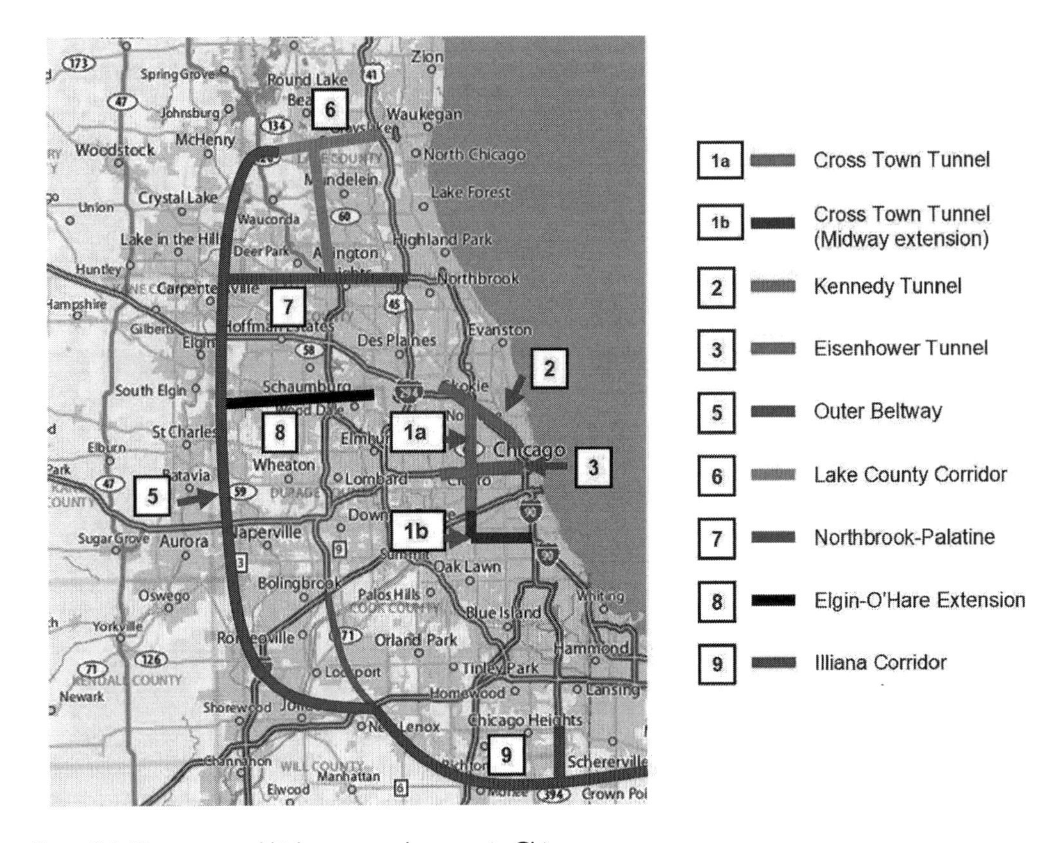

Figure 2.6 The proposed highway tunnel system in Chicago.

jumpers, a US$3.5 billion investment, to critical intersections within the city of Chicago. The proposed arterial queue jumpers are funded through electronic tolling. Combined, these additions would reduce travel delays by about 20% compared to the traffic condition in 2014.

While the integrated scale of the Chicago plan is meaningful, the transportation plans are composed of component parts that draw on current technologies. For example, West Connect is a 21-mile (33-km) highway plan under construction in Sydney, Australia. In addition to improving existing roads, the megaproject includes 16 miles (26 km) of new tunnels that carry traffic below suburban towns. The most notable component from a design perspective may be the Rozell Interchange which connects two main roads and avoids one of the city's most congested corridors. Much of the project is tolled, providing a mechanism for the private sector to fund significant portions of the facilities through a PPP. Parts of the project opened to traffic in 2019 and are expected to be completed in 2023.

2.6 FUNDAMENTAL ELEMENTS OF MEGACITY MOBILITY

Combining the 3D transportation investments with a spiderweb planning framework has the potential to address a range of transportation planning and performance needs. Most importantly, the comprehensive and integrated approach to transportation planning pragmatically addresses the way technology, economic shifts, and wealth are fundamentally transforming travel needs and demand. The conventional hub-and-spoke network presumes an urban form and development model that is less and less relevant as cities become larger and more complex. These trends appear robust across cities in North America, Europe, and Asia (Bertaud et al., 2009; Bertaud, 2018).

A robust and resilient transportation system, however, needs to be more than building out infrastructure. The system needs to be managed effectively to ensure optimal use of facilities with adequate funding to avoid obsolescence. The following chapters will dive deeper into these issues as they explore travel demand management (Chapter 3), multimodal transportation capacity building (Chapters 4 and 5), performance-based mobility management (Chapter 6), funding (Chapter 7), and budget allocation using asset management principles (Chapter 8).

For now, megacities interested in improving mobility and transportation system performance will have to pay careful attention to foundational elements of their transportation systems and networks.

- *Spiderweb transportation planning.* Global megacities will need a complex interlinked and multimodal transportation network that allows travelers and suppliers of goods to navigate to increasingly dispersed points, not just major corridors, at varying times of the day (or night) and, increasingly, every day of the week.
- *3D capacity expansion.* Global megacities will increasingly need to consider elevated and subterranean options for building out and adapting their facilities to meet ever-changing transportation needs, especially to add missing links in a spiderweb network in areas already densely developed.
- *Multimodal integrated expansion with flexible capacity.* Global megacities will need to pay close attention to the multimodal needs of their populations and businesses, recognizing that the demand, utilization, or preference for particular modes is dependent on time, income, reliability, and quality of service. And that technological innovation increasingly is modifying modes and changing the tradeoffs travelers face, the mix of providers of transportation services, and the policy constraints facing government. More productive and efficient transportation systems will seamlessly integrate these modes to travelers and optimize the use of mode based on their own travel goals and objectives.
- *Comprehensive travel demand management.* Optimizing the transportation network will necessitate effectively communicating the costs and benefits of using different aspects of the transportation system in real time or proactively. The comprehensive and integrated use of pricing to manage an entire range of transportation facilities, whether parking, roads, or transit, will be essential to using facilities at optimal capacity, identifying critical needs within the transportation network, financing new facilities, and sustainably managing transportation assets. At the same time, recognizing that pricing set in a market through a decentralized process inherently incorporates a vast array of information that is not possible to reproduce centrally with a planning-based price. Equity issues should be addressed not by modifying transportation prices and thus undermining their effect on decisions but with policies aimed directly at individuals (Ardila-Gomez and Ortegon-Sanchez, 2015).
- *Multilevel investments (arterial, regional, intercity).* Transportation systems will need to integrate and provide access points to multiple levels of transportation facilities. Traditional engineering approaches to transportation facilities develop focus on specific uses and purposes, such as regional travel through limited access highways or local roads to serve neighborhoods and blocks. Infrastructure in megacities requires an integrated and comprehensive approach bridging private systems and services with planners and recognizing the ways the different components of the system influence the performance of others.

- *Integrating freight transportation needs and priorities*. While much transportation planning in megacities focuses on passenger travel, freight and commercial traffic needs are critical to their local and regional economies. Megacities will often be challenged with attempts to address freight and passenger transportation needs through singular facilities, such as bridges, highways, or rail tracks. Finding new ways to finance dedicated facilities for the specific requirements of freight traffic will often be necessary to improve system performance and public safety. The separation of these facilities, such as truck-only toll facilities and BRT with exclusive travelways, using 3D approaches and others, will also have the benefit of identifying sustainable revenue streams and more effective management. Connecting freight rail to rubber-tire facilities is important for the first and last mile of goods movement but it also creates conflicts that must be resolved to improve system performance and safety.
- *Sustainable funding, innovative financing, and performance-based budget allocation using transportation asset management principles.* Megacities must ensure sustainable revenue streams from a variety of methods to ensure facilities are maintained, improved, and built to meet the changing transportation needs of the city economy. Funding systems should focus on returns to mobility, charge system users for their costs, provide clear incentives for users and service providers, and efficiently bring about maximum return on investment in construction and maintenance of infrastructure.

REFERENCES

Aguiléra, A., Wenglenski, S., and Proulhac, L. 2009. Employment suburbanisation, reverse commuting and travel behaviour by residents of the central city in the Paris metropolitan area. *Transportation Research Part A: Policy and Practice* 43(7), 685–691.

Ardila-Gomez, A., and Ortegon-Sanchez, A. 2015. *Sustainable Urban Transport Financing from the Sidewalk to the Subway: Capital, Operations, and Maintenance Financing.* The World Bank, Washington, DC.

Balaker, T., and Staley, S.R. 2006. *The Road More Travelled: Why the Congestion Crisis Matters More Than You Think, and What We Can Do about It.* Rowman and Littlefield, New Kingstown, PA.

Bertaud, A. 2018. *Order without Design: How Markets Shape Cities.* MIT Press, Cambridge, MA.

Bertaud, A., Brueckner, J.K., and Fu, Y. 2009. Managing urban development in Chinese cities. In *Smart Urban Growth in China.* Editors: Song, Y., and Ding, C. Lincoln Institute for Land Policy, Cambridge, MA, 77–108.

Bruegmann, R. 2001. Urban density and sprawl: an historic perspective. In *Smarter Growth: Market–Based Strategies for Land-Use Planning in the 21st Century.* Editors: Holcombe, R.G., and Staley, S.R. Greenwood Press, Westport, CT, 155–178.

Bruegmann, R. 2005. *Urban Sprawl: A Compact History.* The University of Chicago Press, Chicago, IL.

Cherry, C. 2005. China's Urban Transportation System: Issues and Policies Facing Cities. Working Paper. The University of California- Berkeley, Berkeley, CA, USA. [Online]. Available: https://escholarship.org/content/qt1fx5m1ph/qt1fx5m1ph.pdf [Accessed on: March 21, 2021]

Farahani, R.Z., Miandoabchi, E., Szeto, W.Y., and Rashidi, H. 2013. A review of urban transportation network design problems. *European Journal of Operational Research* 229(2), 281–302.

Feigenbaum, B. 2015. *Increasing Mobility in Southern California: A New Approach.* Reason Foundation, Los Angeles, CA [Online]. Available: https://reason.org/wp-content/uploads/2015/11/southern_california_mobility_plan.pdf [Accessed on February 4, 2021].

INRIX. 2019. INRIX Global Traffic Scorecard: Congestion Cost UK economy £6.9 billion in 2019, press release [Online]. Available: https://inrix.com/press-releases/2019-traffic-scorecard-uk/ [Accessed on September 5, 2021].

Knapp, G., and Zhao, X. 2009. Smart growth and urbanization in China: Can an American tonic treat the growing pains of Asia? In *Smart Urban Growth in China*. Editors: Song, Y., and Ding, C. Lincoln Institute for Land Policy, Cambridge, MA, 11–24.

Kotkin, J. 2005. *The New Suburbanism: A Realist's Guide to the American Future*. The Planning Center. Chicago, IL [Online]. Available: http://www.csun.edu/~rdavids/350fall08/350readings/Kotkin_The_New_Suburbanism.pdf [Accessed on February 4, 2021].

Kotkin, J. 2014. City of villages. *City Journal*, Winter Issue [Online]. Available: https://www.city-journal.org/html/city-villages-13621.html [Accessed on February 4, 2021].

Moore, A.T., Staley, S.R., and Li, Z. 2010. Meeting mobility needs in megacities using 3-D transportation planning: Concept and applications. In *Proceedings of the 15th International Conference of Hong Kong Society for Transportation Studies (HKSTS)*. Editors: Sumatee, A., Lam, H.W.W., Ho, H.W., and Siu, B. Hong Kong Society of Transportation Studies, Hong Kong, China, 125–132.

Moore, A.T., Staley, S.R., and Li, Z. 2012. Financing rail transit projects in China: Challenges of fiscal sustainability. In *Proceedings of the 17th International Conference of Hong Kong Society for Transportation Studies (HKSTS)*. Editors: Mak, H.-Y., and Lo, H.K. Hong Kong Society of Transportation Studies, Hong Kong, China, 81–88.

Poole, R.W. 2018. *Rethinking America's Highways: A 21st-Century Vision for Better Infrastructure*. The University of Chicago Press, Chicago, IL.

Poole, R.W., and Swensen, C.R. 2012. Managed arterials: New application of managed lanes concept. *TRB Journal of Transportation Research Record* 2297(1), 66–72.

Safarik, D., Ursini, S., and Wood, A. 2018. The tall, polycentric city: Dubai and the future of vertical urbanism. *CTBUH Journal* 4: *Special 2018 Conference Themed Issue: Polycentric Cities (2018)*, 20–29.

Song, Y., and Ding, C. (eds). 2009. *Smart Urban Growth for China*. Lincoln Institute for Land Policy, Cambridge, MA.

Staley, S.R. 1994. *Planning Rules and Urban Economic Performance: The Case of Hong Kong*. Chinese University Press, Hong Kong, China.

Staley, S.R. 2012a. Mobility & economic productivity in the 21st century city. *Presentation at the ITE Midwest District-TRB Urban Streets Conference*, Chicago, IL.

Staley, S.R. 2012b. Practical strategies for reducing congestion and increasing mobility for Chicago. Policy Study No. 404. Reason Foundation, Washington, DC.

Staley, S.R. 2018. *Planning Rules and Urban Economic Performance: The Case of Hong Kong*. The Chinese University Press, Hong Kong, China.

Staley, S.R., and Moore, A.T. 2009. *Mobility First: A New Vision for Transportation in a Globally Competitive Twenty-First Century*. Rowman & Littlefield, New Kingstown, PA.

TomTom. 2020. Traffic index [Online]. Available: https://www.tomtom.com/traffic-index/ranking [Accessed on February 4, 2021].

Von Criekingen, M., Bachmann, M., Guisset, C., and Lennert, M. 2007. Towards polycentric cities: An integration into the restructuring of intra-metropolitan spatial configurations in Europe. *Belgeo* 1, 31–50.

Wang, Y, Wang, Z, Li, Z, Staley, S.R., Moore, A.T., and Gao, Y. 2013. Study of modal shifts to bus rapid transit in Chinese cities. *ASCE Journal of Transportation Engineering* 139(5), 515–523.

Wang, Y., Li, L., Wang, L., Moore, A.T., Samuel, S.R., and Li, Z. 2014a. Modeling traveler mode choice behavior of a new high–speed rail corridor in China. *Transportation Planning and Technology* 37(5), 466–483.

Wang, J., Mao, X., Li, Z., Moore, A.T., and Staley, S.R. 2014b. Determining the reasonable scale of a toll highway network in China. *Journal of Transportation Engineering* 140(10). doi: 10.1061/(ASCE)TE.1943-5436.0000671.

Wu, W. 2007. Urban infrastructure and financing in China. In *Urbanization in China: Critical Issues in an Era of Rapid Development*. Editors: Song, Y., and Ding, C. Lincoln Institute for Land Policy, Cambridge, MA, 251–270.

Zhang, L., and Levinson, D. 2007. The economics of transportation network growth. In *Essays on Transport Economics*. Editors: Coto-Millán, P., and Inglada, V. Physica-Verlag HD, Heidelberg, Germany, 317–339.

Zhang, H., Mao, B., Goa, Z., and Jiang, Y. 2009. Entrance-exit location model and solution algorithm for Beijing underground expressway network. In *Proceedings of the 14th Annual Meeting of the Kong Society of Transportation Sciences*. Editors: Wang, D., and Li, S.-M. Hong Kong Society of Transportation Studies, Hong Kong, China, 517–524.

Chapter 3

Travel demand management

The pace of transportation capacity expansion has failed to keep up with escalating travel demand. As a result, many urban/metropolitan transportation networks around the globe are operating near or at their capacities. This means more cities are suffering from capacity shortfalls to accommodate more intensive traffic, especially during daily peak periods, leading to more severe recurrent traffic congestion for larger geographic areas and extended time duration.

One possible venue for mitigating recurrent congestion is to manage travel demand in ways that focus on enhancing the efficient movements of people and goods, rather than only on improving mobility for vehicles that facilitate passenger and freight travel. Travel demand management (TDM) represents a complex and complementary set of policies and responses that seeks to reduce the need for physical travel through more efficient land use or transportation substitutes, alter destination locations, adopt flexible arrival times, shift travel modes, promote demand leveling by reducing the number of single-occupancy vehicles (SOVs) trying to simultaneously use congested network links and nodes, change departure times of travel, change travel route, and manage travel lanes, and more (VTPI, 2017).

Depending upon the local conditions, some or all demand management strategies could be selected for implementation. A properly designed mix of strategies will mean that the goals of passenger and freight trips can be met within the capacity constraints of the network and its facilities while allowing for an appropriate level of growth in that capacity as well. Demand management strategies reach maximized effectiveness if they are implemented in conjunction with active traffic management strategies in which the network is actively operated, not just passive preserving and adding infrastructure (Cracknell, 2000; Kasipillai and Chan, 2008; FHWA, 2011; Sammer, 2016; Ferguson, 2018).

Essential to demand management is to understand that markets inherently seek the most efficient ways to meet the needs of people travel and goods delivery. A mobility market would seek to minimize the use of land for development, energy consumption and emission, travel time, and direct costs of people travel and goods delivery, as well as maximizing the safety of people travel and goods movements. Since megacity mobility takes place in a transportation network that is not entirely, or in most places much at all of, a free and flexible market, demand management strategies can be used to emulate many market forces to reach similar efficiencies. The effort, however, is constrained by persistent information problems without the prices signals and decentralized decision process of markets.

The pricing mechanisms used for transportation systems and services (discussed in more detail in Chapter 7) can bridge some of this gap. Pricing a roadway or light-rail line to keep it uncongested during peak demand, with dynamic pricing that adjusts in real time, creates a functional market for the use of that road or rail line and over time. Pricing strategies could lead to internalizing various travel for many location and travel decisions on that route and shift the network to use resources more efficiently. That is a very direct example, but an

DOI: 10.1201/9780429345432-3

important one, as megacities should move to use smart pricing of transportation as much as possible while dealing with equity issues in more direct ways, as discussed in Chapter 7. Other transportation demand strategies should strive for similar frameworks.

3.1 INFLUENCE OF LAND USE ON MOBILITY

Historically, high-density urban areas with mixed land use served by better transit systems in the United States, like urban areas in other places, are correlated with lower levels of dependence on private vehicles for travel and higher use of non-auto travel modes. A number of researchers have proposed the use of land-use measures of residential density, diversity, design, destination accessibility, and distance to transit (referred to as 5D measures) that characterize the built environment to assess influence of land use on the use of transportation (Cervero and Kockelman, 1997; Ewing and Cervero, 2001, 2011; Zhang, 2004). Much of the dialogue and discussion over travel behavior, land use, and travel mode has been siloed as either/or with an emphasis on reducing SOV travel and privately owned automobiles. At higher densities, the reasoning goes, more alternatives combined with great pedestrian accessibility to neighborhood services and retail opportunities will reduce automobile travel. A key element missing in this discussion is the travel benefits provided by self-directed and spontaneous travel decisions made possible by automobile or auto-based travel. Thus, while research shows that the built environment can influence travel behavior when alternatives to the automobile are available, the effects can be modest except at very high residential and employment densities.

Residential density is defined as a ratio of people or dwelling units divided by land area. Studies using aggregate and individuals' travel data for cities and metropolitan areas have found that an increase in residential density could lead to a reduction in vehicle miles of travel (VMT) and an increase in the use of alternative travel modes (Newman and Kenworthy, 1989, 1999; van de Coevering and Schwanen, 2006; Kim and Brownstone, 2013). But whether these reductions in trips by automobile are beneficial on average is unclear. If higher densities and traffic congestion reduce travel and, as a result, convenient access to services, places, and goods people want to consume, overall social welfare could fall (Balaker and Staley, 2006). The presumption by many planners is that the services will be placed closer to the consumers and users of the services. Thus, "mixed" land uses are preferred, particularly at the terminal points or major nodes along a transportation corridor.

Diversity refers to mixed-use development in presence of multiple types of land use functions, such as those for residential, office, and retail purposes within one mid- or high-rise building, or a complex. With a variety of residential, employment, and recreational opportunities offered within the mixed-use development, coupled with opportunities at nearby locations, the auto ownership of residents in the mixed-use development tends to be lower because of limited parking provisions and high parking fees. These higher fees and the more limited allocation of land to parking often reflect the higher demand for land in these places for alternative, non-automobile uses. Also, a significant portion of otherwise auto trips would shift to walking, biking, and scooting, as well as transit riding. The captive auto trips are likely with shorter trip lengths. The combined effect would exhibit significant reductions in VMT (Kockelman, 1997; Bento et al., 2005; Ewing and Cervero, 2010). The design characteristics of a built environment such as elements of the urban street network (such as complete streets to be described in Chapter 4), network connectivity, pedestrian walkways, and density and opportunities of pedestrian crossings could potentially encourage walking of local residents and promote non-auto trips (Frank and Engelke, 2005; Ewing and Cervero, 2010).

Destination accessibility measures the potential for reaching a number of office, shopping, and recreational destinations or personal places of interest in a given time duration. In general, destination accessibility is higher in more central locations as opposed to relatively lower-density suburban developments. The higher the destination accessibility, the lower the dependence on auto travel (Ewing et al., 2011).

In the presence of adequate pedestrian walkway and bikeway facilities, keeping the distance of a residential location or employment center to the transit stop or station within the feasible walking, biking, or scooting distance will likely attract potential transit riders for the use of transit services and even induce new transit riders. This will likely decrease the dependence on the use of private vehicles, although research is unclear on whether the magnitude of these effects is significant enough to alter general travel patterns (Frank and Engelke, 2005; Ewing and Cervero, 2010; Cervero, 2013). The quality of the alternative transportation services and facilities is important. For example, high levels of bike commuting and use in the Netherlands and Denmark are attributed to high-quality dedicated bikeways that can manage high volumes of bike traffic, separated from automobile and truck traffic, to ensure safety and reliability.

Additional land-use measures such as network connectivity, job-housing balance, and level of mixed land use have also been utilized in empirical studies to evaluate influence of land use on travel behavior and mobility. Practically, an urban area with better network connectivity, more balance between job and housing locations, and a higher level of mixed-land development will help reduce the trip length and could potentially lead to decreases in the auto dependence for travel. As a closed loop, the changing travel behavior could influence transportation investments in reshaping the transportation system, which will have a fundamental role in local accessibility of places, residential locations, and economic activities and in turn affect land use in the long run.

However, planners and policymakers must be careful to avoid imposing presupposed expectations on travel behavior promoting transportation-compatible land use and supporting facilities. Research and experience in most cities reveal that travel mode choices are determined by the preferences and priorities of the individual traveler. These choices are varied and dynamic as priorities shift and evolve based on individual circumstances. Living within walking distance of employment may be a major initial determinant for living in a particular neighborhood. However, if the commuters' job changes to another location, the decision about whether to commute by transit or automobile will depend on the relative importance of staying in their neighborhoods versus the time and cost of commuting to the new location. Similarly, the housing and commuting preferences of single professionals differ significantly from those households with small children, which in turn differ from those with older children. Moreover, land use changes typically unfold over decades and organically, even under rigid planning systems. Household needs and preferences change much more frequently. The task of transportation planners is to ensure that transportation system supports the desires, needs, and preferences of the users, not preconceived notions about how people should behave or prioritize specific travel nodes or patterns.

3.2 PHYSICAL TRAVEL MANAGEMENT

Urban travel decisions are mainly affected by attributes concerning vehicle ownership, trip frequency and length, transit availability and service quality, ridesharing availability, cycling, scooting, and walking-related non-motorized travel conditions closely relevant to land use factors (Al Mamun and Lownes, 2011; Amirkiaee and Evangelopoulos, 2018; Asirin and Azhar, 2018). In addition, flexible working arrangements including flextime, compressed workweek, and telecommuting or telework influence the total physical travel and travel intensity. These characteristics are in significant flux during the COVID-19 Pandemic, and it's unclear how employers and employees will reset the balance between telework, remote work, and traditional work commercial building settings.

Vehicle ownership and use decisions are shaped in part by land use factors such as density, transportation network connectivity, parking supply and management, mobility management, and integrated land use and transportation planning considerations, as well as implementation of connected and automated/autonomous vehicles (CAVs) that will significantly alter the mobility landscape for automobiles, trucks, and transportation infrastructure (to be covered in Chapter 4).

Increased connectivity among roadways can directly reduce vehicle travel. Parking supply refers to parking spaces available per building unit, floor area, or land area. Parking management pertains to how parking spaces are managed and priced. Mobility management aims to encourage travel activities accomplished by multiple travel modes. Planning considerations seek to integrate land use and transportation planning to achieve environmentally sensitive, economically viable, community-oriented, and sustainable urban development. Land-use policies could reduce barriers to the use of alternative travel modes other than auto mode that will lead to reductions in physical travel by autos (Farinloye et al., 2019). They typically include removing disincentives to and allowing mixed land use for housing, commercial, and institutional purposes in proximity, clustering of jobs and other activities in urban core areas or central business districts and other efficient locations, and multimodal accessible designs for sites, roadways, and urban streets. Avoiding overreliance on auto travel requires ensuring that demand for transit services and for microtransit or non-motorized travel including walking, biking, and scooting is easily met within land-use patterns; appropriate infrastructure is available; and effective intersection traffic control and adequate access management are implemented. On a regional scale, it is crucial to allow development and density to locate jobs and housing in proximity for shorter commuting and other travel without reducing accessibility or mobility.

3.3 DESTINATION LOCATION AND ARRIVAL TIME MANAGEMENT

In terms of physical travel, each trip is characterized by origin point, destination location, and travel time—which includes a required or expected arrival time and a planned time window for departure to ensure on-time arrival, which in turn includes an estimate of expected travel time uncertainty or variability. For a given trip, the origin-destination travel time or travel impedance is affected by the sequential choices of destination location, arrival time, travel mode, departure time, travel route, and travel lane. Also, travel impedance exhibits varying levels of sensitivity to physical travel by trip purpose that is mainly classified into home-based work (HBW), home-based other (HBO), and non-home-based (NHB) trips. Therefore, strategies for destination and arrival time management need to focus on influencing the choices of destination locations and arrival times of specifically purposed trips.

For work-related trips, flexible working arrangements can influence the extent of physical travel. Typical strategies include flexible working hours, compressed workweek, teleworking or telecommuting, and staggered working times. Flexible work hours are also called flextime, which refers to a working schedule of a government agency or private entity that allows employees to adjust the start and stop times in accordance with the core hours of presence of all employees established by the management to accommodate personal obligations and avoid peak traffic hours.

The compressed workweek adopts a schedule such as 10 hours per day and 4 working days per week to replace the traditional arrangement of 8 hours per day and 5 days per week schedule. It not only reduces the travel frequency of each responsive employee by two or more commuting trips per week but also leads to savings of travel time of each trip because the earlier start time and later end time help shift the travel to non-peak periods (Zong et al., 2013; Duddu and Pulugurtha, 2015; Su et al., 2020).

Teleworking or telecommuting refers to a work situation of allowing the employee to work away from the traditional workplace while keeping smooth contacts with colleagues with communications technologies facilitated by the employer. The common forms of teleworking include working at home by remotely accessing the workplace's electronic server, working in a business park that allows for remote access to the workplace, and working from a shared workspace at the client' office or a permanent location, which is also called hosteling or free addressing (Turnbull et al., 1995).

Under a scheme of staggered working times, employees or groups of employees of a public or private entity start and finish work at slightly different times within prescribed limits. The starting and finishing times could be chosen by the employee or assigned by the employer. A typical example of overlapping shifts would be to have one shift from 7:30 to 15:00 and a second from 10:30 to 18:00. Aside from lunch breaks, this working time arrangement would provide all employees with 4.5 core hours of presence per day between 10:30 and 15:00. This arrangement may be particularly important if there are peaks in demand within the period of core hours.

As applicable, one or more flexible working strategies could be publicly promoted and effectively implemented. This will potentially lead to drastic reductions in daily work-based trips and trip intensities especially in peak periods. Therefore, a portion of non-home-based trips such as lunch trips will be reduced. This will eventually result in reductions in total travel demand and demand intensities in daily peak periods and the adjacent-to-peak period around midday.

3.4 TRAVEL MODE MANAGEMENT

For passenger travel, primary travel modes include auto; transit—bus, bus rapid transit, and fixed-guideway transit; ridesharing modes—carpool, vanpool, ride hailing (only one rider served by a ridesharing vehicle), and other ridesharing services; biking; scooting; and walking. For freight travel, typical travel modes are trucking, freight rail, cargo drone where trucking mode is predominant within an urban/metropolitan area. When choosing among competing passenger travel modes, travelers consider attributes generally including out-of-vehicle travel time, in-vehicle travel time, inter- and intra-modal transfers, and out-of-pocket costs such as vehicle operating costs, tolls, parking fees, transit fares, ridesharing charges, and bike/scooter renting fees. These combine to form a value for each traveler that lets them pick a travel mode based on how all those factors tradeoff for them for a given trip. Those decisions tend to be highly influenced by factors such as traveler's income level and auto ownership (Ben-Akiva and Lerman,1985; Verplanken et al., 1997; Bamberg et al., 2003; Chan and Shaheen, 2012; Aziz et al., 2018).

Policies that affect those value variables will influence travelers' choice of travel mode for a given trip (Buehler, 2011; Maduwanthi et al., 2015; Feng et al., 2017; He and Thøgersen, 2017; Farinloye et al., 2019). However, there are also many cultural and behavioral factors that influence travelers' choice that are difficult to disentangle and understand how they will interact with policies (Lanzini and Khan, 2017). Given how many variables shape travel mode choice decisions, it is not surprising that policies designed to influence them frequently fail to work as hoped and often may be more costly than warranted. This challenge only grows as the mode choice options travelers enjoy grow and change, for example, with mobility-as-a-service and automated vehicles (Durand et al., 2018; Hawkins and Habib, 2019). There is a strong case to be made for focusing policies on improving the travel options preferred by travelers, and finding ways to improve their efficiency will have more positive effects on mobility outcomes.

3.5 DEMAND LEVELING MANAGEMENT

Of all trips occurring in a weekday, a predominant portion of the daily total trips will be accommodated by auto, carpool, vanpool, and other ridesharing services facilitated by private vehicles, such as automobiles, vans, taxicabs, and so forth. Some of those trips are SOV trips with a single private vehicle carrying only one person, namely, driving alone. Demand leveling strives for reducing the number of SOV trips and single-occupancy VMT. Reducing the number of SOV trips and related VMTs on an average weekday can improve the efficiency of mobility and limit traffic congestion, excessive vehicle operating costs, and harmful air pollution.

Demand leveling and trip reduction policies generally provide commuting incentives to reduce SOV trips and VMTs via park-and-ride services, vanpool programs, shared use mobility programs such as casual carpooling where drivers arrive and pick up enough passengers to meet the high-occupancy vehicle (HOV) lane eligibility requirements to avoid a toll, travel along the HOV lane, and drop off passengers at an agreed-upon location, transit programs, bicycle, scooter, and pedestrian amenities, parking management, and outreach and education programs (Gärling et al., 2002; Gärling and Schuitema, 2007; VTPI, 2019).

Strategies to reduce SOV trips and VMT can also be promulgated by employers, usually with some policy incentives or requirements, and primarily include vehicle parking charges; traveler information on available ridesharing and transit alternatives; service to provide ride matching, shuttles, and guaranteed ride home; subsidies for carpool, vanpool, and transit; and cash benefits offered in lieu of accepting free parking (Shoup, 2017). Employer-based strategies tend to work best in high-concentration industries with clustering that provide a density for alternatives to SOV making them easier to adopt, for example, high-tech industry, medical complexes, university campuses, and military bases.

3.6 DEPARTURE TIME AND TRAVEL ROUTE MANAGEMENT

Automobile travel is the dominant mode in the United States and Europe and is a rising share of travel in most regions of the globe as it tends to rise with wealth (ECA, 2020; Sumit et al., 2020; USCB, 2020). Congestion from auto travel can be influenced through measures that affect departure times and route choices of individual auto trips with pre-trip and en route traveler information provisions to potentially avoid congested time period and/or corridor. Auto travelers may modify their departure times if the utility gained from the time saving of the total trip time offsets the utility lost from the extra time spent for earlier departure or longer waiting time for delayed departure. Practically, deploying HOT lanes for facilities experiencing or expected to experience recurrent congestion in peak periods can influence departure times of auto, carpool, vanpool, and other ridesharing travelers to the edges of the peak period, create route changes to parallel facilities, or even trigger shifts to transit or other less costly modes (Thorhauge et al., 2016; Poole, 2020; TTI, 2021).

3.7 TRAVEL LANE MANAGEMENT

Once a traveler's choices are made on the departure time, travel route, or a given facility, demand management could impose final influences on how travel lanes of a given facility would be used by travelers. In normal traffic conditions, the adoption of HOV and HOT lanes offers auto, carpool, vanpool, and other ridesharing travelers the choice to experience reliable travel times in exchange for increased occupancy or paying a fee on a given facility. The use of HOT lanes offers choices and does not force anyone to change their travel behavior (Yang and Huang, 1999). In traffic conditions induced by incidents owing to vehicle crashes, vehicle disablements, travelway debris, temporary workzones, or inclement weather, dynamic travel lane control and temporary use of paved shoulders both upstream and downstream of the incident, coupled with overhead messages of speed control, displays of lane closure, merging, and diversion, and real-time monitoring of traffic condition and updating of traveler information that would minimize the entire duration of incident detection, response, clearance, and recovery become essential.

TDM strategies targeting physical travel and a series of choices of destination location and arrival time, travel mode, departure time, route or facility, travel lane, coupled with demand leveling throughout the trip chain will help control the total travel demand, demand intensity, vehicle volumes, and vehicular traffic intensity that will eventually help achieve a reliable movement of people and goods.

3.8 CASES IN ACTION

3.8.1 Bay Area travel demand leveling program, California

The Bay Area fair value commuting (FVC) demonstration aims to reduce the share of SOV commuting in the Bay Area by implementing the FVC solution set, which aims to solve many of the above problems. Stanford's commuter program provides a conceptual starting point for FVC. Stanford reduced SOV from 75% to 50% (transit share increased from 8% to 31%), eliminating the need for a new US$107 million parking structure (BAMC, 2018). This project will demonstrate two key concepts:

An integrated commuter wallet software platform will try to provide commuters with the greatest convenience to plan, compare, and pay for other modes of transportation. Back-end systems will be coordinated to provide seamless commuter rewards and benefits to employees.

Second, a tip or cash system will be shown. A "tip" system will simultaneously assess the SOV royalties (allocate "fair value") and reallocate the income received to fund incentives to use other modes of transportation to establish a self-sustaining commuter program. The cash-out system is an incentive-based program in which incentives are paid to non-SOV employees. Although pure cash disbursement plans cannot solve the "expense" part, cash disbursement plans are likely to lead to a significant reduction in car use by participating employees (Figure 3.1).

3.8.2 Sustainable urban mobility in Stockholm, Sweden

The Urban Mobility Index (UMI) is a measure for assessing the maturity, innovation, and effectiveness of mobility solutions adopted by cities around the globe. According to the Future of Mobility Report 3.0 by Van Audenhove et al. (2018), the UMI of Stockholm, Sweden, scored 57.1%, which was very close to the first-place city, Singapore, with a score

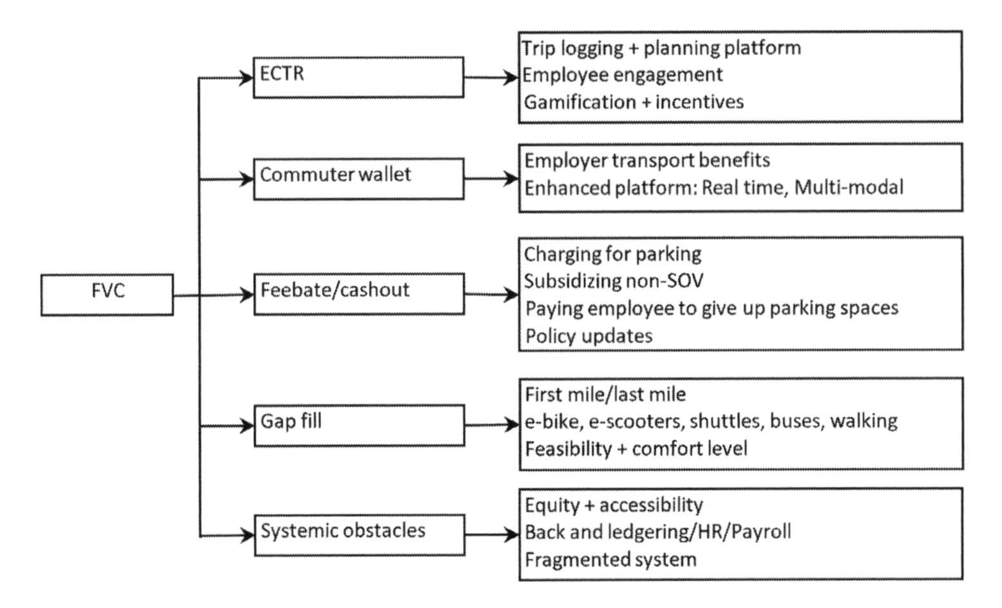

Figure 3.1 The Bay Area fair value commuting program for travel demand leveling in California.

of 59.3%. Sustainable mobility in Stockholm is backed by multiple regulations and policies implemented over the last several decades (Twisse, 2019).

- In 1996, Stockholm City Administration introduced regulations for an environmental zone in the city's central area limiting the access of vehicles weighing more than 3.5 tons. Exceptions were for vehicles retrofitted with an approved emissions control device to control diesel particulate matters (PMs), hydrocarbons (HC), and nitrogen oxides (NO_X) to improve air quality.
- In January 2006, Stockholm City Administration initiated an inner-city congestion charging scheme extending beyond the environmental zone to enable individual travelers to evaluate the relative costs of their journeys based on the charges that they would incur. They could then determine the tradeoffs relative to their choices of trip destinations, arrival times, travel modes, and so forth. Meanwhile, public transportation services were expanded to facilitate travel of visitors and residents. A 7-month trial period was used for travelers to experience how the scheme would work in practice. A referendum was held after the trial period where the travelers voted in favor of formally implementing the scheme that has been in operation since August 2007.
- In 2012, Stockholm City Traffic Administration published the Stockholm Urban Mobility Strategy sought to reduce congestion even further by 2030 while reducing transport emissions in the area by 30%. To this end, regulations were imposed on diesel heavy goods vehicles (HGVs) and buses to encourage the use of alternative fuel and ultra-low-emission vehicles. The strategy also contains plans for promoting TDM by adopting teleworking arrangements such as online working and meeting technologies and ridesharing, investing in public transportation and bike facilities, stimulating the purchase of environmentally friend cars, and increasing parking charges.
- In 2016, Stockholm City Administration increased the congestion charge for entering the inner-city zone by 75% at peak times between 06:30 and 18:30 on weekdays and by 10% during off-peak hours. The regulations were extended to Essinge bypass (E4/E20), which is a heavily congested motorway located west of Stockholm city center.

The congestion charge rates vary by time of the day. Taxis charges are capped per vehicle in 24-hour period. Exemptions are given to the evening, at weekends, public holidays, and the day before a public holiday, as well as buses, emergency vehicles, motorcycles, and diplomatic and military vehicles.

- In December 2016, the Strategy for a Fossil-fuel Free Stockholm by 2040 was published as a collective effort of Stockholm City Development Administration, Stadshub AB, City Traffic Administration, and City Planning Administration. The strategy outlined plans to further regulate vehicular access in combination with complementary schemes to reduce road traffic overall. Specifically, the environmental zone would successively ban older HGVs and buses from entering the inner city over time. The policy was coupled with a Website which provided information on approved suppliers of exhaust emission upgrades and details the fines should a regulated vehicle enter the zone. Further, HGVs and buses over 12 m long, 3.5 m wide, and 3.5 tons are only allowed on certain roads within the city. In Stockholm's Old Town, the length limit is shortened to 8 m and a height limit of 3.5 m is added. Payments can be made via direct debit, by using bank transfers, or the scheme's online virtual shops.
- In 2018, the city of Stockholm issued Urban Vehicle Access Regulations (UVAR) to continue with its endeavors to improve air quality and reduce its impact on climate change and promote alternative travel modes within the city center for congestion mitigation. Key features include (i) regulating the access of a low emission zone by heavy vehicles using older or more polluting technologies and fuel types to help improve air quality in the inner city; (ii) charging vehicles entering the city center and using main access roads extending beyond the low-emission zone limits; (iii) limiting large and heavy vehicles entering Stockholm by type, length, width, height, and weight, and time of the day; and (iv) imposing tighter regulations on vehicles entering the Old Town by prohibiting motorized vehicle entries to the historical center except for 6:00 am to 11:00 am.

According to the Study on UVAR by European Commission's Directorate-General for Mobility and Transport (Ricci et al., 2017), the multiple rounds of regulations aimed to promote sustainable mobility in Stockholm have led to reductions in queueing time by 30%–50%, vehicle volumes by 20% and 25%, and vehicle emissions such as carbon monoxide by 14%, PM10 by 13%, and volatile organic compounds by 13% (Figure 3.2).

3.8.3 Smarter travel choices from better travel information in Reading, Berkshire, UK

Reading in southern England is strategically located offering good access to London and the UK's main international airport, Heathrow. While Reading has a highly successful bus transit network, the Reading Borough Council (RBC) relies heavily on it to manage its road network and provide information to the travelers in support of sustainable travel. The current Real-Time Passenger Information (RTPI) system disseminates real-time passenger transportation information at bus stops and on buses. Additional information on incidents and parking provisions is displayed via variable message signs (Kuiper, 2015). The above information is accessible through an RBC travel Website.

However, the RTPI system lacks information provision on road journey times, congestion, and alternative routes. Given the massive increase in mobile device usage and the challenges in servicing the range of platforms available, developing the system to provide the desired coverage sufficiently was not viable. Alternatively, RBC has developed its Open Data Services (ODS) that links directly to the core Intelligent Transportation Systems (ITS) with the significant amount of static and real-time transportation data presented in either

Figure 3.2 Sustainable urban mobility in Stockholm, Sweden. (Source: Shutterstock (2014) with permission.)

XML or JSON format. The data format of the Reading ODS is aligned with that of the London ODS to allow London apps to be easily modified to use Reading open data. The cloud-based service could be scaled to demand in real-time and is extremely reliable. This allows Web or app developers to easily access and re-use the data under open license to develop high-quality applications to better inform travelers of the transportation network status in real time to enable them to make smarter choices of travel modes, departure times, and routes that would minimize travelers' dissatisfaction.

The first app *trvlRDG* is already available. Other apps and services have also been developed using the open data such as the bus app, *tvptravel*, developed for the Thames Valley Park business park. Through engagement with developers and further development of the ODS to include additional data sets, it will drive the development of high-quality travel apps to provide a wide range of real-time multimodal traveler information on bus travel, car parking, incident and congestion, roadworks, and road journey time enabling travelers to make a more informed choice as to how they move around Reading and the greater London areas. Also, the feedback of ODS developments and experience of data applications for traveler information provisions enable transportation agencies in the two areas to further improve their user services (Figure 3.3).

3.8.4 Urban transportation development and management in Singapore

Singapore is an island city-state of 640 km^2 and a population of 5.7 million in 2019. Since its independence in 1965, Singapore has rapidly grown into one of the four newly industrializing economies in Asia with an average growth rate of 7% per year in the last three decades (Palliyani and Lee, 2017). The vibrant economy, coupled with a large population and confined land area, makes transportation development and management more challenging than many urban areas elsewhere. This is attributable to the adoption of an integrated approach for land use and multimodal transportation planning, as well as effective management of

Figure 3.3 Smarter travel choices from better traveler information provisions in Reading, Berkshire, UK. (Source: RBC (2021).)

travel demand and transportation capacity. Initiated in 1972, the integrated urban land use and transportation master planning contains four key strategies (Müller, 2020):

- Integrating urban development with transportation planning by closely relating the expansion of urban transportation facilities with other measures for desired patterns of land use to minimize the need for travel.
- Developing a comprehensive added capacity road network in which the total road lengths has been increasing from 1,761 km in 1965 in the year of independence, to 2,173 km in 1975, 2,645 km in 1985, and 2,989 km in 1993 to 3,356 km in 2019.
- Aggressively limiting private auto ownership through a quota scheme implemented in 1990 and managing demand of road usage to alleviate traffic congestion.
- Providing quality public transportation choices via development of a mass rapid transit system and an island-wide public bus network.

The traffic restraint policy for controlling ownership and usage of private vehicles is facilitated by implementing measures of fiscal disincentive, area licensing, electronic road pricing, and quota schemes. The fiscal instruments for ownership control were instituted in the early 1970s. Such instruments include an import duty, which is levied by the Customs and Excise Department of the Ministry of Finance on each vehicle brought into the country based on its Open Market Value (OMV). When the vehicle is registered, the Land Transport Authority of the Ministry of Communications (MOC) collects a registration fee, an additional registration fee (ARF), and annual road tax.

In June 1975, the Singapore government introduced a cordon pricing-based area licensing scheme for automobiles entering the CBD during peak hours. All vehicles except for ambulances, fire engines, police vehicles, and public buses that had purchased daily or monthly licenses with visible stickers mounted on windscreens could enter the enforced cordon zone in the restricted hours. Violators are liable to fines issued through mails. Over the years, the restricted zone has been adjusted several times to keep abreast of the CBD expansion. The enforced cordon-pricing hours currently from 7:30 am to 6:30 pm on weekdays and 7:30 am to 3:00 pm on Saturdays, mode of operation, and rates have also been adjusted in response to changes of income socioeconomic and traffic conditions.

In July 1989, the Singapore government announced the policy of implementing electronic road pricing. Owing to the lack of off-the-shelf, ready-to-use electronic toll collection system to automatically charge vehicles traveling in a multi-lane environment, the electronic road pricing system was not put into operation almost 8 years later. By May 1997, road pricing schemes were operational for four expressways and parkways leading into the city. Drivers were mandated to display valid area licenses to pass through any of the scheme gantries during morning peak between 7:30 am and 9:30 am from Monday to Saturday and afternoon peak between 4:30 pm and 7:00 pm, excluding Sundays and public holidays. With the extension of the road pricing scheme, operational peak hours of the cordon pricing-based area licensing scheme have been revised accordingly. Ultimately, the electronic road pricing evolves into an automated version of the area licensing scheme that varies user charge rates by time and location according to traffic conditions, which serves as a means of effective traffic management. The holistic consideration of enforcement hours for area licensing scheme to control vehicles entering and exiting the city center and road pricing scheme to regulate vehicles using freeways and parkways has led to reduction in traffic intensity by 45%, increase of average travel speed by 20%, and decrease of vehicle crashes by 25% (Yuan, 2018).

On May 1, 1990, the MOC began to implement the quota scheme that classified private vehicles into seven categories depending on engine capacity and intended vehicle usage, each category was allocated to an annual quota, and a certificate of entitlement (COE) was required to own a new private vehicle.

Concurrent to actively managing private vehicle ownership and usage, the Singapore government has been instrumental in making funds available for the transportation sector for road building programs and significant expansion of rapid rail and urban bus systems in conjunction with adoption of modern technologies for multimodal transportation management. In addition, the government has taken responsibilities to continually monitor and adjust policies and measures for planning, maintaining, improving, and managing roads, streets, and public transportation systems, as well as their respective uses to ensure they remain effective in providing best travel services at a minimum cost. Sustainable mobility is essential to support economic prosperity of the city (Figure 3.4).

Figure 3.4 Urban transportation development and management in Singapore. (Source: Ling (2020).)

REFERENCES

Al Mamun, M., and Lownes, N.E. 2011. A composite index of public transit accessibility. *Journal of Public Transportation* 14(2), 4.

Amirkiaee, S.Y., and Evangelopoulos, N. 2018. Why do people rideshare? An experimental study. *Transportation Research Part F: Traffic Psychology and Behaviour* 55, 9–24.

Asirin, A., and Azhar, D. 2018. Ride-sharing business model for sustainability in developing country: Case study Nebengers, Indonesia. *IOP Conference Series: Earth and Environmental Science* 158, 012053.

Aziz, H.A., Nagle, N.N., Morton, A.M., Hilliard, M.R., White, D.A., and Stewart, R.N. 2018. Exploring the impact of walk-bike infrastructure, safety perception, and built-environment on active transportation mode choice: A random parameter model using New York City commuter data. *Transportation* 45(5), 1207–1229.

Balaker, T., and Staley, S.R. 2006. *The Road More Travelled: Why the Congestion Crisis Matters More Than You Think, and What We Can Do About It*. Rowman & Littlefield, New Kingstown, PA, USA.

Bamberg, S., Ajzen, I., and Schmidt, P. 2003. Choice of travel mode in the theory of planned behavior: The roles of past behavior, habit, and reasoned action. *Basic and Applied Social Psychology* 25(3), 175–187.

BAMC. 2018. Accelerating Progress: Strong Partnerships Move the Bay Area Forward: Annual Report to Congress. Bay Area Metro Center, San Francisco, CA, USA. [Online]. Available: https://mtc.ca.gov/sites/default/files/2018_Report_to_Congress_MTC-ABAG.pdf [Accessed on March 15, 2021]

Ben-Akiva, M., and Lerman, S.R. 1985. *Discrete Choice Analysis: Theory and Application to Travel Demand*. MIT Press, Cambridge, MA.

Bento, A.M., Cropper, M.L., Mobarak, A.M., and Vinha, K. 2005. The effects of urban spatial structure on travel demand in the United States. *Review of Economics and Statistics* 87(3), 466–478.

Buehler, R. 2011. Determinants of transport mode choice: A comparison of Germany and the USA. *Journal of Transport Geography* 19, 644–657.

Cervero, R. 2013. Bus Rapid Transit (BRT): An efficient and competitive mode of public transport. Institute of Urban and Regional Development, the University of California at Berkeley, Berkeley, CA [Online]. Available: https://escholarship.org/uc/item/4sn2f5wc [Accessed on February 14, 2021].

Cervero, R., and Kockelman, K.M. 1997. Travel demand and the 3Ds: Density, diversity, and design. *Transportation Research Part D: Transport and Environment* 2(3), 199–219.

Chan, N.D., and Shaheen, S.A. 2012. Ridesharing in North America: Past, present, and future. *Transport Reviews* 32(1), 93–112.

Cracknell, J.A. 2000. *Experience in Urban Traffic Management and Demand Management in Developing Countries*. The World Bank, Washington, DC.

Duddu, V.R., and Pulugurtha, S.S. 2015. Assessing the effect of compressed work week strategy on transportation network performance measures. *Journal of the Transportation Research Forum* 54(2), 57–70.

Durand, A., Harms, L., Hoogendoorn-Lanser, S., and Zijlstra, T. 2018. *Mobility-as-a-Service and Changes in Travel Preferences and Travel Behaviour: A Literature Review*. KiM Netherlands Institute for Transport Policy Analysis. The Hague, Netherlands.

ECA. 2020. Sustainable Urban Mobility in the EU, Special Report 06-2020. European Court of Auditors, Luxembourg, Luxembourg.

Ewing, R., and Cervero, R. 2001. Travel and the built environment: A synthesis. *TRB Journal of Transportation Research Record* 1780(1), 87–114.

Ewing, R., and Cervero, R. 2010. Travel and the built environment: A meta-analysis. *Journal of the American Planning Association* 76(3), 265–294.

Ewing, R., Meakins, G., Bjarnson, G., and Hilton, H. 2011. Transportation and land use. In *Book: Making Healthy Places*. Editors: Dannenberg, A.L., Frumkin, H., and Jackson, R.J. Island Press, Washington, DC, 149–169.

Farinloye, T., Mogaji, E., Aririguzoh, S., Kieu, T.A. 2019. Qualitatively exploring the effect of change in the residential environment on travel behaviour, *Travel Behaviour and Society* 17, 26–35.
Feng, J., Dijst, M., Wissink, B., and Prillwitz, J. 2017. Changing travel behaviour in urban China: Evidence from Nanjing 2008–2011. *Transport Policy*, 53, 1–10.
Ferguson, E. 2018. *Travel Demand Management and Public Policy*. Routledge, Abingdon, UK.
FHWA. 2011. Integrating Active Traffic and Travel Demand Management: A Holistic Approach to Congestion Management. Report No. FHWA–PL–11–001. Federal Highway Administration, U.S. Department of Transportation, Washington, DC, USA.
Frank, L.D., and Engelke, P. 2005. Multiple impacts of the built environment on public health: Walkable places and the exposure to air pollution. *International Regional Science Review* 28(-2), 193–216.
Gärling, T., and Schuitema, G. 2007. Travel demand management targeting reduced private car use: effectiveness, public acceptability and political feasibility. *Journal of Social Issues* 63(1), 139–153.
Gärling, T., Eek, D., Loukopoulos, P., Fujii, S., Johansson–Stenman, O., Kitamura, R., Pendyala, R., and Vilhelmson, B. 2002. A conceptual analysis of the impact of travel demand management on private car use. *Transport Policy* 9(1), 59–70.
Hawkins, J., and Habib, K.N. 2019. Integrated models of land use and transportation for the autonomous vehicle revolution. *Transport Reviews* 39(1), 66–83.
He, S.Y., and Thøgersen, J. 2017. The impact of attitudes and perceptions on travel mode choice and car ownership in a Chinese megacity: The case of Guangzhou. *Research in Transportation Economics* 62, 57–67.
Kasipillai, J., and Chan, P. 2008. Travel demand management: Lessons for Malaysia. *Journal of Public Transportation* 11(3), 41–55.
Kim, J., and Brownstone, D. 2013. The impact of residential density on vehicle usage and fuel consumption: Evidence from national samples. *Energy Economics* 40, 196–206.
Kockelman, K.M. 1997. Travel behavior as function of accessibility, land use mixing, and land use balance: Evidence from San Francisco Bay Area. *TRB Journal of Transportation Research Record* 1607(1), 116–125.
Kuiper, D. 2015. Providing access to data to create better travel information in Reading (UK). *The European Local Transport Information Service (ElTiS)*, Brussels, Belgium, EU [Online]. Available: https://www.eltis.org/discover/case-studies/providing-access-data-create-better-travel-information-reading-uk [Accessed on: March 21, 2021].
Lanzini, P., and Khan, S.A. 2017. Shedding light on the psychological and behavioral determinants of travel mode choice: A meta-analysis. *Transportation Research Part F: Traffic Psychology and Behaviour* 48, 13–27.
Ling, S. 2020. LTA to reduce ERP rates at most gantries in the Island, Starting from 6 April. The Online Citizen, Singapore [Online]. Available: https://www.theonlinecitizen.com/2020/03/31/lta-to-reduce-erp-rates-at-most-gantries-in-the-island-starting-from-6-april/ [Accessed on February 10, 2021].
Maduwanthi, M., Marasinghe, A., Rajapakse, R.P.C.J., and Dharmawansa, A.D. 2015. Factors influencing to travel behavior on transport mode choice– A case of Colombo metropolitan area in Sri Lanka. *International Journal of Affective Engineering* 15(2), 63–72. doi: 10.5057/ijae.IJAE-D-15-00044.
Müller, J. 2020. Number of private hire cars in Singapore 2012–2019. Statista, Hamburg, Germany [Online]. Available: https://www.statista.com/statistics/953848/number-of-private-hire-cars-singapore/#:~:text=In%202019%2C%20there%20were%20around,ride%2Dhailing%20market%20in%20Singapore [Accessed on March 21, 2021].
Newman, P., and Kenworthy, J. 1989. *Cities and Automobile Dependence: An International Source Book*. Gower Publishing, Brookfield, VT.
Newman, P., and Kenworthy, J. 1999. *Sustainability and Cities: Overcoming Automobile Dependence*. Island Press, Washington, DC.
Palliyani, S., and Lee, D-H. 2017. Sustainable transport policy- An evaluation of Singapore's past, present and future. *Journal of Infrastructure, Policy and Development* 1(1), 112–128.

Poole, R.W. 2020. The impact of HOV and HOT lanes on congestion in the United States, OECD International Transport Forum Discussion Paper. International Transport Forum, Paris, France.

RBC. 2021. Travel by bus. Reading Borough Council (RBC), Reading, Berkshire, UK [Online]. Available: http://www.reading-travelinfo.co.uk/bus.aspx [Accessed on February 10, 2021].

Ricci, A. Gaggi, S., Enei, R., Tomassini, M., and Fioretto M. (ISINNOVA), Gargani, F., Di Stefano, A., and Gaspari, E. (PwC) with contributions from Archer, G., Kearns, S., McDonald, M., Nussio, F., Trapuzzano, A., and Tretvik, T. 2017. Study on Urban Vehicle Access Regulations, Final Report. Directorate-General for Mobility and Transport, European Commission, Brussels, Belgium, EU.

Sammer, G. 2016. *Travel Demand Management and Road User Pricing: Success, Failure and Feasibility*. Taylor & Francis Group, Boca Raton, FL.

Shoup, D. 2017. *The High Cost of Free Parking*, Routledge, New York, NY, USA.

Shutterstock, S.-F. 2014. Stockholm, Sweden - July 31: Gamla Stan, the old part of Stockholm, Sweden on July 31, 2014. Shutterstock, New York [Online]. Available: https://www.shutterstock.com/image-photo/stockholm-sweden-july-31-gamla-stan-233934952 [Accessed on February 10, 2021].

Su, Q., and Wang, D.Z. 2020. On the commute travel pattern with compressed work schedule. *Transportation Research Part A: Policy and Practice* 136, 334–356.

Sumit, B., Anna, G., Methody, G., Estel, M., Mihaela, M., Zion, S., Josh, T., Tamara, P., Carrie, S., and Dagmawit, W. 2020. Country-level commuting mode share, Wonder, May 15th, 2020.

Thorhauge, M., Cherchi, E., and Rich, J. 2016. How flexible is flexible? Accounting for the effect of rescheduling possibilities in choice of departure time for work trips. *Transportation Research Part A: Policy and Practice* 86, 177–193.

TTI. 2021. Managed (HOV-HOT) Lanes, Technical Strategies Series 2021. Texas Transportation Institute, Texas A & M University, College Station, TX, USA.

Turnbull, K.F., Higgins, L., Puckett, D., and Lewis, C. 1995. Potential of Telecommuting for Travel Demand Management, Research Report 1446-1. Texas Transportation Institute, Texas A & M University, College Station, TX, USA.

Twisse, F. 2019. Stockholm: Achieving sustainable mobility using urban vehicle access regulations. The European Local Transport Information Service (ElTiS), Brussels, Belgium, EU [Online]. Available: https://www.eltis.org/discover/case-studies/stockholm-achieving-sustainable-mobility-using-urban-vehicle-access [Accessed on: March 21, 2021].

USCB. 2020. 2019 American Community Survey, U.S. Census Bureau, U.S. Department of Commerce, Washington, D.C., USA.

Van Audenhove, F-J., Korn, A., Steylemans, N., Smith, A., Rominger, G., Bettati, A., Zintel, M., and Haon, S. 2018. The Future of Mobility 3.0, Arthur D. Little Web Report. [Online]. Available: https://www.adlittle.com/en/insights/viewpoints/future-mobility-30 [Accessed on: March 21, 2021]

Van de Coevering, P., and Schwanen, T. 2006. Re-evaluating the impact of urban form on travel patterns in Europe and North America. *Transport Policy* 13(3), 229–239.

Verplanken, B., Aarts, H., and Van Knippenberg, A. 1997. Habit, information acquisition, and the process of making travel mode choices. *European Journal of Social Psychology* 27(5), 539–560.

VTPI. 2017. Why Manage Transportation Demand? Victoria Transport Policy Institute, Victoria, BC, Canada. [Online]. Available: https://www.vtpi.org/tdm/tdm51.htm [Accessed on March 21, 2021]

VTPI. 2019. Commute Trip Reduction (CTR): Programs That Encourage Employees to Use Efficient Commute Options, *TDM Encyclopaedia*. Victoria Transport Policy Institute, Victoria, BC, Canada.

Yang, H., and Huang, H.J. 1999. Carpooling and congestion pricing in a multilane highway with high-occupancy-vehicle lanes. *Transportation Research Part A: Policy and Practice* 33(2), 139–155.

Yuan, L.L. 2018. A Case Study on Urban Transportation Development and Management in Singapore. Working Paper. National University of Singapore, Singapore. [Online]. Available: https://www.

scribd.com/document/457492232/A-Case-Study-On-Urban-Transportation-Development-And-Management-In-Singapore-pdf [Accessed on March 21, 2021]
Zhang, M. 2004. The role of land use in travel mode choice: Evidence from Boston and Hong Kong. *Journal of the American Planning Association* 70(3), 344–360.
Zong, F., Juan, Z., and Jia, H. 2013. Examination of staggered shifts impacts on travel behavior: A case study of Beijing, China. *Transport* 28(2), 175–185.

Chapter 4

Building out 3D highway transportation with flexible capacity

4.1 GENERAL

Most nations are experiencing rapid growth in auto travel and on-road goods movements, and even where transit and other non-auto modes remain dominant, growth in population and/or wealth drives increased needs for infrastructure and facilities to support mobility. In megacities expanding transportation capacity by adding more travel lanes or routes to the road network, parking spaces, expanding rail transit facilities, bus routes, stops and stations, ridesharing curb spaces, bike and microtransit lanes and parking areas, and even better sidewalks and road crossings for pedestrians becomes ever more difficult or even infeasible owing to prohibitively high costs of facility delivery and scarcity of usable land spaces. On the other hand, the viewpoint of capacity expansion being equivalent to widening or adding to existing facilities or building new ones in crowded spaces has essentially overlooked the fact that a transportation network is 3D in nature where solutions of capacity expansion could be sought perpendicularly to the 2D plane of surface streets either going down to build tunnels or moving up to construct elevated structures (Elbert et al., 2020). The related capacity expansion solutions could achieve enhanced efficiency and effectiveness in mobility improvements when they are deployed in conjunction with demand management measures and use mechanisms such as traveler information systems and pricing in response to dynamically varying travel demand with due cognizance of trip purpose and attributes affecting choices of trip destination location, destination arrival time, travel mode, departure time, travel route, and facility or travel lane.

To be sure, elevated or underground transportation facilities are often more expensive than surface level one, though the more densely developed the megacity and the more precious surface land is, the more 3D makes sense. At the same time, technological advancements are making tunnels and elevated structures more cost-effective (Samuel, 1999; Efron and Read, 2012; NRC, 2013; Michaels, 2016). Thus, while capacity expansions are often necessary to keep pace with economic growth, ensure connectivity, and improve mobility, they are also much more feasible given the state of technologies needed to build, manage, and finance these new facilities (see details in Chapter 7).

Given the size of the topic, however, this chapter focuses primarily on highway capacity expansion that facilitates automobiles and freight vehicular traffic. Chapter 5 discusses mass transit and multimodal capacity expansion in more depth. To some extent, this binary distinction between transit and vehicular traffic is arbitrary and simplistic. Megacities will need to consider and accommodate all modes of passenger travel and freight shipments, finding the right balance for particular cities, regions, and stages of development. Nevertheless, rather than paring down the discussion, the material has been spread out over two chapters. However, the main takeaway is that public officials, planners, and decision makers in megacities should focus on comprehensive and integrated multimodal transportation systems and networks.

DOI: 10.1201/9780429345432-4

4.1.1 3D spiderweb transportation networks

Travel demand in terms of person and goods movement trips generated in an urban/metropolitan area or city center consists of four types, with origin-destination locations of individual trips anchored external-internal, internal-external, internal-internal, and external-external to the area. Taking passenger trips for example, each type of trips may be split by trip purpose of home-based work (HBW), home-based other (HBO), and non-home-based (NHB) trips. The individual trips exhibit temporal and spatial, length, and travel time dynamics. To accommodate trip dynamics, different types of transportation facilities, travel modes, and vehicles are supplied to achieve safe and efficient movements. Higher functional classifications of highways such as Interstates and freeways are designed to accommodate uninterrupted traffic flows, which are appropriate to handle all types of trips with relatively longer distances. Commuter rails are well suited for HBW trips between suburbs and the central business district. Freeways and freight rails are appropriate for good shipments. Arterials and collectors in an urban street network are designed for accommodating interrupted traffic flows which are highly correlated with internal–internal trips with comparatively short travel distances, which also accommodate shorter bus transit, bike, and microtransit trips.

The total capacity of a transportation system is governed by capacities of subsystems that accommodate different types of traffic flows—uninterrupted and interrupted—associated with various travel modes with flow intensities dynamically varying by time and location. To ensure adequate use of potentially available total capacity, the capacity of each subsystem needs to be used efficiently and meanwhile the capacity utilization of subsystems needs to be well coordinated. Taking the highway system for example, when the capacity of the freeway system becomes insufficient in serving for the peak period traffic, the capacities of expressway and arterial systems could accommodate a portion of traffic. However, expressways and arterials are designed mainly for serving interrupted flows. To maximize the capacity utilization of expressway and arterial systems, the number of traffic movement conflicts and duration of delays to individual vehicles using at-grade intersections need to be minimized. Possible solutions are to adopt measures for more effective traffic control for signalized intersections, convert designs of at-grade intersections to unconventional at-grade intersections and unconventional overpasses or interchanges, and implement pricing schemes for some facilities. Inherent in this already in megacities is some extent of a 3D configuration of the transportation network blended by limited access highways, expressways, arterials, collectors, and local roads carrying private vehicles and bus fleets and partly overlying each other in places, and also with fixed guideway transit, with light rail tending to be elevated while metros/subways largely go underground, as well as bike, pedestrian, and microtransit facilities with over and underpasses, tunnels, and the like. Therefore, the multimodal transportation system is essentially an incomplete interconnected 3D spiderweb system with subsystems complementary to each other to serve for uninterrupted and interrupted traffic flows generated by passenger and freight demand.

4.1.2 Multimodal integration

In practice, multimodal transportation networks have long meant facilities serving personal auto, bus, and rail transit travel. These networks are expected to accommodate and facilitate the expanded use of walking and bicycling. Now they are expected to adjust to include ridesharing and microtransit. An extra layer of complexity but also opportunity comes with adding goods movements to the multimodal mix.

For a trip originated at a specific location, its occurrence involves a series of choices concerning destination location, arrival time, travel mode, departure time, route, and facility/travel lane. Central to sustaining mobility is to reach an efficient balance between vehicle volumes from personal trips and goods shipments and available capacity of a transportation system by time and location so that mobility performance of the entire system is maintained at the highest possible level. The first tier of multimodal integration is the integration between auto and transit modes, possibly including other modes concerned with private vehicles, such as carpool, vanpool, and other ridesharing modes. The second tier of multimodal integration aims to integrate transit sub-modes, including bus, bus rapid transit, and fixed guideway transit. The two-tier integrations begin from integrated land use and multimodal transportation planning to integrated planning, design, and operations of auto and transit modes, including transit sub-modes, and possibly other modes involved with the use of private vehicles (Rondinelli and Berry, 2000). The third tier of multimodal integration adds microtransit and emerging technologies and perhaps trucks, freight rail, and cargo drones for goods movements to some extent. It is easy to see how 3D thinking can be applied to multimodal facilities to help integration, with underground transit, parking, and perhaps pedestrian and goods movement flows; ground-level movement of all types, and elevated pedestrian, bike, scooter, and microtransit; parking structures; and even elevated roads or transit, etc. You could easily invert or mix and match this example—thinking 3D dramatically expands what is possible.

4.2 CONVENTIONAL OPTIONS OF HIGHWAY CAPACITY EXPANSION

The conventional options of highway capacity expansion mainly include adding travel lanes, building new highways, widening a portion of an existing roadway, and improving designs of highway crossings. These types of improvements are implemented by public agencies or via public-private partnerships (Adams, 2001).

4.2.1 Adding new travel lanes or building new roads

Adding new lanes to an existing road or building new roads can potentially reduce congestion and provide alternate routes for highway users. In general, these capacity expansion options require significant investments and involve extensive construction durations, which are commonly applied to reduce severe recurrent congestion along major highway corridors. They are most practical for deployments in urban corridors without the needs for acquisition of additional right-of-way and jointly with other mobility improvement strategies that manage traffic and provide alternative travel options by considering the current and future demand levels and traffic diversion effects after adding new travel lanes or building new roadway facilities.

4.2.2 Roadway widening

The availability of travel lanes of a roadway segment influences traffic mobility and safety. In some cases, it may be feasible to add capacity within existing right-of-way by expanding the road footprint, or through creative reconfiguring of space within the existing footprint (DeCorla-Souza, 2009). There are also bottleneck situations for just a short distance where a geometric design improvement by adding auxiliary lanes could potentially help mitigate these constraints. The auxiliary lanes can be used for speed changing, turning, weaving, truck climbing, maneuvering of entering and exiting vehicles, and other purposes supplementary to through traffic movements, which will balance the traffic load and maintain a more uniform roadway level of service.

4.2.3 Grade separation improvements

In some cases, grade separations designed between freeways and other principal arterials that accommodate high-traffic volumes would limit traffic diverging and converging between them. Improving designs of grade separations to system interchanges offer enhanced connectivity and accessibility of the overall roadway system in handling high-volume uninterrupted flow. The primary constraints of grade separation improvements are associated with available budget and right-of-way for construction.

4.2.4 Case in action: U.S. Interstate 2.0

With just 2.5% of the nation's lane miles of highways, the U.S. Interstate highway system carries 25% of total vehicle miles of travel, which plays a vital role in serving the country to travel, trade, and economic development. Moreover, Interstate highways provide important crosstown links in virtually all large U.S. cities. All pavements of the first generation of the Interstate highways built in the 1960s exceeded their design useful service lives about a decade ago. Reconstruction of Interstate pavements has begun back then and will continue toward the end of the next decade. Meanwhile, the continuing population increase and economic prosperity have led to steady growth in travel demand and traffic volumes. This imposes new challenges to pavement reconstruction for the initial Interstate highways by adding more travel lanes and redesigning interchanges along some Interstate corridors to ensure achieving a desirable level of mobility. Massive investments are needed for reconstruction and expansion of the first generation of the U.S. Interstate highway system to the new system called the U.S. Interstate 2.0 (Poole, 2013, 2018).

Historically, the primary funding source for the U.S. highway system is based on fuel taxes. Steady increases in vehicle fuel efficiency, combined with a lack of a rational approach for adjusting fuel tax rates over time, opposition by highway users, and a paucity of political support, have resulted in underfunding the network. The current highway finance system does not generate sufficient financial resources to cover the expenses of pavement maintenance and repair treatments, let alone the investments needed for reconstruction or system expansion and improvements.

To slow down and fill in the ever-increasing funding gaps, the transportation community—public policy, engineering, transportation planners, and other professionals and stakeholders—generally agrees to phase out fuel taxes. Fuel taxes are unsustainable revenue sources as transportation vehicles and fleets adopt technologies less reliant on liquid-based fossil fuels such as oil. This shift is likely to accelerate under policies designed to limit and reduce carbon dioxide emissions to meet climate change goals. Instead, these carbon-based taxes should be replaced with a hybrid of mileage- and weight-based electronic tolls dynamically adjusted by time and location of travel. These tolls would be set to fully recover the agency costs of construction, maintenance, and repair treatment in highway facility service life cycle (Li, 2018). They would also cover user costs of marginal over average user costs imposed by additional travel to ensure efficient utilization of the capacity-expanded Interstate system. Phased field deployments of the technology are preferred to ensure smooth transitions from fuel taxes and electronic tolls. A phased implementation would also further confirm the cost recovery potential for options of tolling added capacity, testing the effects of variable toll rates for lanes with and without this pricing technology, and how rates affect travel on existing and added travel lanes.

4.3 ELEVATED FREEWAYS

A freeway is a controlled-access highway with the primary function of accommodating uninterrupted mainline traffic flows with vehicle entries and exits handled by distantly spaced interchanges along its alignment. Intuitively, an elevated freeway is a freeway that is raised above grade as a single long pier bridge for its entire length. Elevated freeways are more expensive to build than already costly at-grade freeways and are only used when it would be infeasible for at-grade construction of conventional freeways owing to land scarcity, budget, or time constraints to eliminate impeding structures, blockage of existing crossings, hilly terrain, pedestrian, or wildlife conflict avoidance, and environmental impacts mitigation (Kutz, 2004; Wolshon and Pande, 2016).

4.3.1 Elevated crosstown expressways

Elevated crosstown expressways are essentially elevated freeways deployed as crosstown routes or as the center city "ring road" in an urban/metropolitan area. Examples include Shanghai's extensive elevated expressway network, the Metro Manila Skyway System, the Hanshin Expressway in Osaka/Kobe/Kyoto, and elevated expressways in a number of U.S. cities including Chicago, Los Angeles, Philadelphia, and Tampa.

4.3.2 Case in action: Tampa Bay crosstown expressway

The Tampa Bay Crosstown Expressway System in Florida was planned during 1950s–1970s. Initially, it included components of north, south, and west crosstown expressways throughout the Tampa area and only the southern component was built due to financial problems, land acquisition, and community revolts. The Southern Crosstown Expressway, renamed Lee Roy Selmon Crosstown Expressway in 1999 in honor of former Tampa Bay Buccaneers hall-of-fame football player Lee Roy Selmon, is a 14.168-mile (22.801-km) all-electronic, limited access toll road in Hillsborough County, Florida, built in stages opening between 1976 and 1987, and in 2006. It features the world's first reversible, all-electronic, and elevated express lanes which operate westbound from 6:00 am to 10:00 am, westbound/eastbound transitional operation from 10:00 am to 3:00 pm, and eastbound operation from 3:00 pm to 6:00 am the next morning. The expressway uses open-road tolling via SunPass transponders with Toll-By-Plate license plate recognition and billing for travelers without SunPass transponders, which mails toll bills to the vehicle owners without accounts with penalties being imposed after 30 days. At present, field experiments are underway for placements of automated vehicles and bus toll lanes along the expressway.

The design and construction of the elevated tolled lanes are noteworthy. The Tampa-Hillsborough County Expressway Authority (THCEA) built the elevated lanes above the median of their Crosstown Expressway toll road, a route developed for very congested peak direction travel in the morning and evening peak hours, optimal conditions for a reversible facility. As seen in Figure 4.1, the travel lanes were built with an innovative design and construction method using a very small footprint of 2 m in the median of the existing road to support significant new capacity—18 m wide with three lanes plus a breakdown lane on each side. The precast segments were built off-site and then put in place in sequence (Poole, 2006).

Figure 4.1 Illustration of Lee Roy Selmon expressway flyovers in Tampa, Florida. (Source: THEA (2016).)

4.4 TRANSPORTATION TUNNELS

A transportation tunnel is a horizontal underground passage that is used to avoid an obstruction to provide a more direct route for passenger travel and freight movements with reduced travel time. It primarily consists of the tunnel structure and portals, lighting system, ventilation system, traffic operations control system, and safety and security system. Transportation tunnels are generally classified by the cross-section shape, lining, invert type, and construction method (Li et al., 2008). Historically, they have been used extensively in large cities for light and heavy rail subways and to a lesser extent for highways and other travel modes. In built-up megacities, the main obstruction is existing surface development, where tunnels provide a means to expand capacity for all modes as density increases. As more megacities emerge and demands for greater mobility increase, tunnels are an ever more important part of the 3D spiderweb urban transportation networks (Figure 4.2 and Table 4.1).

4.4.1 Importance of tunnels

In urban areas where open land has generally been exhausted, construction of road tunnels has been increasingly considered as an option to expand mobility; cope with ever-growing traffic to mitigate congestion that has been a major challenge to transportation planners, engineers, and decision makers; and increase travel mode options (TRB, 2010, 2016; FHWA, 2018). The National Research Council in the United States in 2013 published a book which provides a comprehensive analysis of the role of underground facilities for sustainable urban development, pointing out that limiting our city infrastructure to the surface alone will force many less sustainable options to emerge (NRC, 2013). Indeed, the USDOT states:

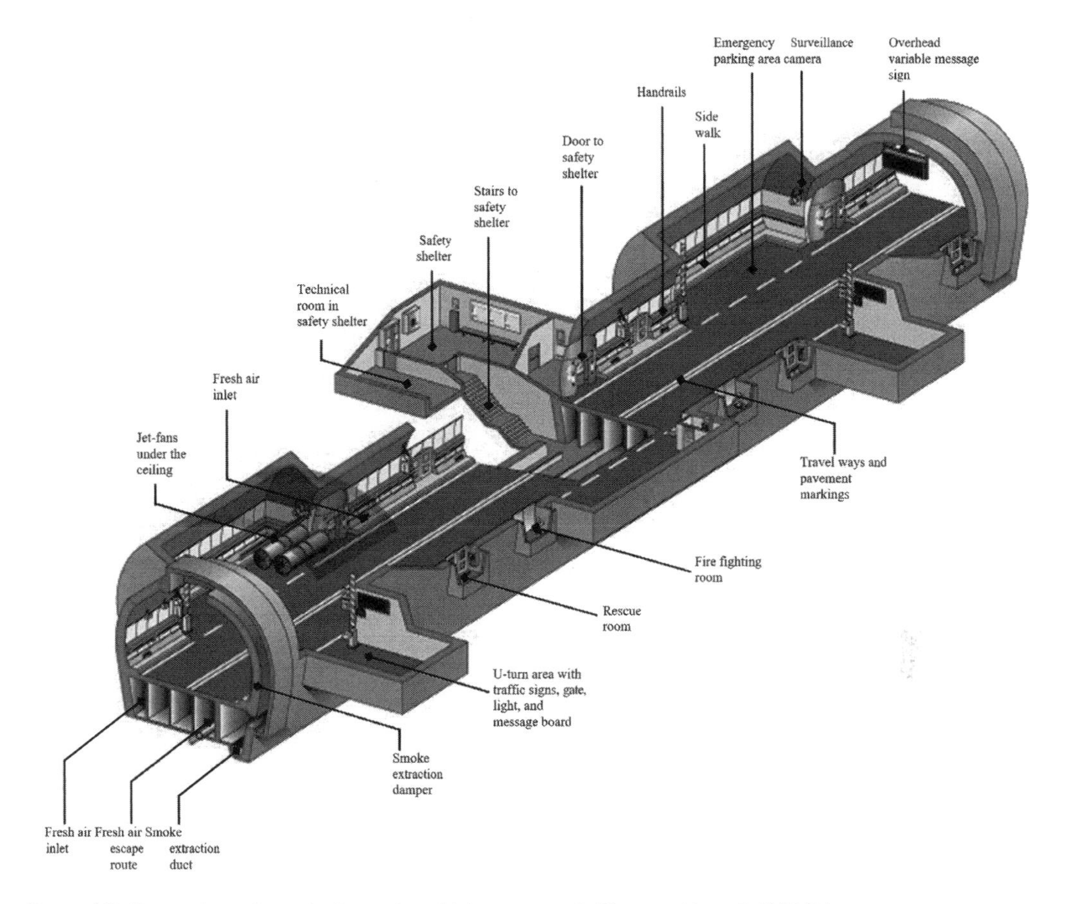

Figure 4.2 Illustration of a typical two-lane highway tunnel. (Source: Li et al. (2008).)

Table 4.1 Summary of ten longest highway tunnels in the world

Country	*Tunnel*	*Length (mile)*	*Width (ft)*	*Lanes*	*Cost (million USD)*	*Construction period*
Norway	Laerdal	15.24	32.8	2	114	1995–2000
China	Zhongnanshan (two tubes)	11.22	35.8	2/tube	410	2002–2007
Switzerland	St. Gotthard	10.52	25.6	2	626	1969–1980
Austria	Arlberg	8.69	45.9	4	21.2	1974–1978
Taiwan, China	Hsuehshan (two tubes)	8.05	24.6	2/tube	580	1993–2006
France—Italy	Fréjus	8.02	32.8	2	443	1974–1980
France—Italy	Mont-Blanc	7.22	28.2	2	58	1957–1965
Norway	Gudvanga	7.11	19.7	2	69	Opened in 1991
Norway	Folgefonn	6.93	31.2	2	89	1997–2001
Japan	Kanetsu (north)	6.80	23.0	2	No info	1980–1985
	Kanetsu (south)	6.88	23.0	2	No info	1986–1991

Source: Li et al. (2008).

Tunnel structures offer significant economic, social, and environmental advantages over the other more commonly employed highway structure alternatives; this is because the land above the transportation corridors remain available for economic development and community or recreational use; and when appropriately considered, the benefits from building a tunnel might outweigh the costs saved by building cheaper alternative highway structures. For example, when neighborhoods are not separated by long liner transportation arteries, local communities remain intact while heavy, and sometimes hazardous, goods vehicles (trucks) flow below ground where this traffic is grade separated from local roadways and sidewalks which greatly improves safety for those living and working in the area. Furthermore, when viaducts are replaced with tunnels, and when depressed roadways are capped to form a tunnel, long linear parks and recreational trails can be developed for creating green space, which will generate additional value for nearby properties, thus potentially resulting in increased revenue streams for local municipalities while sales taxes will also increase as commerce better thrives due to the near elimination of unpleasant loud noises and unhealthy vehicle emissions. This sentiment is echoed by researchers:

In the coming years, the world's population will increase in the urban areas. So, the urban centers will have to adapt, in order to guarantee that their future population will have the necessary and sustainable growth. Due to the constraints for surface construction, also connected to environmental issues, this population growth will imply a greater use of the underground. With this optimal growth, the population of the cities will have better mobility, quality of life, and economic and social sustainability (Tender et al., 2017). And by tunnel entrepreneurs like Elon Musk:

If you think of tunnels going 10, 20, 30 layers deep (or more), it is obvious that going 3D down will encompass the needs of any city's transport of arbitrary size. [I]t isn't possible for everyone to be using a "2D road network" but if you want to create more space, you have to go up or down—with down being [the] preferred option (Heathman and Simon-Lewis, 2017).

4.4.2 Feasibility of tunneling

The context of whether tunneling is appropriate depends on physical feasibility, environmental impacts, and financial affordability, weighed against anticipated benefits to mobility and economic activity.

4.4.2.1 Physical feasibility

One of the primary purposes of physical feasibility assessment is to evaluate the viability and suitability of constructing highway tunnels in dense urban areas based on engineering and construction practices. The assessment mainly considers geologic and geotechnical, hydrological, and seismic conditions and the ability of the tunnel option to satisfy traffic demand, geological and geotechnical conditions, highway geometric design standards, and construction methods. Advances in precision guidance, vibration monitoring, and chemical softening of rock and soil all allow tunneling where it was previously not feasible—"Miami recently dug a traffic tunnel beneath a busy waterway by eating through a mix of saturated ground and porous coral rock that was previously defied affordable excavation" (Michaels, 2016) (Figure 4.3).

4.4.2.2 Environmental impacts

The focus of environmental impacts assessment is to identify and address the potential tunnel issues and impacts associated with construction and operation of a highway tunnel in

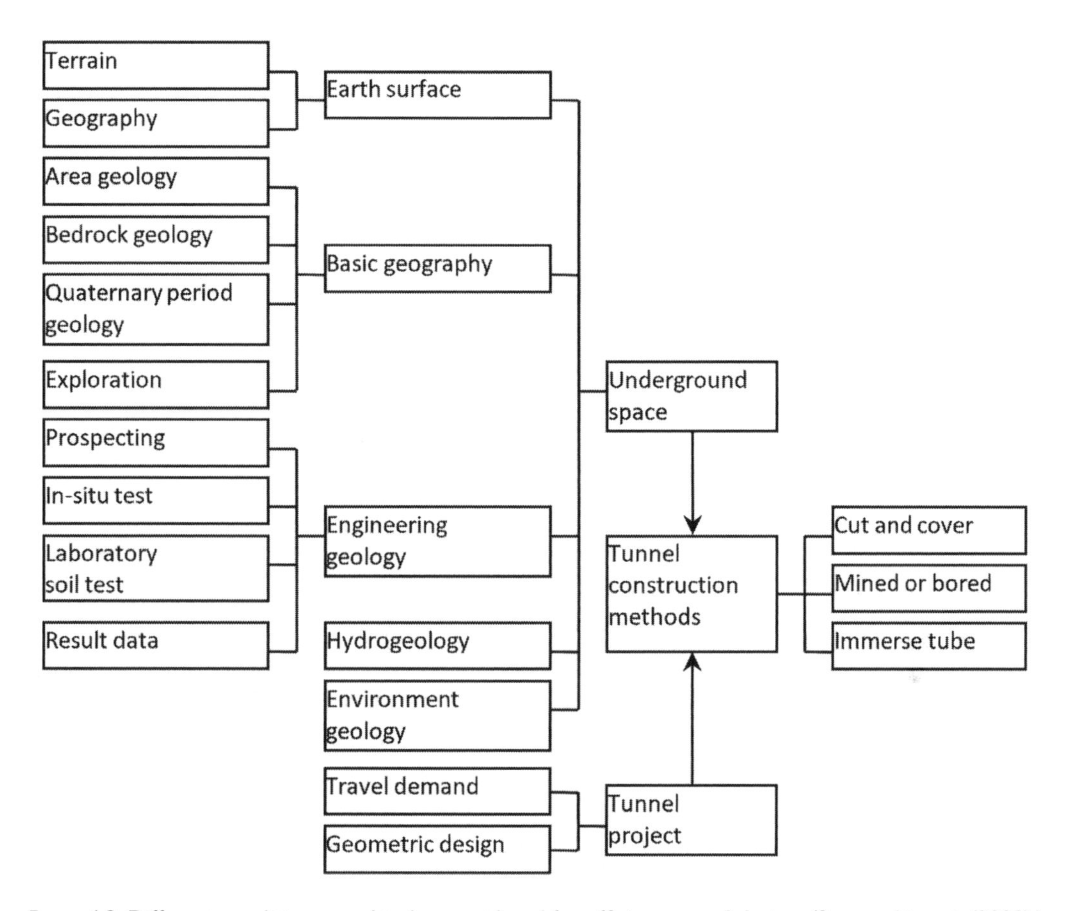

Figure 4.3 Different conditions need to be considered for efficient tunnel design. (Source: Li et al. (2008).)

aspects of noise, air quality, historic properties, esthetics, archaeology, hazardous waste, soil disposal, and stormwater. These assessment contents are consistent with those considered for surface or elevated highway construction projects. However, the assessment for surface or elevated highway construction needs to consider additional impacts on the ecological environment (vegetation, soil erosion, surface water quality, and wild animals) and social economy caused by the project (Table 4.2).

4.4.2.3 Financial feasibility

Admittedly, tunneling is more expensive than surface construction in open terrain, but growing cities no longer have the latter, so tunnels are really the alternative to no growing infrastructure at all. Institutional settings do matter for the costs of tunnel infrastructure. One analysis points out that tunnel costs (in U.S. dollars) vary widely by global region (Munfah and Nicholas, 2020):

- *New York*: US$1.5–US$2.5 billion per mile
- *Other parts of United States and Australia*: US$600–US$900 million per mile
- *Europe, Middle East*: US$250–US$500 million per mile
- *India, China, and Southeast Asia*: US$100–US$200 million per mile

Table 4.2 Tunnel-induced environmental issues and mitigation measures

Environmental condition	*Existing condition*	*Anticipated future condition*	*Possible mitigation measure*
Land use planning	Include local and regional plans in corridor study	Change and redevelopment of community space	Avoid unplanned development and have local support to optimize potential land use benefits
Topography, geomorphology, geology, and soils	Topography affects landscape and reflects subsurface conditions	Possible settlement, erosion, and contaminated soils	Address each problem in construction management
Groundwater quality	Depth and extent of existing aquifers	Effects of groundwater-level fluctuation due to dewatering and long-term drawdown	Monitor level and quality of groundwater
Surface water quality	Waterways and wetland along the corridor	Increase surface run-off, seepage, and spillage	Provide new drainage systems
Air quality	Ambient air quality	Impact due to tunnel construction and operation, dust, and emissions	Provide sufficient ventilation and air quality control systems
Noise and vibration	Noise monitoring prior to construction	Activities associated with construction	Significant reduction after construction
Social conditions	Mostly urban dwellers with diversities	Residential and commercial facilities may be resettled	Urban regeneration with support from local entities
Flora, fauna, cultural heritage, landmark	Floras and fauna endangered species, historic heritage	May need to be removed	Temporary displacement and replacement or reinforcement
Economic analysis	Impact on business along the corridor	Effects of temporary closure due to construction activities	Perform tunnel benefit–cost analysis
Hazard and risk	Nearby communities, use of sensitive land for spoil placement, and users	Hazards due to construction, including operation of vehicles and equipment, material delivery	Implement traffic management strategies
Waste management	No waste from tunnel prior to construction	Potential to generate liquid and solid wastes during construction	Action for waste minimization and management

Source: Li et al. (2008).

In Table 4.3, that same analysis points out many cost drivers for tunnel projects and how they align with total costs in each region. A study by Booz Allen Hamilton (BAH) for the Reason Foundation that surveyed 19 recent tunnel projects throughout the world revealed that the potential costs of tunneling vary from US$643 million per lane-mile (in Hong Kong) to US$17 million per lane-mile (in Shanghai). The average cost per lane-mile was US$145 million per lane-mile, and the median was US$93 million per lane-mile. Other research finds similar results and provides analysis of ways to manage the costs of tunnel projects (Efron and Read, 2012; Benardos et al., 2013). Table 4.4 exhibits information on the costs of some large urban tunnels.

Compared with typical highway projects, road tunnels are subject to a much larger range of cost variations and much higher average costs. At the same time, innovation in the tunneling industry is driving down costs rapidly. Elon Musk's Boring Company was created explicitly to seek ways to dramatically reduce tunneling costs and expand their use for urban mobility (Heathman and Simon-Lewis, 2017). That would build on a trend, as

Table 4.3 Comparative influence of tunnel cost drivers by region

Cost drivers by region	*New York*	*Other U.S. regions*	*Europe*	*China, India, and Southeast Asia*
Geology and geotechnical challenges	H	H	H	M
Labor cost and requirements	VH	H	M	VL
Materials and plant	H	H	H	M
Construction risk and contingencies	VH	H	M	L
Soft costs	VH	H	M	L
Safety and environmental issues	VH	VH	M	L
Market structure and situation	VH	H	M	L
Government approvals process and permitting	VH	VH	M	L
Stakeholders and community issues	VH	H	M	VL
Project procurement type	VH	H	M	L
Client sophistication and knowledge in underground construction	H	H	M	M
Owner's risk sharing	VH	H	M	M
Labor and other laws	VH	H	H	L
Political influence on infrastructure projects	H	H	L	L

Source: Munfah and Nicholas (2020).

VH, very high; H, high; M, medium; L, low; and VL, very low.

Table 4.4 List of recent large urban tunnels in the world

City	*Tunnel*	*Tunneling methods*	*Length (mile)*	*Width (ft)*	*Lanes*	*Cost (US $ billion)*	*Completion*
Brisbane, Australia	North-South Bypass (two tubes)	Combination of cut-and-cover, roadheaders, and TBM	4.2	28.5	2/tube	1.6	2010
Chongqing, China	Light Rail Daping station (two tubes)	Cut-and-cover	0.9	65.6	2/tube	0.28	2007
Hong Kong, China	Kowloon Southern link	Mixshield TBM	1.4	28.2	2	1.8	2009
Kuala Lumpur, Malaysia	SMART-storm water SMART-highway (double-decker)	Slurry shield TBM	6.0 2.5	43.3	– 2/decked	0.52	2007
Madrid, Spain	Madrid Calle 30 project (two tubes)	Combination of cut-and-cover and earth pressure balanced TBM	34.8	49.9	3/tube	5.0	2007
Paris, France	A86 West project (double-decker)	Earth pressure balanced and slurry shield TBM	6.2	34.1	2/decked	2.3	2009
Shanghai, China	Changjiang under river tunnel project (two tubes)	Mixshield TBM	4.5	50.6	3/tube	1.5	2010

Source: Li et al. (2008).

boring machines have made huge progress—in one case in New York manually digging a 120-ft section of tunnel cost almost US$1 million per foot compared to US$19,000 per foot in nearby sections with a boring machine (Michaels, 2016). As part of financial assessment, alternative financial strategies using traditional and non-traditional funding sources to offset the cost disadvantages can be adopted to increase the economic feasibility of tunnel projects. The public and private sectors can be considered. Based on these revenue sources and the assumed level of contribution from each source, financial affordability can be determined.

4.4.3 Cases in action

4.4.3.1 Paris A86 West tunnels

As the final link of the 80 km A86 ring road around the greater Paris, the A86 West tunnels were built to relieve traffic congestion and improve traffic connectivity between the suburbs of Paris. At a cost of 2.23 billion Euros, the A86 West tunnels consist of two toll tunnels: the first one is an innovative 10 km long, double-deck tunnel featured with three lanes for light vehicles, and the second one is a 7.5 km long, single-deck tunnel featured with two lanes designed along more traditional lines to be used by all vehicles. The A86 West tunnels significantly reduce the travel time between Malmaison and Versailles from 45 to 10 minutes (Figure 4.4).

4.4.3.2 Sydney M5 East Freeway tunnels

The M5 East Freeway links the M5 Motorway at Beverly Hills with General Holmes Drive at Sydney Airport and on to the Eastern Distributor. It is 9 km long and includes twin 4-km tunnels and a 550-m tunnel under the Cooks River. The M5 East Freeway forms a part of Sydney's Orbital Network. It greatly improves access between southwestern Sydney, the city, and major industrial and commercial areas of South Sydney. It has significantly reduced traffic congestion in residential areas, improved traffic flow, and removed heavy vehicles from key local roads, with safety, noise, and air pollution benefits for local communities (Figure 4.5).

4.4.3.3 Istanbul Bosporus multimodal crossings

As Istanbul grows toward megacity status, integrating travel between the east and west sides of the Bosporus Strait grows from important to vital. Already the narrowest navigable strait for international shipping, the waterway is a crucial link between the Black Sea and Sea of Marmara (and Dardanelles and Aegean Sea). Tunnels will play an ever more important role in expanding mobility in the city without the surface land limitations and shipping interference of bridges and existing extensive ferry services, as seen in Figure 4.6.

Turkey has for decades been incorporating tunnels into infrastructure for crossing the Bosporus divide between Europe and Asia, with plans for more in the future (TunnelTalk, 2015). Table 4.2 shows how they added tunnels to the current and under construction bridge crossings, including the innovative Marmaray immersed tube railway crossing (opened in 2013) and two roadway tunnels totaling six lanes in each direction. Currently, planning and surveying work are underway for a three-deck multimodal tunnel with one deck for each direction of vehicle traffic and another for rail transit (Figure 4.7 and Table 4.5).

Figure 4.4 Illustration of the Paris A86 West tunnels on the A86 ring road, Paris, France.

4.4.3.4 Shanghai Yangtze River Tunnel–Bridge crossing

At a cost of US$1.8 billion, the 25.5 km long Shanghai Yangtze River Tunnel–Bridge crossing is comprised of 8.9 km of the world's largest diameter tunnel (15.43 m) from Shanghai Pudong to Changxing Island, 10.3 km of roadway in building the world's largest cable-stayed bridge (with a navigable hole separation of 730 m) from the Yangtze River estuary to Chongming Island, and 6.3 km of connecting roads.

Since opening to traffic on October 31, 2009, the travel time from the mainland of Shanghai to Changxing Island is reduced from about 30 to 10 minutes and from Changxing to Chongming Island is reduced from 1 hour to 20 minutes. The Shanghai Yangtze River Tunnel–Bridge crossing not only closely connects the mainland of Shanghai to Chongming Island, but it is stepping up north to connect with Qidong, Jiangsu Province, via a new bridge crossing the upper north channel of the Yangtze River bridge. This eventually allows

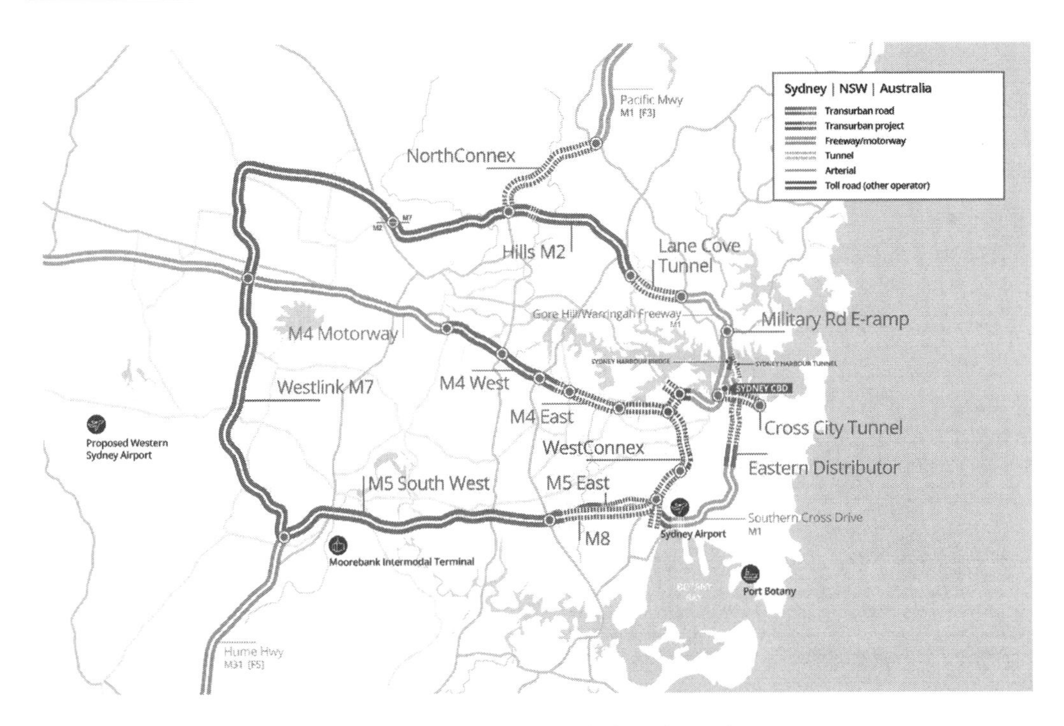

Figure 4.5 Illustration of the M5 East expressway tunnels, Sydney, Australia.

Figure 4.6 Ferries in Istanbul, Turkey. (Source: Shutterstock (2019) with permission.)

Figure 4.7 Concept for three-deck multimodal Bosporus tunnel in Istanbul, Turkey. (Source: TunnelTalk (2015).)

Table 4.5 Existing and planned fixed links across the Bosporus, Istanbul, Turkey

Link and construction date	*Design and construction company*	*Capacity*	*Investment (US$ in billion)*
First bridge (1970–1973)	Freeman Fox & Partners Enka Construction & Industry Co Cleveland Bridge & Engineering Co Hochtief AG	Four lanes per direction	0.20
Second bridge (1986–1988)	Freeman Fox & Partners IHI and Mitsubishi (Japan), Impregilo (Italy), and STFA (Turkey)	Four lanes per direction	0.13
Melen water tunnel by tunnel boring machine (TBM) (breakthrough 2012)	OAO Mosmetrostroy (Russia)	Bored tunnel	0.665
Marmaray immersed tube railway (2004–2013)	Parsons Brincherhoff (USA) lead Taisei (Japan)-Gama-Nurol (Turkey)	Teo rail tracks	4.00
Third road and rail bridge (2012–2016)	İçtaş (Turkey)-Astaldi (Italy) Concession Hyundai-SK E&C (Korea) Construction	Four lanes and one rail track per direction	2.50
Eurasia bored highway tunnel (2010–2016)	Parsons Brincherhoff (USA) lead Yepi Merkezi (Turkey)-SK E&C (Korea)	Two lanes per direction	1.245
Triple-deck bored highway and metro tunnel (planned)	TBD	Two lanes and one metro track per direction	TBD

Source: TunnelTalk (2015).

Figure 4.8 Illustration of Shanghai Yangtze River tunnel-bridge system, Shanghai, China.

the Shanghai Yangtze River Tunnel-Bridge crossing to be connected with the Shanghai-Hangzhou and Shanghai-Nanjing expressway network that significantly promotes the Yangtze River Delta economic integration (Figure 4.8).

4.4.3.5 Chongqing Jiefangbei underground circle

The 3 km circle is designed 20 m deep, with a 5.5 m vertical clearance and 7–9 m width to support one-way, two-lane traffic. Phase I construction of the two-phase project began in June 2010. Multiple exits and entrances are provided to access 20,000 underground parking spaces. The project has drastically increased the utilization of underground parking facilities in the Jiefangbei business circle of the city's CBD and reduced unauthorized on-street parking, leading to improved efficiency of utilizing the surface street capacity to mitigate traffic congestion (Figure 4.9).

4.5 REDESIGNING AT-GRADE INTERSECTIONS

Essential to redesigning of at-grade intersections is to improve left-turn traffic operations by isolating conflict movements to achieve mobility and safety improvements. Two broad categories of designs may be employed, which are unconventional at-grade intersection designs and unconventional overpass and interchange designs (Li et al., 2010).

4.5.1 Unconventional at-grade intersection designs

4.5.1.1 Doublewide intersections

In this at-grade intersection design, each direction of the arterial highway is separated into two sets of lanes upstream of the signalized cross street. At some distance downstream of the intersection, a simple two-phase signal controls the merge of the two sets of lanes back into the original number of arterials through movement lanes.

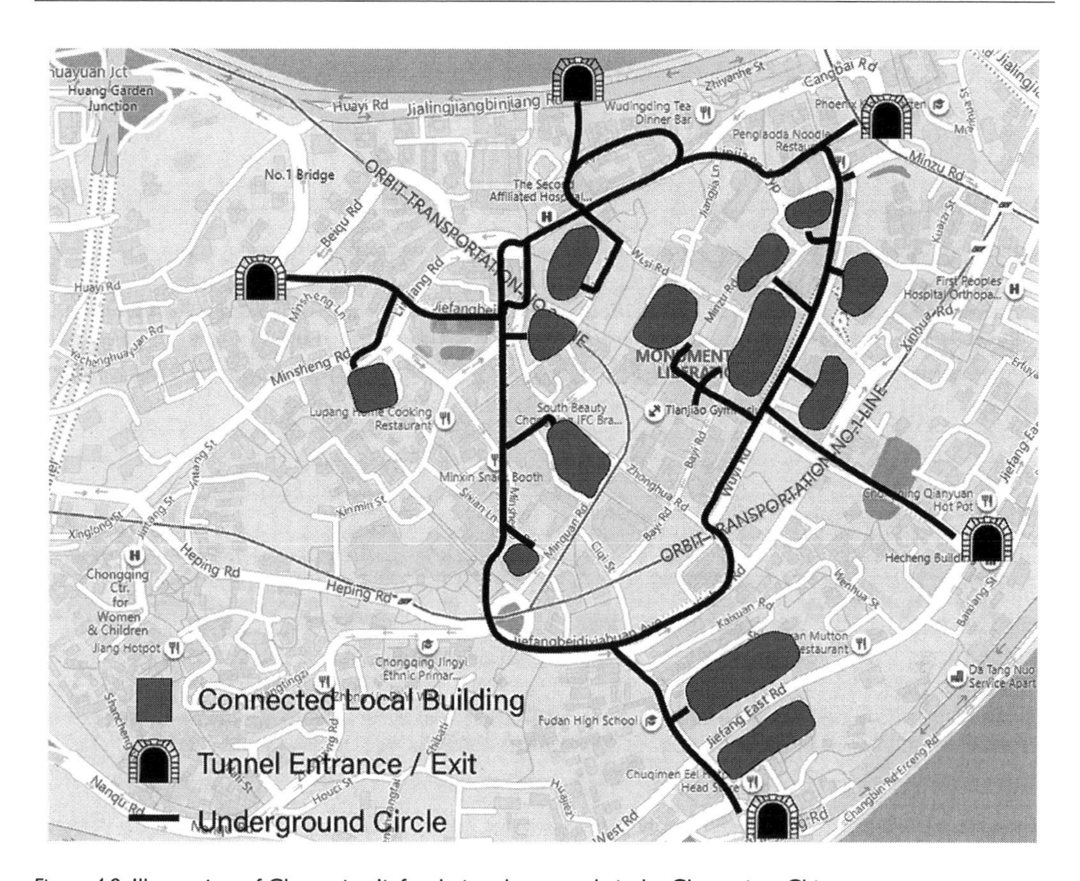

Figure 4.9 Illustration of Chongqing Jiefangbei underground circle, Chongqing, China.

4.5.1.2 Continuous flow intersections

In this at-grade intersection design, left-turning vehicles begin their turn several 100 ft upstream of the main intersection at a signalized "crossover" intersection for protected left turns and move into separated lanes to the right of the opposing through movements. The protected left turns are completed simultaneously with through movements. This design only requires a simple two-phase intersection signal-timing plan for the main intersection traffic control.

4.5.1.3 Median U-turn intersections

In this at-grade intersection design, left turns are made indirectly using directional U-turn crossovers closest to the intersection. A vehicle traveling on the arterial highway that desires to make a left turn at the cross street must travel through the cross-street intersection, make a U-turn at the first median U-turn crossover, and then turn right onto the cross street. Similarly, left-turn vehicles on the cross street must turn right onto the arterial and use the median U-turn crossover to finish left turn onto the arterial highway.

4.5.1.4 Super street intersections

The super street design is similar to the median U-turn design concept, but it features a break in cross-street traffic that allows the signals on opposite directions of the arterial highway to operate independently. Left turns from the arterial can make direct left turns onto the cross street. However, cross street through and left-turn movements must use the directional U-turn crossovers and then turn right onto the cross street (Figure 4.10).

4.5.2 Unconventional overpass, queue jumper, and interchange designs

4.5.2.1 Center-turn overpasses

In this overpass design, left-turn traffic is separated from arterial and cross street through and right-turn movements where all left-turn vehicles are accommodated by an elevated intersection within the medians. Both the elevated and at-grade intersections use two-phase traffic signals. Left-turn traffic descends from the elevated intersection and merges into through traffic lanes on the arterial and cross street.

4.5.2.2 Queue jumpers

Queue jumpers are grade separations added to at-grade intersections along major arterials to allow vehicles to bypass the intersections. Vehicles not turning at an intersection can use the through movement travel lanes provided by the grade separations to continue without interruption. Grade separations may be designed as underpasses in denser, more-urbanized locations to ensure they are visually unobtrusive. By contrast, either underpasses or overpasses may be considered for lower-density, suburban locations. Major arterials with queue jumpers could be operated as managed arterials for use by buses free of charge and by toll-paying vehicles with toll rates varying by time of the day to help improve people mobility. In addition, flyover connectors could be added between managed arterials and freeways with managed lanes or HOT lanes to facilitate uninterrupted vehicle entries and exits that would lead to maximized people mobility (Poole et al., 2012).

4.5.2.3 Tight diamond interchanges

In this interchange design, free-flow through movements for the arterial are provided by using exit and entry ramps tight to the arterial roadway outer edges. Two elevated ramp intersections are used for the arterial left-turn and right-turn movements as well as all cross-street movements. Ramp intersections are closely spaced and are coordinated using special phasing patterns to eliminate vehicle storage lanes.

4.5.2.4 Single point interchanges

In this interchange design, free-flow through movements for the arterial are provided by creating an elevated, signalized intersection to handle the arterial left-turn movements as well as cross-street left-turn and through movements. All right turns are made at unsignalized ramps separated from the elevated intersection.

4.5.2.5 Echelon interchanges

In this interchange design, one-half of two approaches on both the arterial and intersecting cross streets, respectively, is elevated on structures as they intersect, while the other half of the two approaches on both the arterial and cross street intersect at grade. The interchange contains a symmetrical but offset pair of intersections separated by grade, operated by two-phase signals as they meet the one-way streets.

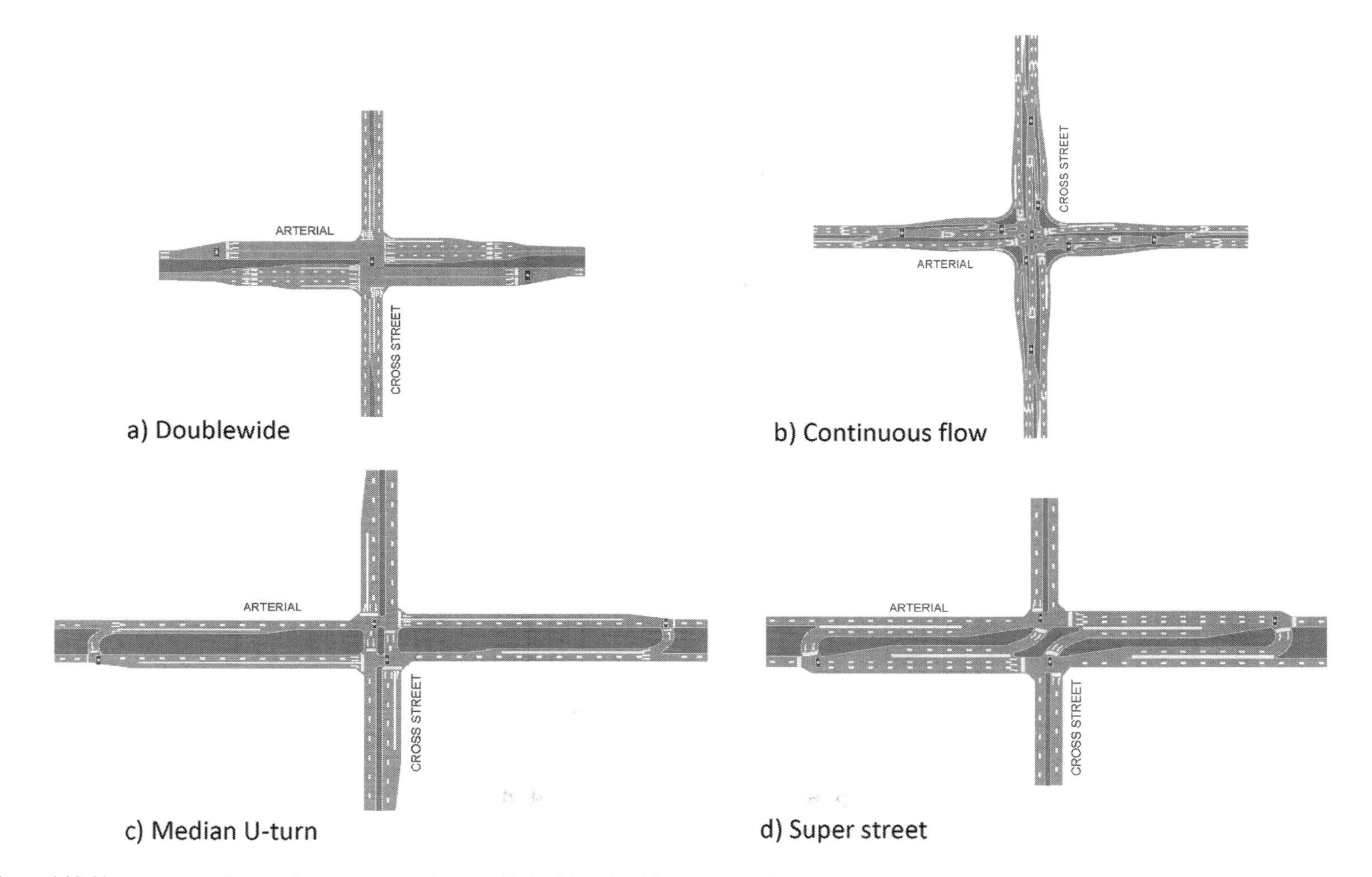

Figure 4.10 Unconventional at-grade intersection designs: (a) doublewide, (b) continuous flow, (c) median U-turn, and (d) superstreet.

4.5.2.6 Median U-turn diamond interchanges

In this interchange design, free-flow through movements for the arterial are provided by using directional, signalized crossovers downstream of the main intersection for all left turns. The arterial and cross-street turning movements are diverted onto one-way frontage roads on the outer edge of depressed through movement lanes and arterial off ramps prior to reaching the main intersection. Signals of the main intersection and left-turn movements are coordinated to ensure efficient channelization of arterial turning movements and all cross-street movements (Figure 4.11).

4.5.3 Cases in action

4.5.3.1 Young Circle in Hollywood, Florida

The Young Circle in Hollywood, Florida is one of the busiest and most confusing gateways in South Florida initially designed with a one-way counterclockwise traffic circle in conjunction with seven traffic signals. The daily entering vehicle volume is about 50,000, and on average, 450 vehicle crashes occur yearly. A 20-million, 6-year design improvement is underway to replace all signals by five roundabouts integrated into a supersized two-way traffic circle, with vehicles in the outer lanes traveling counterclockwise consistent with the existing traffic movements and vehicles in the innermost lanes moving in a clockwise direction. Traffic impacts study reveals that the new design would increase the average traffic speed from 9 to 18 mph, reduce congestion at peak hours by 77%, and lead to fewer crashes (Figure 4.12).

4.5.3.2 Lujiazui pedestrian circle/vehicular roundabout in Shanghai, China

Several years ago, a new pedestrian circle/vehicular roundabout was unveiled in Lujiazui area of the Pudong new district of Shanghai, China. This massive facility is an example of accommodating roadway and non-roadway travel. Pedestrians avoid vehicular traffic via elevated bridges and pedestrian circle at the roundabout terminus of Lujiazui Road. The pedestrian circle sits above the surface street by nearly 20 ft, can accommodate up to 15 people walking side by side, and is equipped with numerous escalator stairway entrances and exits. It provides accessibility to the Oriental Pearl Tower that connect employees of the nearby financial area and domestic and international tourists for activities of shopping, catering, and sightseeing (Figure 4.13).

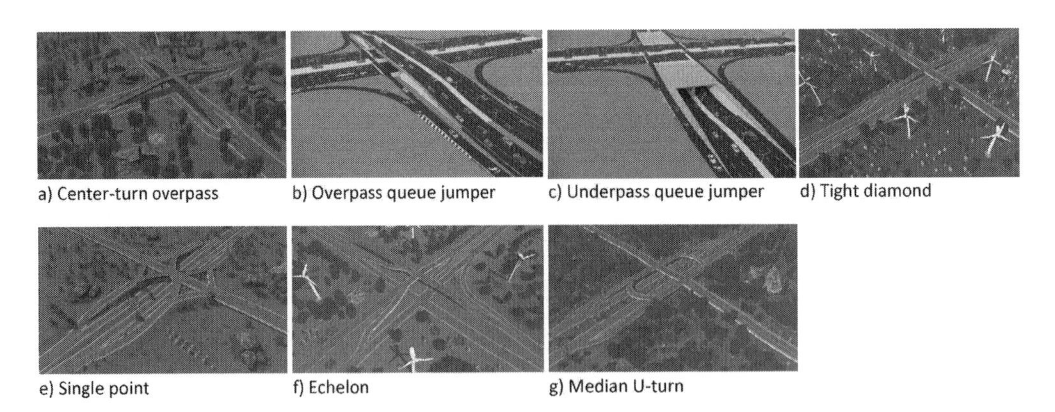

Figure 4.11 Unconventional overpass, queue jumper, and interchange designs: (a) center-turn overpass, (b) overpass queue jumper, (c) underpass queue jumper, (d) tight diamond, (e) single point, (f) echelon, and (g) median U-turn.

a) Current configuration

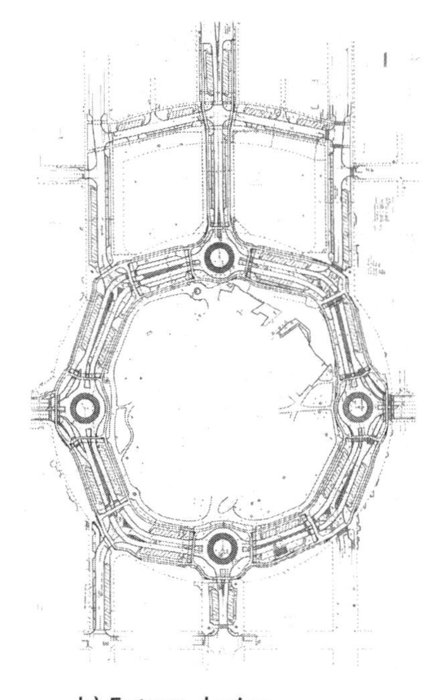

b) Future design

Figure 4.12 Innovative design of young circle in Hollywood, Florida. (a) Current configuration. (*Source:* Shutterstock (2016) with permission.) (b) future design. (*Source:* Toole Design (2019).)

Figure 4.13 Lujiazui pedestrian circle/vehicular traffic roundabout in Shanghai, China.

4.6 COMPLETE STREETS

Too often, planners presume that transit and road traffic are discrete options and are incapable of being optimized or managed concurrently or simultaneously. While Chapter 5 discusses transit planning and implementation more fully, the concept of "complete streets" shows that travel modes can be considered as part of integrated transportation development and management.

4.6.1 Basic elements

The concept of complete streets is a transportation policy and design approach that requires streets to be planned, designed, operated, and maintained for safe, convenient, and comfortable access and use for travelers of all ages and abilities by various travel modes including walking, scooting, biking, transit, and auto, as well as trucking for delivery purposes (Newman and Kenworthy, 1999; Ritter, 2007). The Complete Streets movement emerged out of a general belief among some segments of the urban planning communities that new land development in towns and cities tended to ignore.

In the United States, Oregon was the first state that enacted the complete streets policy dating back to 1971, which required that new or rebuilt roads accommodate bicycles and pedestrians, and also calling on state and local governments to fund pedestrian and bicycle facilities in the public right-of-way. Subsequently, more than 16 state legislatures have adopted complete streets laws (ORDOT, 2016). The specific design elements of complete streets vary based on community context and project goals, which largely include (LaPlante and McCann, 2008):

- Pedestrian infrastructure that includes sidewalks; traditional and raised crosswalks; median crossing islands; Americans with Disabilities Act (ADA) of 1990 compliant facilities, including audible cues for people with low vision, pushbuttons reachable by people in wheelchairs, and curb cuts; and curb extensions.
- Traffic calming measures to lower vehicle speeds of and define the edges of vehicle travel lanes, including a road diet, center medians, shorter curb corner radii, elimination of free-flow right-turn lanes, angled, face-out parking, street trees, planter strips, and ground cover.
- Bicycle accommodations that cover protected or dedicated bicycle lanes, neighborhood greenways, wide-paved shoulders, and bicycle parking.
- Public transit provisions that are concerned with bus rapid transit, bus pullouts, transit signal priority, bus shelters, and dedicated bus lanes.

Deploying complete streets could potentially achieve health, safety, and environmental impact-related benefits. For instance, field data-driven studies by Powell et al. (2003), Dumbaugh and Li (2010), and Pucher et al. (2010) revealed that complete streets could promote walking and bicycling by providing safer places to achieve physical activity through transportation. Reynolds et al. (2009) found that converting the traditionally designed urban streets mainly for auto movements to complete streets could reduce auto miles of travel, vehicle-related crashes, and safety risks of pedestrians and bike riders when well-designed bicycle-specific infrastructure is included, as well as improving air quality due to reduced vehicle air emissions.

However, these are tradeoffs, improving pedestrian, bike, and scooter mobility by reducing auto mobility, and given the shares of those modes of travel in most developed economies, that could mean reducing mobility for many while improving it for few. In cities with more balance between auto and non-motorized travel, and other microtransit modes as they grow, complete streets can improve overall mobility. Even in developed cities with large auto mobility shares, complete streets can be helpful if applied thoughtfully, leaving the major arterials primarily to autos and improving pedestrian, bike, and scooter infrastructure on parallel but less heavily traveled roads.

4.6.2 Case in action: complete streets in Saint Paul, Minnesota

The city of Saint Paul, Minnesota, is an innovative leader in the implementation of complete streets. Funded by a USDOT TIGER II Grant for capital investments in surface transportation infrastructure, the city developed a street design manual to implement its complete streets policies. Numerous pilot projects are underway that include public workshops to prioritize potential street improvements, especially transit accommodations and pedestrian walkways (Figure 4.14).

4.7 OPTIMAL CONTROL OF SIGNALIZED INTERSECTIONS

4.7.1 General

Traffic signal control systems and associated methods vary in space and time. Broadly, they could be grouped into two categories based on their target: (i) isolated signal control; and (ii) coordinated area signal control.

For isolated intersection signal control, signalized intersections are located and operated independently. This type of control is selected when traffic arrivals are basically unaffected

Figure 4.14 Illustration of complete streets in Saint Paul, Minnesota. (Source: Lemon (2014).)

by any adjacent signalized intersection. Isolated signal control method is more common in suburban/rural areas where density of signalized intersections is low. An isolated system can be pre-timed or actuated.

When signalized intersections are more closely spaced, coordinated control is often implemented. Operations at one signalized intersection are then influenced by operations at one or more adjacent signalized intersections. All signalized intersections in a coordinated system are controlled by a computerized signal control system, and the system could be pre-timed or traffic responsive.

With pre-timed control, signal timing plans are calculated and optimized offline, using software such as SYNCHRO (Trafficware, 2019) and TRANSYT (Binning et al., 2019). Historic traffic data are used to generate optimal signal timing plans, and signalized intersections are coordinated by proper offsets. In contrast, traffic-responsive coordinated systems rely on certain traffic detection methods, such as loop detector, camera, and GPS, on intersection approaches to provide real or near real-time traffic data to calculate optimum signal settings timely. The improved traffic mobility demonstrated with traffic-responsive control has led to the development of a number of systems, including TANSTP (Traffic Adaptive Network Signal Timing Program) (Kessemann et al., 1973), SCATS (Sydney Coordinated Adaptive Traffic System) (Sims and Dobinson, 1980; Lowrie, 1982), SCOOT (Split Cycle Offset Optimisation Technique) (Hunt, 1981; Robertson and Bretherton, 1991), UTOPIA (Urban Traffic OPtimisation by Integrated Automation) (Mauro and DiTaranto, 1981), OPAC (Optimized Policy for Adaptive Control) (Gartner, 1983), PRODYN (Henry et al., 1983), CRONOS (Boillot et al., 1992), RHODES (Real-time Hierarchical Optimizing Distributed Effective System) (Head et al., 1992; Mirchandani and Head, 2001), RT-TRACS (Real-Time Traffic Adaptive Control System) (Gartner et al., 1994), BALANCE (Friedrich et al., 1995), MOTION (Kruse, 1998), TUC (Traffic-responsive Urban Control) (Diakaki, 1999), ACS-Lite (Adaptive Control Software-Lite) (Luyanda et al., 2003), InSync (Chandra et al., 2011; Chandra and Gregory, 2012), and SMIFAATC (Systems and Methods Involving Features of Adaptive and/or Autonomous Traffic Control) (Robinson, 2014).

4.7.2 SCOOT adaptive traffic signal control system

The SCOOT adaptive traffic signal control system was developed in the United Kingdom and is the most widely deployed adaptive signal control system in existence. SCOOT uses both stop line and advance detectors, typically 150–1,000 ft (50–300 m) upstream of the stop line or exit loops which are located downstream of the intersection. The advance detectors provide a count of the vehicles approaching at each intersection. This gives the system a high-resolution picture of traffic flows and a count of the number of vehicles in each queue, several seconds before they touch the stop line, allowing SCOOT to perform a short-term traffic prediction and time for communications between the traffic signal controller and the central computer. SCOOT also provides queue length detection and estimation. Under the SCOOT system, green waves can be dynamically delayed on a "just-in-time" basis based on the arrival of vehicles at the upstream detector, which allows extra time to be allocated to the previous green phase, where warranted by heavy traffic conditions. SCOOT controls the exact green time of every phase on a traffic controller by sending "hold" and "force-off" commands to the controller.

The SCOOT model utilizes three optimizers: splits, offsets, and cycle. At every intersection and for every phase, the split optimizer will make a decision as to whether to make the change earlier, later, or as due, prior to the phase change. The split optimizer implements the decision, which affects the phase change time by only a few seconds to minimize the degree of saturation for the approaches to the intersection. During a predetermined phase in each cycle, and for every intersection in the system, the offset optimizer makes a decision to alter, all the offsets by a fixed amount. The offset optimizer uses information stored in cyclic flow profiles and compares the sum of the performance measures on all the adjacent links for the scheduled offset and the possible changed offsets.

A SCOOT system is split into cycle time "regions" that have pre-determined minimum and maximum cycle times. The cycle optimizer can vary the cycle time of each "region" in small intervals in an attempt to ensure that the most heavily loaded node in the system is operating at 90% saturation. If all stop bars are operating below 90% saturation, then the cycle optimizer will make incremental reductions in cycle time (DETR, 1999).

4.8 NEW TRUCK ROUTE CAPACITY

The need for new truck route capacity dwells on reliable forecasts of future freight demand, mode split mainly between truck and rail, and various types of added and redistributed truck traffic utilizing different classes of highways in the metropolitan area, in conjunction with added truck traffic induced by intermodal transshipments. Travel characteristics of freight movements differ greatly from those of passenger trips. Traditionally, freight demand forecasting for a metropolitan area is carried out using a variational process of the typical four-step passenger demand forecasting that contains freight trip generation, distribution, and assignment steps without accounting for freight mode split between truck and freight rail that is largely excluded from urban transportation. Further, the exclusion of freight-mode split step makes it hard for the traditional model to rationalize choices of different types of trucks that handle freight shipments, leading to missing linkage between freight travel demand and truck traffic for current and future time periods. In recent years, commodity-based freight models have been increasingly thought of as structurally superior in that by treating freight demand as commodities. They could associate production and distribution of commodities directly with the economy of a metropolitan area and readily translate commodity flows with different freight modes and different types of trucks. One major hurdle for commodity-based freight model development is willingness of all commodity producers, purchasers, and shippers contributing to the economy of the metropolitan area potentially extending far beyond its geographic coverage to share fine-grained, highly sensitive data for model construction, calibration, and validation (Gonzalez-Feliu et al., 2014). One compromised solution is to establish connections of aggregated prediction of commodity flows at major freight generators such as ports, railyards, terminals, air cargo operations, warehouses, and distribution facilities in the metropolitan area to the regional travel demand forecasting model containing multimodal freight components. With supplementary data on total truck movements, trips, and routes collected from facility-specific gate and carrier's driver surveys, as well as cordon and classification counts made available, the demand forecasting model could then carry out commodity-based freight trip generation, and activity-based trip distribution, mode split, and truck traffic forecasts (SteadieSeifi et al., 2014).

4.8.1 Truckways

The forecasted truck flows can be utilized to identify primary truck routes along various classes of highways and urban streets in the metropolitan area. Based on the minimum acceptable mobility standard measured by the qualitative measure in terms of level of service or quantitative measures such as route-based travel time and travel time reliability index, the addition of required travel lanes for all truck routes could be determined. As appropriate, a mixture of shared-use and exclusive truck lanes may be adopted for the added travel lanes. For the truck-only lanes predominantly accommodating heavy duty single-unit and combination trucks for long-distance shipments, they might be deployed as truck-only toll lanes or tolled truck ways, typically with barrier separations with the remaining travel lanes (Poole and Samuel, 2004; Poole, 2007).

For truck-only lanes or truck ways accommodating heavy duty trucks, higher axle loads of the trucks will cause much greater extents of pavement damages. The repetitive truck use will lead to accelerated deterioration of pavement conditions and, in turn, require more frequent and costly maintenance and repair treatments for pavement condition restoration or even trigger pavement reconstruction much earlier than the timeline of the pavement design service life. Further, heavy duty trucks will cause similar impacts on the deterioration and preservation of highway bridges, elevated highways, and travelways of tunnels. Imposing tolls on trucks to recover excessive costs of pavement, bridge, elevated highway, and tunnel construction, maintenance, and repairs will help preserve the conditions of truck-only lanes or truckways to the best possible level. To ensure efficiency and equity of toll rates imposed on different types of trucks, data details of truck-related axle loads and classifications should be utilized to determine the fair share of load and non-load-related facility damages and cost recovery distributions across the heavy-duty trucks expected in the traffic stream. Improved ride qualities in conjunction with smoother truck operations along dedicated travel lanes will result in better fuel efficiency, less tire and brake wear, reduced air emissions and pollution, safer truck traffic, and reduced fatigue to truck operators.

4.8.2 Case in action: tolled truckways in Georgia

Over a decade ago, the state of Georgia in the United States began the screening process to identify tolled truckways in the freeway system based on criteria of freeway level of service, truck volumes, truck percentage of total freeway flow, truck-related crashes, and truckers' willingness to pay. Various toll rates were assessed to achieve maximized toll revenues under different circumstances of an acceptable level of service, a minimum level of tolled truck only

Figure 4.15 Illustration of tolled truckways in Georgia. (Source: RS&H (2020).)

lane utilization, and a truck diversion rate from parallel roads to tolled truckways. Alternative tolled truckway designs were considered in association with the level of through truck trips, placement of tolled truck lanes inside or outside of the directional travel lanes, relocation of HOV lanes, and acquisition of additional right-of-way (Chu, 2007) (Figure 4.15).

4.9 CONNECTED AND AUTOMATED/AUTONOMOUS VEHICLES

The invention of automobile by Karl Benz in 1886 has greatly expanded mobility landscape for human beings. Over the years, drastic improvements of vehicle performance in fuel economy, safety, and dynamics have been made. One of the innovations is to introduce self-driving vehicles, coupled with intelligent highways, that could be traced back to 1950s (Science Digest, 1958). With the continuing technological advancements in sensing, telecommunication, high-performance computing, and information dissemination, the idea of making self-driving vehicles for public use to improve safety and mobility, as well as mitigating adverse impacts on the environment, has been rigorously explored in the last decade, with leading effects made in developed countries such as Japan, EU nations, and the United States (Whelan, 1995; Makino, 2006; Faber et al., 2014; USDOT, 2014; Hawkins and Habib, 2019).

In the context of self-driving vehicles, connected and automated/autonomous vehicles (CAVs) stand out as two separate but related technological advancements. Connected vehicles (CVs) are those that can communicate with other vehicles, infrastructure, and devices through wireless network technology. Automated/autonomous vehicles (AVs) are also known as driverless vehicles equipped with technology that enables them to operate with little to no human assistance. A vehicle can be connected but not automated, automated but not connected, neither, or both.

4.9.1 Connected vehicles

For CVs, data collection, exchange, and processing are achieved by using on-board units, roadside equipment, and communications technologies. The on-board units (including vehicle sensors) collect, transmit, and receive vehicle status data such as vehicle speed, location, brake status, and airbag activation to and from other vehicles and nearby infrastructure. The roadside equipment mounted along roadsides and at interchanges and intersections support data exchange between vehicles and infrastructure. Communications technologies such as cellular communications, wireless communications, and dedicated short-range communications (DSRCs) that are not affected by severe weather conditions render vehicle-to-vehicle (V2V), vehicle-to-infrastructure (V2I), and vehicle-to-others (V2D) interactions. For this reason, the applications of CVs could be split into the above three categories. The V2V applications allow each vehicle exchanges data with other vehicles in the vicinity. Typical V2V safety applications include emergency brake-light warning, forward-collision warning, intersection movement assistance, blind spot and lane change warning, no passing warning, and control loss warning. The V2I applications involve data exchanges between vehicle's on-board equipment and infrastructure's roadside equipment, which could support dynamic mobility applications involving various types of vehicles, real-time environment-friendly applications, and dynamic transit applications specifically. For V2D applications, a vehicle turning right may be able to send an alert to a bicyclist's cell phone or device on the bike and avoid a potential collision. In consideration of the need for investments in V2I equipment, V2V and V2D applications appear to be more likely in the not-too-distant future (NHTSA, 2014; RITA, 2014).

4.9.2 Automated/autonomous vehicles

AVs rely on a variety of sensors such as cameras, radar, lidar (image sensing), sonar, GPS, odometry, inertial measurement units, and computer vision to perceive their surroundings. Once an environment has been scanned and obstacles and relevant signage detected, the vehicle-mounted advanced control system interprets sensory data, reacts as the situation dictates, and performs steering control, accelerating, or braking as required (Taeihagh and Lim, 2019). The first self-sufficient automated/autonomous cars appeared in the United States in 1984 with Carnegie Mellon University's Navlab and ALV projects and in Europe in 1987 with Mercedes-Benz and Bundeswehr University Munich's Eureka Prometheus Project (Wallace et al., 1985; Kanade, 1986; Schmidhuber, 2009). As of 2019, 29 U.S. states have passed laws permitting automated/autonomous cars (NCSL, 2019). Currently, the United Kingdom, the European Union (EU), and Japan are on track to regulate automated/autonomous cars.

The term "automated" connotes control or operation by a machine, while "autonomous" refers to acting alone or independently. Since most of the vehicle concepts have a person in the driver's seat, utilize a communications connection to other vehicles and highway infrastructure, and do not independently select either destinations or routes for reaching them, the term "automated" appears to be more accurate (Wood et al., 2012). According to the European Union Regulation 2019/2144 (EU, 2019), an "automated vehicle" means a motor vehicle designed and constructed to move autonomously for certain periods of time without continuous driver supervision but in respect of which driver intervention is still expected or required; and a "fully automated vehicle" refers to a motor vehicle that has been designed and constructed to move autonomously without any driver supervision.

Table 4.6 summarizes driving automation levels defined by Society of Automotive Engineers (SAE) International (SAE International, 2018). For zero to two driving automation levels, a human driver is needed to monitor the driving environment for. From level 3 automation and beyond, an automated driving system directly monitors the driving environment, and at level 5 with full driving automation, the system also commands

Table 4.6 Driving automation levels defined by SAE International (2018)

Automation level		*Definition*	*Execution of steering and acceleration/ deceleration*	*Monitoring of driving environment*	*Fallback performance of dynamic driving task*	*Driving modes*
Level 0	No automation	Full-time performance by the human driver of all aspects of the dynamic driving task, even when enhanced by warning or intervention systems	Human driver	Human driver	Human driver	n/a
Level 1	Driver assistance	Driving mode-specific execution by a driver assistance system of steering or acceleration/deceleration using information about the driving environment and with the expectation that the human driver performs all remaining aspects of the dynamic driving task	Human driver and system	Human driver	Human driver	Some
Level 2	Partial automation	Driving mode-specific execution by one or more driver assistance systems of both steering and acceleration/deceleration using information about the driving environment and with the expectation that the human driver performs all remaining aspects of the dynamic driving task	Automated driving system	Human driver	Human driver	Some
Level 3	Conditional automation	Driving mode-specific performance by an automated driving system of all aspects of the dynamic driving task with the expectation that the human driver will respond appropriately to a request to intervene	Automated driving system	Automated driving system	Human driver	Some
Level 4	High automation	Driving mode-specific performance by an automated driving system of all aspects of the dynamic driving task, even if a human driver does not respond appropriately to a request to intervene	Automated driving system	Automated driving system	Automated driving system	Some
Level 5	Full automation	Full-time performance by an automated driving system of all aspects of the dynamic driving task under all roadway and environmental conditions that can be managed by a human driver	Automated driving system	Automated driving system	Automated driving system	All

The gray shading indicates those fields where the human driver is the primary actor. Distinct from the unshaded fields where the automated driving system is the primary actor.

all safety-critical driving functions. On this front, some significant technical and human factor-related challenges need to be overcome. As noted by Schwarz et al. (2013), technical challenges of automated vehicles include sensing, and localization and mapping for identifying the location of a vehicle and detecting the highway infrastructure, sensing and object detection for identifying static and moving objects on travelways, and path building and decision-making for O-D route planning. On the human factor-related challenges, situations of a driver not actively involved with vehicle control, driver performance when vehicle control is transitioned back to the driver, and trust in automation when evolving from lower-level automation to higher-level automation need to be properly handled. The implementation of self-driving vehicles with full automation will cause a wide range of impacts on the automobile industry, health risks, welfare, urban planning, traffic management, vehicle insurance, labor market, energy and the environment, self-parking, parking space, privacy, terrorist attack risks, shared mobility providers, vehicle repairs, emergency response and military forces, interior design and entertainment, telecommunications, hospitality industry, and airline services. Policy, regulation, technical, and human factor issues need to be thoroughly addressed before large-scale deployments of connected and fully automated vehicles are feasible.

4.9.3 Cases in action

4.9.3.1 Self-driving cars by Waymo in California

Derived from its mission of a new way forward in mobility, Waymo based in Mountain View, California, is formerly known as the Google self-driving car project launched in January 2009 with the goal of fully automated long-distance driving over uninterrupted flow highways. In 2012, the project began to test a fleet of Toyota Prius and Lexus RX450h vehicles with safety drivers accompanying passengers on urban streets in North America. A Toyota Prius modified with Google's experimental driverless technology was licensed by the Nevada Department of Motor Vehicles (DMV) in May 2012 as the first license issued in the United States for a self-driven car (Slosson, 2012). In 2015, the "Firefly" vehicle with custom sensors, computers, steering and braking technology, and without a steering wheel, pedals, and a safety driver was tested on urban streets in Austin, Texas. In 2017, Waymo announced the first vehicle built on a mass-production platform and began a public self-driving vehicle trial in Phoenix, Arizona. In 2018, this trial became Waymo One, a commercial self-driving service, with safety drivers in all cars. In 2019, Waymo announced plans for vehicle assembly in Detroit by using vehicle assembler Magna to turn Jaguar I-PACE and Chrysler Pacifica Hybrid minivans into Waymo Level 4 automation vehicles (Sage, 2019). In July 2020, the company announced an exclusive partnership with auto manufacturer Volvo to integrate Waymo's self-driving technology into Volvo's vehicles (Bell, 2020).

4.9.3.2 Autopilot/full self-driving by Tesla in California

Tesla Autopilot is a suite of advanced driver-assistance system features offered by Tesla, Palo Alto, California, that has lane centering, traffic-aware cruise control, self-parking, automatic lane changes, semi-autonomous navigation on limited access freeways, and the ability to summon the car from a garage or parking spot. In all of these features, the driver is responsible and the car requires constant supervision. The company claims the features reduce accidents caused by driver negligence and fatigue from long-term driving (Williams, 2019). In late 2020, Tesla released a beta version software supporting full self-driving by augmenting the base Autopilot capabilities to a small group of testers in the United States (Stoklosa, 2020).

4.10 CONCLUSION

In sum, expanding highway capacity is both feasible and necessary in rapidly growing megacities. As the size and scope of human activities and commodity needs expansion, access to a wider variety and range of travel modes and sub-modes becomes essential in order to maximize connectivity, access, and mobility. Given the organic nature of urban development, however, urban/transportation planners, engineers, and decision makers will need to think beyond traditional investments that rely primarily on 2D surface expansions or simply inserting new segments to address specific bottlenecks. On the contrary, thinking of 3D multimodal transportation requires taking a holistic and strategic approach to capacity building in a large-scale transportation network with regional implications. This approach is particularly important when considering the interface and interactions between people and goods movements.

REFERENCES

Adams, T.M., Koncz, N.A., and Vonderohe, A.P. 2001. Guidelines for the Implementation of Multimodal Transportation Location Referencing Systems. NCHRP Report 460. Transportation Research Board, National Academies Press, Washington, DC, USA.

Bell, A. 2020. Volvo Cars, Waymo partner to build self-driving vehicles. *Reuters*, 2020-06-25 Issue. London, UK [Online]. Available: https://www.reuters.com/article/us-waymo-volvo-autonomous-idUSKBN23W2V0 [Accessed on May 14, 2021].

Benardos, A., Paraskevopoulou, C., and Diederichs, M.S. 2013. Assessing and benchmarking the construction cost of tunnels. Presented in the Conference of the Canadian Geotechnical Symposium GeoMontreal on Geoscience for Sustainability, Canadian Geotechnical Society, Montreal, Canada.

Binning, J.C., Burtenshaw, G., and Crabtree, M. 2019. *TRANSYT 16 User Guide*. Transport Research Laboratory, Wokingham, Berkshire, UK.

Boillot, F., Blosseville, J.M., Lesort, J.B., Motyka, V., Papageorgiou, M., and Sellam, S. 1992. Optimal signal control of urban traffic networks. *In Proceedings of the 6th IEEE International Conference on Road Traffic Monitoring and Control*, Institution of Electrical Engineers, London, UK, Publication No. 355, 75–79.

Chandra, R.J., and Gregory, C. 2012. InSync adaptive traffic signal technology: Real-time artificial intelligence delivering real-world results. White Paper, Rhythm Engineering, LLC, Lenexa, KS.

Chandra, R.J., Bley, J.W., Penrod, S.S., and Parker, A.S. 2011. Adaptive control systems and methods. Patent No. US 8,050,854 B1. Rhythm Engineering, LLC, Lenexa, KS.

Chu, H.C. 2007. Implementing truck-only toll lanes at the state, regional, and corridor levels: Development of a planning methodology. Ph.D. Dissertation, Georgia Institute of Technology, Atlanta, GA.

DeCorla-Souza, P. 2009. Congestion pricing with lane reconfigurations to add highway capacity. *Public Roads* 72(5). Report No. FHWA-HRT-09-003. Federal Highway Administration, U.S. Department of Transportation, Washington, DC.

DETR. 1999. SCOOT urban control system, Traffic Advisory Leaflet 7/99. Department of the Environment, Transport and the Regions, London, UK.

Diakaki, C. 1999. Integrated control of traffic flow in corridor networks. Ph.D. Dissertation, Technical University of Crete, Chania, Greece.

Dumbaugh, E., and Li, W. 2010. Designing for the safety of pedestrians, cyclists, and motorists in urban environments. *Journal of the American Planning Association* 77(1), 69–88.

Efron, N., and Read, M. 2012. Analyzing international tunnel costs. Worcester Polytechnic Institute, Worcester, MA [Online]. Available: https://web.wpi.edu/Pubs/E-project/Available/E-project-043012-122558/unrestricted/Analysing_International_Tunnelling_Costs_Public_Report.pdf [Accessed on March 15, 2021].

Elbert, R., Müller, J.P., and Rentschler, J. 2020. Tactical network planning and design in multimodal transportation: A systematic literature review. *Research in Transportation Business & Management.* doi: 10.1016/j.rtbm.2020.100462.

EU. 2019. Regulation (EU) 2019/2144 of the European Parliament and of the Council of 27 November 2019. The European Parliament and the Council, European Union, Brussels, Belgium, EU [Online]. Available: https://eur-lex.europa.eu/eli/reg/2019/2144/oj [Accessed on: May 13, 2021].

Faber, F., Ballingall, S., Burke, M., and Williams, J. 2014. Policy implications of C-ITS core functions. *2014 Australian Institute of Traffic Planning and Management (AITPM) National Conference*, Adelaide, South Australia [Online]. Available: http://citeseerx.ist.psu.edu/viewdoc/download;jsessionid=D57F483C5D20DAD9373428C0D30B3B07?doi=10.1.1.679.3320&rep=rep1&type=pdf [Accessed on May 13, 2021].

FHWA. 2018. Tunnels. Federal Highway Administration, U.S. Department of Transportation, Washington, DC [Online]. Available: https://www.fhwa.dot.gov/bridge/tunnel/ [Accessed on: March 15, 2021].

Friedrich, B., Sachse, T., Hoops, M., Jendryschik, W., and Reichert, G. 1995. Balance and varia methods for traffic adaptive control. In *Proceedings of the Second World Congress on the Intelligent Transport Systems*, Yokohama, Japan, 2356–2361.

Gartner, N.H. 1983. OPAC: A demand-responsive strategy for traffic signal control. *TRB Journal of Transportation Research Record* 906, 75–81.

Gartner, N.H., Stamatiadis, C., and Tarnoff, P.J. 1994. Development of real time, traffic-adaptive control strategies for IVHS. In *Proceedings of the First World Congress on Applications of Transport Telematics and Intelligent Vehicle-Highway Systems.* Towards an intelligent transport system, Vol. 2, Artech House Publishing, Boston, MA, 423–430.

Gonzalez–Feliu, J., Semet, F., and Routhier, J.L. 2014. *Sustainable Urban Logistics: Concepts, Methods and Information Systems.* Springer, Berlin, Germany.

Hawkins, J., and Habib, K.N. 2019. Integrated models of land use and transportation for the autonomous vehicle revolution. *Transport Reviews* 39(1), 66–83.

Head, K.L., Mirchandani, P., and Shepherd, D. 1992. A hierarchical framework for real-time traffic control. *TRB Journal of Transportation Research Record* 1360, 82–88.

Heathman, A., and Simon-Lewis, A. 2017. Elon Musk's Boring Company has completed the first 'segment' of a tunnel in LA. *Wired*, 2017-06-29 Issue. Condé Nast Britain, London, UK [Online]. Available: https://www.wired.co.uk/article/elon-musk-tunnel-boring [Accessed on March 21, 2021].

Henry, J.J., Farges, J.L., and Tufal, J. 1983. The PRODYN real time traffic algorithm. In *Proceedings of the IFAC Symposium, International Federation of Automatic Control in Transportation Systems*, Baden-Baden, Baden-Württemberg, Germany, 305–310.

Hunt, P.B. 1981. SCOOT: A traffic responsive method of coordinating signals, Report 1014. Transport and Road Research Laboratory, Crowthorne, Berkshire, UK.

Kanade, T. 1986. Autonomous land vehicle project at CMU. *Proceedings of the 1986 ACM Fourteenth Annual Conference on Computer Science*, Cincinnati, OH, 71–80. doi: 10.1145/324634.325197.

Kessemann, R., Bolling, L., Cooper, D.L., Ritter, J., and Bravo, E. 1973. Urban Traffic Control System and Bus Priority System (UTCS/BPS): Traffic Adaptive Network Signal Timing Program (TANSTP). Software Description. Report No. DOT-FH-11-7594. Federal Highway Administration, U.S. Department of Transportation.

Kruse, G. 1998. MOTION: Signal control for urban road networks. *Presented at the Transportation Research Board (TRB) Midyear Traffic Signal Committee Meeting*, Adaptive Traffic Signal Control Workshop, Pacific Grove, CA.

Kutz, M. 2004. *Handbook of Transportation Engineering.* McGraw–Hill, New York.

LaPlante, J., and McCann, B. 2008. Complete streets: We can get there from here. *ITE Journal* 78(-5), 24–28.

Lemon, J. 2014. The transit tourist: Minneapolis – St. Paul, MI. *The Source*, 2014-07-22 Issue. METRO, Los Angeles, CA [Online]. Available: https://thesource.metro.net/2014/07/22/the-transit-tourist-minneapolis-st-paul-minn/ [Accessed on May 14, 2021].

Li, Z. 2018. *Transportation Asset Management: Methodology and Applications.* CRC Press, Boca Raton, FL.

Li, Z.C., Lam, W.H., and Sumalee, A. 2008. Modeling impact of transit operator fleet size under various market regimes with uncertainty in network. *TRB Journal of Transportation Research Record* 2063, 18–27.

Li, Z., Shen, J., Budiman, J., and Sutchiewcharn, N. 2010. Prototype designs of a continuous traffic flow system for urban intersection congestion mitigation. In A Report of Findings of the Robert W. Galvin Congestion Project. Illinois Institute of Technology, Chicago, IL.

Lowrie, P.R. 1982. The Sydney co-ordinated adaptive traffic system: Principles, methodology, algorithms. *Proceedings of International Conference on Road Traffic Signaling.* Institution of Electrical Engineers, London, UK, No. 207, 67–70.

Luyanda, F., Gettman, D., Head, L., Shelby, S., Bullock, D., and Mirchandani, P. 2003. ACS-Lite algorithmic architecture: Applying adaptive control system technology to closed-loop traffic signal control systems. *TRB Journal of Transportation Research Record* 1856, 175–184.

Makino, H. 2006. SmartWay project cooperative vehicle highway systems. *Presentation at the Vehicle Highway Automation Committee Meeting of the 85th Annual Transportation Research Board Meeting.* The National Academies Press, Washington, DC.

Mauro, V., and DiTaranto, C. 1981. "Utopia," control, computers, communications in transportation. In *Selected Papers from the IFAC Symposium. International Federation of Automatic Control (IFAC) Congress*, Kyoto, Japan, 245–252.

Michaels, D. 2016. The high-tech, low-cost world of tunnel building, *Market Watch*, May 10, 2016 [Online]. Available: https://www.marketwatch.com/story/the-high-tech-low-cost-world-of-tunnel-building-2016-05-10 [Accessed on March 21, 2021].

Mirchandani, P., and Head, K.L. 2001. A real-time traffic signal control system: Architecture, algorithms, and analysis. *Transportation Research Part C: Emerging Technologies* 9(6), 415–432.

Munfah, N. and Nicholas, P. 2020. Why tunnels in the US cost much more than anywhere else in the world. *Tunnel Business Magazine*, August 18, 2020 [Online]. Available: https://tunnelingonline.com/why-tunnels-in-the-us-cost-much-more-than-anywhere-else-in-the-world/ [Accessed on March 21, 2021].

NCSL. 2019. Autonomous vehicles: Self-driving vehicles enacted legislation. *National Conference of State Legislatures*, Washington, DC [Online]. Available: https://www.ncsl.org/research/transportation/autonomous-vehicles-self-driving-vehicles-enacted-legislation.aspx [Accessed on May 13, 2021].

Newman, P., and Kenworthy, J. 1999. *Sustainability and Cities: Overcoming Automobile Dependence.* Island Press, Washington, DC.

NHTSA. 2014. Preliminary statement of policy concerning automated vehicles. National Highway Traffic Safety Administration, U.S. Department of Transportation, Washington, DC [Online]. Available: https://www.nhtsa.gov/staticfiles/rulemaking/pdf/Automated_Vehicles_Policy.pdf [Accessed on: May 13, 2021].

NRC. 2013. Lifecycle sustainability, costs, and benefits of underground infrastructure development. In *Underground Engineering for Sustainable Urban Development.* National Research Council, National Academies Press, Washington, DC, 246 pp.

ORDOT. 2016. *State Pedestrian and Bicycle Funding Programs Manual.* Oregon Department of Transportation, Salem, OR [Online]. Available: https://www.oregon.gov/odot/Programs/TDD%20Documents/StatePBFundingProgramsManual_v4.pdf [Accessed on: March 21, 2021].

Poole, R.W. 2006. Elevated express lanes add much-needed capacity. Surface Transportation Innovations No. 34. Reason Foundation, Washington, DC [Online]. Available: https://reason.org/transportation-news/surface-transportation-innovat-32/#feature4 [Accessed on March 21, 2021].

Poole, R.W. 2007. The case for truck-only toll lanes. *Public Works Management and Policy* 11(4), 244–249.

Poole, R.W. 2013. Interestate 2.0: Modernizing the interstate highway system via toll finance. Policy Study No. 423. Reason Foundation, Washington, DC [Online]. Available: https://reason.org/policy-study/modernizing-the-interstate-highway/ [Accessed on March 21, 2021].

Poole, R.W. 2018. *Rethinking America's Highways: A 21st Century Vision for Better Infrastructure.* The University of Chicago Press, Chicago, IL.

Poole, R.W., and Samuel, P. 2004. Corridors for toll Truckways: Suggested locations for pilot projects. Policy Study 316. Reason Foundation, Washington, DC [Online]. Available: https://reason.org/policy-study/corridors-for-toll-truckways/ [Accessed on March 21, 2021].

Poole, R.W., Rubin, T.A., and Swenson, C. 2012. Increasing mobility in Southeast Florida: A new approach based on pricing and bus rapid transit. Policy Study 400. Reason Foundation, Washington, DC [Online]. Available: https://reason.org/wp-content/uploads/files/mobility_southeast_florida.pdf [Accessed on March 21, 2021].

Powell, K.E., Martin, L., and Chowdhury, P.P. 2003. Places to walk: Convenience and regular physical activity. *American Journal of Public Health* 93, 1519–1521.

Pucher, J., Buehler, R., Bassett, D.R., and Dannenberg, A.L. 2010. Walking and cycling to health: A comparative analysis of city, state, and international data. *American Journal of Public Health* 100(10), 1986–1992.

Reynolds, C.C., Harris, M.A., Teschke, K., and Cripton, P.A. 2009. The impact of transportation infrastructure on bicycling injuries and crashes: A review of the literature. *Environmental Health* 8(47). doi: 10.1186/1476-069X-8-47.

Ricci, A. Gaggi, S., Enei, R., Tomassini, M., and Fioretto M. (ISINNOVA), Gargani, F., Di Stefano, A., and Gaspari, E. (PwC) with contributions from Archer, G., Kearns, S., McDonald, M., Nussio, F., Trapuzzano, A., and Tretvik, T. 2017. Study on urban vehicle access regulations, Final Report. Directorate-General for Mobility and Transport, European Commission, Brussels, Belgium, EU.

RITA. 2014. Vehicle-to-Vehicle (V2V) communications for safety. Research and Innovative Technology Administration, U.S. Department of Transportation, Washington, DC [Online]. Available: https://www.its.dot.gov/research_archives/connected_vehicle/dot_cvbrochure.htm [Accessed on: May 13, 2021].

Ritter, J. 2007. 'Complete streets' program gives more room for pedestrians, cyclists. *USA Today*, 2007-07-29 Issue [Online]. Available: https://bikepgh.org/2007/07/30/complete-streets-program-gives-more-room-for-pedestrians-cyclists/ [Accessed on March 21, 2021].

Robertson, D.I., and Bretherton, R.D. 1991. Optimizing networks of traffic signals in real time-The SCOOT method. *IEEE Transactions on Vehicular Technology* 40(1), 11–15.

Robinson, K.B. 2014. Systems and methods involving features of adaptive and/or autonomous traffic control (SMIFAATC). Patent No. US 8,825,350 B1, Newcastle, CA.

Rondinelli, D., and Berry, M. 2000. Multimodal transportation, logistics, and the environment: managing interactions in a global economy. *European Management Journal* 18(4), 398–410.

RS&H. 2020. Courtesy image of Georgia's truck-only lanes. RS&H Company, Jacksonville, FL.

SAE International. 2018. SAE International from SAE J3016™ taxonomy and definitions for terms related to driving automation systems for on-road motor vehicles (2018-06-05). Society of Automotive Engineers, Warrendale, PA [Online]. Available: https://www.sae.org/standards/content/j3016_201806/ [Accessed on: March 31, 2021].

Sage, A. 2019. Waymo picks detroit factory for self-driving fleet, to be operational by mid-2019. *Reuters*, 2019-04-23 Issues. London, UK [Online]. Available: https://www.reuters.com/article/us-waymo-selfdriving-idUSKCN1RZ1IZ [Accessed on May 14, 2021].

Samuel, P. 1999. How to 'build our way out of congestion'. Policy Study Report No. 250. Reason Foundation, Washington, DC [Online]. Available: https://reason.org/policy-study/how-to-build-our-way-out-of-co/ [Accessed on March 21, 2021].

Schmidhuber, J. 2009. Prof. Schmidhuber's highlights of robot car history. Dalle Molle Institute for Artificial Intelligence Research, Manno, Switzerland [Online]. Available: https://people.idsia.ch/~juergen/robotcars.html [Accessed on May 13, 2021].

Schwarz, C., Thomas, G., Nelson, K., McCrary, M., Schlarmann, N., and Powell, M. 2013. Towards autonomous vehicles. Report No. MATC-UI: 25-1121-0003-117. Mid-America Transportation Center, The University of Nebraska-Lincoln, Lincoln.

Science Digest. 1958. Electronic highway of the future. *Science Digest*, April 1958 Issue. Hearst Magazines, Des Moines, IA [Online]. Available: http://blog.modernmechanix.com/electronic-highway-of-the-future/ [Accessed on May 13, 2021].

Shutterstock, M.-F. 2016. Aerial image of Hollywood Young Circle Arts Park and Commons construction site, in Hollywood, FL, USA on December 21, 2016. Shutterstock, New York [Online]. Available: https://www.shutterstock.com/image-photo/hollywood-december-21-aerial-image-young-542225908 [Accessed on February 10, 2021].

Shutterstock, D.-L./ 2019. The Bosporus strait with sea traffic, ships and boats, and the Bosporus bridge in Istanbul, Turkey. Shutterstock, New York [Online]. Available: https://www.shutterstock.com/image-photo/bosporus-strait-sea-traffic-ships-boats-1534255307 [Accessed on February 10, 2021].

Sims, A., and Dobinson, K. 1980. The Sydney Coordinated Adaptive Traffic (SCAT) system philosophy and benefits. *IEEE Transactions on Vehicular Technology* 29(2), 130–137.

Slosson, M. 2012. Google gets first self-driven car license in Nevada. *Reuters*, 2012-05-18 Issue. London, UK [Online]. Available: https://www.reuters.com/article/uk-usa-nevada-google/google-gets-first-self-driven-car-license-in-nevada-idUSLNE84701320120508 [Accessed on May 14, 2021].

SteadieSeifi, M., Dellaert, N.P., Nuijten, W., Van Woensel, T., and Raoufi, R. 2014. Multimodal freight transportation planning: A literature review. *European Journal of Operational Research* 233(1), 1–15.

Stoklosa, A. 2020. Tesla puts "Beta" version of full self-driving capability in hands of select few. *Motor Trend*, 2020-10-22 Issue. El Segundo, CA [Online]. Available: https://www.motortrend.com/news/tesla-full-self-driving-beta-capability-testing/ [Accessed on May 14, 2021].

Su, Q., and Wang, D.Z. 2020. On the commute travel pattern with compressed work schedule. *Transportation Research Part A: Policy and Practice* 136, 334–356.

Sumit, B., Anna, G., Methody, G., Estel, M., Mihaela, M., Zion, S., Josh, T., Tamara, P., Carrie, S., and Dagmawit, W. 2020. Country-level commuting mode share. *Wonder*, May 15th, 2020.

Taeihagh, A., and Lim, H.S.M. 2019. Governing autonomous vehicles: Emerging responses for safety, liability, privacy, cybersecurity, and industry risks. *Transport Reviews* 39(1), 103–128.

Tender, M.L., Couto, J.P., and Bragança, L. 2017. The role of underground construction for the mobility, quality of life and economic and social sustainability of urban regions. *REM- International Engineering Journal* 70(3), 265–271.

THEA. 2016. Courtesy photo of Selmon expressway viaducts. In Transportation Authority Monitoring and Oversight: Fiscal Year 2015 Report by Florida Transportation Commission. Tampa Hillsborough Expressway Authority, Tampa, FL.

Toole Design. 2019. Courtesy image of conceptual design of Young Circle in Hollywood, FL, USA. Toole Design Group, Silver Spring, MD [Online]. Available: https://www.sun-sentinel.com/local/broward/hollywood/fl-ne-young-circle-redesign-hollywood-20190502-story.html [Accessed on May 14, 2021].

Trafficware. 2019. *Synchro Studio 11 User Guide*. Trafficware, LTD, Sugar Land, TX.

TRB. 2010. *Highway Capacity Manual*, 5th Edition. Transportation Research Board, National Academies Press, Washington, DC.

TRB. 2016. *Highway Capacity Manual; A Guide for Multimodal Mobility Analysis*, 6th Edition. Transportation Research Board, National Academies Press, Washington, DC.

TTI. 2021. Managed (HOV-HOT) lanes, Technical Strategies Series 2021. Texas Transportation Institute, Texas A&M University, College Station, TX.

TunnelTalk. 2015. Boundaries busted for highway under the Bosporus, TunnelTalk [Online]. Available: https://www.tunneltalk.com/Turkey-24Sep15-Eurasia-highway-tunnel-crossing-of-the-Bosphorus-in-Istanbul.php [Accessed on March 21, 2021].

USDOT. 2014. Connected vehicle TEST bed: Testing connected vehicle technologies in a real-world environment. Intelligent Transportation Systems Joint Program Office, U.S. Department of Transportation, Washington, DC [Online]. Available: https://www.its.dot.gov/research_archives/connected_vehicle/dot_cvbrochure.htm [Accessed on: May 13, 2021].

Wallace, R., Stentz, A., Thorpe, C., Moravec, H., Whittaker, W., and Kanade, T. 1985. First results in robot road-following. *Proceedings of the 9th International Joint Conference on Artificial Intelligence*, Los Angeles, CA, 1089–1095 [Online]. Available: https://www.ijcai.org/Proceedings/85-2/Papers/086.pdf [Accessed on May 13, 2021].

Whelan, R. 1995. *Smart Highways, Smart Cars*. Artech House Publishers, Boston, MA.

Williams, E. 2019. Does Tesla's Autosteer make cars less safe? *Hackaday*, 2019-03-04 Issue. Pasadena, CA [Online]. Available: https://hackaday.com/2019/03/04/does-teslas-autosteer-make-cars-less-safe/ [Accessed on May 14, 2021].

Wolshon, B., and Pande, A. 2016. *Traffic Engineering Handbook*, 7th Edition. John Wiley & Sons, Hoboken, NJ.

Wood, S.P., Chang, J., Healy, T., and Wood, J. 2012. The potential regulatory challenges of increasingly autonomous motor vehicles. *52nd Santa Clara Law Review* 4(9), 1423–1502.

Chapter 5

Building out transit and multimodal transportation

Multimodal transportation networks mean integrating transportation facilities that serve people travelling by auto, transit, ridesharing, micromobility, and non-motorized modes and goods shipments typically by truck, freight rail, and cargo-drone. Taking passenger travel for example, the segments of origin-destination (O-D) paths associated with individual travelers "link" modes at various stages of their trips with differences in the point of origin, location of destination, duration, and length. Individual travelers make choices of destination location, arrival time, travel mode, departure time, travel path, and facility use based on accessibility, convenience, out-of-pocket costs, and trip time. In megacities, access to a wide range of travel modes is central to sustaining mobility. This chapter focuses on capacity provision and multi-tier modal integration as it relates to transit and freight.

Central to successful transit is an adequate transit network, top-quality transit vehicles, and effective measures for system operations and management. Adequate transit route coverage avoids excessive transfer requirements and wait times between routes and inadequate service capacity, both of which discourage use. Poor-quality transit vehicles with low standards of on-time service and riding comfort discourages use (Wunsch, 1996; Cordahi et al., 2018). Transit is a service, and good service and a customer-centered culture help provide an attractive service. As does ensuring safety, including safe vehicle handling without aggressive driving maneuvers and violence between operators, sustained high and consistent operating speeds, frequent service, avoiding transit-induced traffic blockage, and mitigating transit-related noise and emission impacts. Sensible fare and subsidy policies are also crucial.

Multi-tier integration begins by taking advantage of integrated land use and multimodal transportation planning to integrate planning, design, construction, and operations of auto and transit modes, including transit sub-modes, and possibly other modes involved with the use of private vehicles (Rondinelli and Berry, 2000). The next tier of multimodal integration adds microtransit, non-motorized modes, emerging technologies, and perhaps trucks, freight rail, and cargo drones for goods movements to some extent. It is easy to see how 3D thinking can be applied to multimodal transportation facilities to help integration, with underground transit, parking, and perhaps pedestrian and goods movement flows; ground-level movement of all types; and elevated pedestrian, bike, scooter, and microtransit, parking structures; and even elevated roads or transit, etc. Through inversion, mixing modes, or maximizing access, the 3D transportation framework can dramatically expand what is possible.

DOI: 10.1201/9780429345432-5

5.1 TRANSIT CAPACITY PROVISION

5.1.1 Transit performance benchmarks

Benchmarks of transit capacity provision are concerned with transit demand, network configuration, fleet provision, staff productivity, operations management, and financial performance (Fielding, 1992). For transit demand, benchmarks may be focused on the role and importance of transit in terms of total travel demand, transit mode share and daily ridership, transit sub-mode shares and daily ridership, average daily trips per transit vehicle, and average ridership per transit vehicle trip (Lee and Vuchic, 2005; Vuchic, 2005). Concerning network configuration, ensuring adequate transit network connectivity and accessibility is key to attract potential transit users (Hadas and Ceder, 2010; Al Mamun and Lownes, 2011; Foth et al., 2013). For fleet provision, emphasis may be given to total fleet size and distribution of vehicle type, passenger capacity per vehicle, amenities, and age (Li et al., 2008).

For operations management, the productivity of staff including drivers, conductors, dispatchers, maintenance crew, and administrative and managerial personnel directly affects the quality of service to transit riders, efficiency of passenger loading to individual transit vehicles, vehicle bridging, and overall transit capacity utilization. To assess vehicle-driver assignment and utilization performance, desirable benchmarks may include ratios of different categories of staff to licensed vehicles in a transit agency; annual average days worked per driver; annual average days per vehicle available for service; average percentage of available vehicles actually used per day; percentage of peak-only vehicles available for service and actually used per day; average daily mileage driven per driver or per vehicle; average mileage per incident triggered by crash; vehicle disablement; temporary work zone for route, line, guideway repair, and maintenance; average daily passengers served per driver or per vehicle; daily average load factor; daily peak-hour maximum load factor; average daily mileage with no passenger served per driver or per vehicle; average daily passenger-miles of travel per driver or per vehicle; and average daily passenger-hours of travel per driver or per vehicle.

For financial performance, key issues are affordability from both the transit rider's and operator's perspectives. Affordability of fares by transit riders is essential. On the operator's side, the benchmark should be established to assess the ratio of cost recovery between revenues of farebox earnings and other types of incomes such as advertisement fees and subsidies and costs of transit operations including staff salaries and wages, office expenses, building maintenance, vehicle storage, vehicle operations, vehicle maintenance and repairs, and fixed asset preservation.

These benchmarks are tied in with performance-based mobility management that helps to ensure that adequate resources are allocated to the transit system that would help support high mobility (Sen et al., 2011). The U.S. Federal Transit Administration (FTA) points out that mobility performance metrics should be traveler-centric and help determine how well a system meets the needs of individual travelers, how well the system performs while meeting overall travel demand, and what the system's impact is locally and nationally (FTA, 2020). The World Bank's evaluation of its support of urban transport projects takes a similar focus on mobility performance metrics for judging the success of projects and their cost effectiveness (World Bank, 2017).

5.1.2 Transit network planning and design

Many planners focus on maximizing transit mode shares, which requires attracting potential transit travelers to become actual transit riders. For transit capacity expansion, the connectivity and accessibility of the existing transit system should be assessed to ensure that the planning of the augmented transit network would deliver a well-coordinated network with extensive route coverage, limited transfers between transit sub-modes and routes, and regular service frequencies. Practically, transit route planning always lags changes in transit demand. Plans for added transit capacity must be kept under regular review and revised as necessary using the most current data or even refined proactively using forecasted data to keep abreast of changes in population, total travel demand, transit mode share, and travel patterns. Regular route revisions are also necessary at the operations level (Fielding, 1987; Vuchic, 2005).

During a transit trip, transfers from one transit route to another of the same transit sub-mode, or even across different transit sub-modes, are common. This is particularly likely in a hub-and-spoke transit network where major routes or trunk lines meet at a central focal point such as a transit hub at one of a city's central places. It can also occur when passengers traveling between any two points in the urban area are required to transfer from one route to another. One possible reason for adopting such a network is that some unproductive routes other than the end-to-end trunk routes are not provided. Historically, the coverage of transit routes in some urban areas suffers from disconnects at jurisdictional boundaries when jurisdictions fight for market share. Cooperative arrangements should be sought to optimize routes for the travel patterns of users and share revenues between jurisdictions appropriately.

Poor transit route coverage across the megacity which might cause aggravated inconvenience from transfers, especially for non-radial transit riders, should be avoided. As with roads, an interconnected spiderweb network provides the most mobility and connectivity, facilitating more flexible modes or combining modes. One option for route planning is to add routes linking various central places and circumferential areas as well as feeder routes to trunk routes and major transfer hubs. The combined use of end-to-end trunk routes, feeder routes, and circular routes could help maximize network coverage. The second option for improving the transit coverage is to consider a combined use of straight and circular route sections where the transit route operates in a loop at the end of a straight outward transit route in the CBD area and rejoins the inward route after completing the loop. This could help provide better service coverage than a terminal operation since passengers are able both to board and alight throughout the circuit. The third option is to deploy a straight route in conjunction with fork straight routes or loop routes near one or both ends to serve different terminal points to integrate transit services in central places and sub-regions. The applicability of these options for route coverage improvements is partly influenced by policy and market forces, the nature of the urban street network used by transit vehicles such as buses and streetcars.

Another issue for route planning for an expanded transit network is to limit walking distances of transit riders boarding, alighting, and making transfers. To this end, transit routes will need to consider penetrating further into residential areas with stops, stations, or terminals readily accessible from the potential riders. Emerging technologies in microtransit and personal and shared mobility devices are making this easier. In the presence of parallel urban streets along a transit corridor, tradeoffs need to be made between the options of concentrating all routes along one street and splitting the transit routes to multiple urban streets. Compared with the distributed layout of transit routes, the concentrated layout of transit routes will result in longer walking distances for some transit riders, require higher transit service frequencies, and potentially cause traffic congestion to the transit compatible urban streets but will likely reduce waiting times.

Unfortunately, many conventional transit planners have focused on the surface transportation system as a "zero-sum game," assuming that cars, trucks, and transit compete for the same amount of space (or real estate). A 3D framework for thinking about mass transit investments, however, allows planners and policymakers to break out from this simplistic thinking. Tunnels and elevated transit facilities allow for a more varied and nimble transportation network. Indeed, a 3D network should enable a wider range of access to transit modes and sub-modes and transfers between modes that are uniquely suited to serving the needs and preferences of individual travelers. Some urban centers are inevitably congested due to high employment or shopping densities, or very expensive to individual consumers because of very high congestion charges. Providing easy access to alternative non-automobile modes, whether pedestrian, microtransit, or mass transit with integrated regional transportation planning can reduce travel times as well as increase accessibility to desired destinations. Thus, decision-making and transit investments should pay careful attention to the needs and preferences of the transportation system's users.

5.1.3 Transit vehicles

With a well-planned transit network in place, proper transit vehicles should be selected to match the transit demand with travel intensities that vary significantly by time and location. Unlike highways and roadways, where capacity is most often determined by the size of the facility itself, transit capacity is managed through size and capacity of the vehicles using the network. The choice of transit vehicles is mainly affected by characteristics of vehicle size and type, speed, and flexibility as transit vehicles serving an overall transit system play complementary roles within a specific transit sub-mode and across transit sub-modes.

The choice between the use of multiple small-sized vehicles and a single large vehicle to meet the needs and expectations of transit riders and operators is context sensitive. Factors influencing transit vehicle selection may include transit ridership, service area, service type, terrain, network configuration, route-specific geometric design and traffic control, operating speed, background traffic impacts, safety, and costs of vehicle purchasing, maintenance, and repair (AASHTO, 1978; TCRP, 2002).

5.1.3.1 Bus transit

For bus transit services, various types of buses configured according to body and chassis layouts, mechanical specifications, and in-vehicle amenity feature standards may be adopted. The buses could be single deck or double deck with varying lengths. Double-deck buses have advantages of providing a high seating capacity within a limited space, and occupying less road, terminal, and depot floor space per seated passenger but have disadvantages of increased loading and unloading times, higher manufacturing costs, and demanding for greater vertical clearances that would be limited to bus routes due to overhead obstructions, low bridges, and headroom in depots.

5.1.3.2 Bus rapid transit

The term "bus rapid transit (BRT)" first appeared in the 1937 Chicago public transit plan. During this pre-World War II period, buses were rapidly replacing street railways for high ridership transit corridors in many U.S. urban areas (Levinson et al., 2002). Compared with regular bus transit, it could reach higher operating speeds with reduced speed variations owing to significantly reduced or even no intersection delays facilitated by bus signal priorities (BSPs) given at intersections. BRT can easily achieve the same capacity as fixed guideway systems, given sufficient vehicles and technology to allow close following but have the advantage of being much more flexible. Any vehicle can leave and enter the "express flow" or the BRT route, so the ability to mix express vehicles with vehicles' running routes with more stops is much greater than with fixed guideway, as is the ability to adjust the types and level of service with seasonal changes, changes in travel or development patterns, facility maintenance, etc., without compromising levels of service. Analysis of BRT projects around the globe have found them to compete with fixed guideway systems in performance and are much more cost-effective and flexible (Kumar et al., 2012; Cervero, 2013; Wang et al., 2013).

Rather than the rigid single-deck bus for regular bus operations, the articulated single-deck bus or large-sized bendy bus with the rear body section connected to the main body by an accordion-like joint mechanism allowing for sharp turns and curves appears to be effective for a rapid bus corridor. Owing to longer sizes, articulated buses may be operated along relatively wide and straight bus corridors. They also need specially designed accesses and depot facilities, coupled with experienced operators to ensure safe maneuvers. The more complex construction of articulated buses makes them more costly for purchasing, maintenance, and repair.

5.1.3.3 Fixed guideway transit

Fixed guideway transit typically consists of light rail, heavy rail, commuter rail, and newly evolving automated people mover (APM) modes. Both light and heavy rails are with motive capability, typically driven by electrical power taken from overhead lines. Light rail may not need to run on exclusive right-of-way, which could be deployed by mounting tracks on elevated structures or having the tracks installed on urban streets, rendering light rail vehicles being called streetcars or trolley cars. Unlike light rail, heavy rail is usually operated on exclusive right-of-way, either by mounting tracks on elevated structures or inside underground tunnels called subways or metros. Commuter rail vehicles can be largely grouped into commuter rail non-self-propelled passenger coaches, self-propelled passenger cars that carry passengers, and locomotives that pull or push the passenger coaches or cars. An APM is a type of small-scale automated fixed guideway transit serving relatively small areas such as airports, downtown districts, or theme parks. The APM vehicles could be propelled by conventional on-board electric motors, linear motors, or cable traction.

5.1.4 Transit signal priority

5.1.4.1 Bus signal priority

For bus transit, the existing methods to provide BSP at signalized intersections could be categorized as: (i) link-based; (ii) intersection-based; and (iii) integrated measures and strategies (Hounsell et al., 1996). Examples of link-based measures include with-flow bus lanes, contra-flow bus lanes, bus-only streets, and busways. Bus-moving priority methods at traffic signals are examples of intersection-based measures. Whereas the combination of traffic signal measures with physical measures, queue relocation and virtual bus lanes are examples of integrated measures in which any of these systems alone is not effective.

- *Link-based bus priority.* The most common used link-based bus priority solutions are with-flow bus lanes, which are reserved traffic lanes, usually on the nearside, for the use of buses and may accommodate bicycles (DETR, 1997). A with-flow bus lane allows buses to bypass traffic queues and produce substantial time savings to buses and their passengers.

A contra-flow bus lane is a bus-only lane where buses are allowed to travel against the main direction of traffic flow. This enables buses to avoid unnecessary diversions, to maintain route patterns when new one-way streets are introduced, and to gain better access to business and shopping areas. Contra-flow bus lanes are normally adopted in area-wide one-way traffic systems, where the effect is to create a two-way road with "buses only" allowed in one direction and all other vehicles including buses in the other direction (IHT, 1987, 1997).

Corridors or networks of bus-only sections of the road network constructed specifically for the exclusive use of buses are called busways. They are designed to segregate buses from general traffic that protect them from congestion. For reasons of economy and land requirements, automatically guided or tracked busways may be preferred over those leaning on manual steering (IHT, 1987, 1997).

- *Intersection-based bus priority.* Intersection-based measures vary from an exemption for buses from turning restrictions applied to other vehicles to technology-led adaptive bus priority at traffic signals. The BSP is an important way of bus priority measure in urban areas without implementation of additional infrastructures.
- *Integrated measures and strategies.* Various bus priority options available could be summarized as passive priority, which weights or re-optimizes the signal timings to take account of streams of traffic containing significant bus flows, and active priority, which make the traffic signal responsive to approaching individual buses and give them priorities to pass the intersection (DETR, 1997).

Passive bus priority strategies consist of methods for developing signal timing plans concerning cycle length, green times, and offsets to favor transit along signalized arterials. They do not require any detection and request systems. In general, when transit operations are predictable (e.g., consistent dwell times), transit frequencies are high and traffic volumes are low; passive priority strategies can be an efficient form of BSP. However, in general situation in urban areas, active control systems consistently provide greater benefits to the buses than passive priority systems (ITSA, 2004).

Active bus priority control strategies generally include green extension, red truncation, phase insertion, and queue jump phase. Green extension is a strategy that will extend the green phase for the approaching transit vehicle. This strategy only applies when the signal is able to keep green for the approaching BSP-equipped buses. Green extension is one of the fundamental and most effective strategy. It does not require additional clearance intervals and allows buses to be served in the current green phase. Red truncation, which is also called early green or recall, is a strategy that will truncate the current red phase, or equivalently, start the next green phase earlier. For phase insertion, a special priority phase is inserted within the normal signal sequence. An example of phase intersection would be a leading left-turn-only phase specifically for transit vehicles. Queue jump phase is a special case of actuated transit phases. It allows a transit vehicle to enter the downstream link ahead of the normal traffic stream.

The technique of BSP that assigns different priorities to two types of vehicles, i.e., autos and buses moving through an isolated signalized intersection or multiple intersections along an arterial corridor can be popularized to multiple types of vehicles, such as autos, buses, microtransit, and connected and automated/autonomous vehicles that are gradually adding to the traffic stream. Also, the geographic coverage of signal priorities could be expanded from one or more arterial corridors to all major streets in an urban area. This is expected to maximize the capacity provision and utilization of intersections in an urban street network that will help improve urban mobility greatly.

5.1.4.2 Bus signal priority in SCOOT

Split Cycle Offset Optimisation Technique (SCOOT) system have a number of facilities that can be used to provide signal priority to buses. Passive priority, which does not differentiate between vehicles, can be given to links or routes using split and offset weightings. As all vehicles on the weighted link receive a similar benefit, the level of priority that can be given is limited. Active priority can be given to individual buses: extensions to prevent a bus being stopped at the start of red and recalls to start the bus green earlier than normal. In addition, in SCOOT MC3, intermediate stages between the current stage and the bus stage can be skipped. Differential priority allows different levels of priority to be given to certain buses, e.g., limited priority to late buses and high priority to very late buses, but no priority to those ahead of schedule. All these techniques are controlled by user set parameters to prevent the priority causing undesired excessive delays to other vehicles (DETR, 2000). Figure 5.1 presents the concepts of extension, recall, and skipping strategies in SCOOT. A four-phase signal timing plan is assumed, and the first phase is the green phase for the bus that requested a BSP.

The SCOOT kernel software allows for buses to be detected either by selective vehicle detectors, i.e., using bus loops and transponders on buses, or by an automatic vehicle location (AVL) system. Bus loops, or AVL systems where bus detection points can be specified, have an advantage as they can be placed in optimum positions. The best location for detection usually is a compromise between the need for detection as far upstream as possible and the downstream of any bus stop. This is because SCOOT does not attempt to model the time spent at bus stops and the journey time from downstream of the bus stop to the stop line for each bus needs to be accurately predicted. Depending on site conditions, a bus detection location giving a bus journey time of 10–15 seconds to the stop line is recommended (DETR, 2000; Oliveira-Neto et al., 2009).

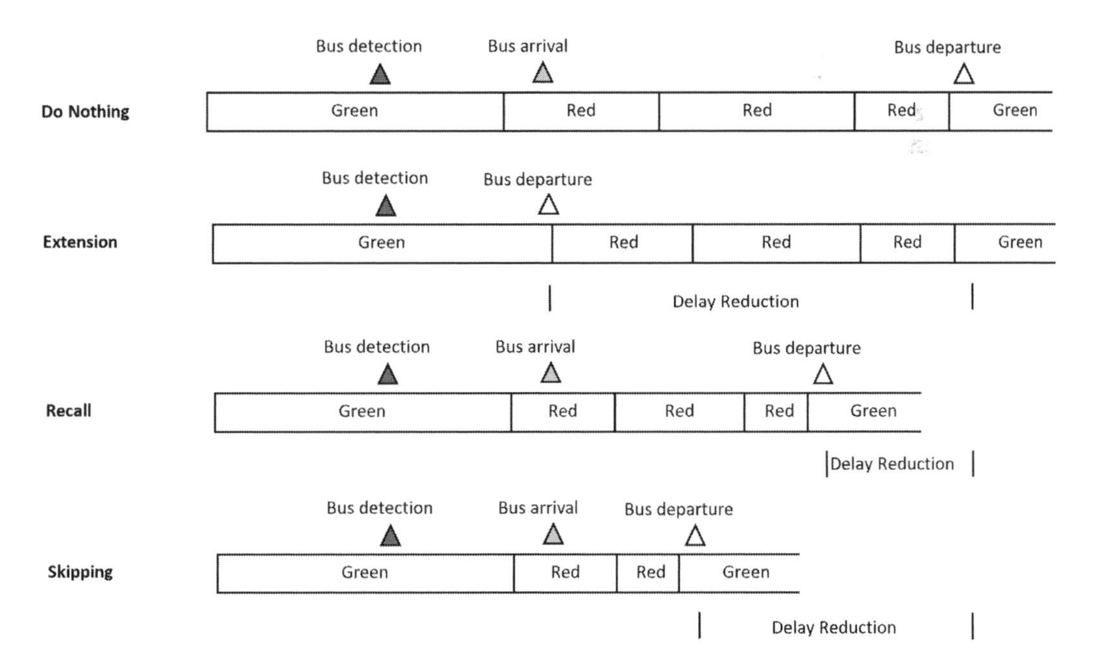

Figure 5.1 Bus signal priority in SCOOT.

5.1.5 Cases in action

5.1.5.1 Downtown Seattle Transit Tunnel

The Downtown Seattle Transit Tunnel, also known as the Metro Bus Tunnel, is a 1.3-mile (2.1-km) long, 18-ft (5.5-m) wide double-track transit tunnels in Seattle, Washington. It is one of two tunnels in the United States shared by buses and light rail trains. (The other facility is being the Mount Washington Transit Tunnel in Pittsburgh, Pennsylvania, and is the only one in the United States with shared stations until Seattle renovated its tunnel.) The construction of the transit tunnel began in March 1987. It was opened in September 1990, rebuilt from 2005 to 2007, and reopened for service in September 2007. The total cost of the tunnel was US$468.7 million in 1984 dollars (equivalent to US$917 million in 2019 dollars). The tunnel was used by buses only before the reconstruction in 2005 and shared by buses and light rail from 2009 onward. The operating speed was set at 30 mph (48 km/h).

Bus routes from King County Metro and Sound Transit Express leave the tunnel north via Interstate 5, south via the SODO Busway, or east via Interstate 90. Owned by King County Metro and shared with Sound Transit through a joint-operating agreement signed in 2002. The transit tunnel is part of the "Third Avenue Transit Spine," the busiest transit corridor in Seattle, which has a theoretical capacity of 40 trains per hour per direction with a minimum of 90-second headways, carrying 22,000 passengers per hour per direction. As of 2012, the transit tunnel carried 52,600 daily riders with ~20% riders taking the light rail service (TunnelTalk, 2009; SoundTransit, 2020) (Figure 5.2).

5.1.5.2 Los Angeles Metro Busway system

The Los Angeles Metro Busway system is a limited-stop BRT system with exclusive or semi-exclusive roadways known locally as busways or transitways running in the medians of Interstate highways. There are currently three primary corridors in operation, including

Figure 5.2 Illustration of the Downtown Seattle Transit Tunnel, Washington. (Source: TunnelTalk (2009).)

the 11-mile long El Monte Busway which is an express bus corridor running west along the median of Interstate 10 (San Bernardino Freeway), the 11-mile long Harbor Transitway that runs in the median of Interstate 110 (Harbor Freeway), and the 18-mile long G Line (evolved from the 14-mile Orange Line) Busway that uses a dedicated, exclusive right-of-way for the entirety of its route with stations located at ~1-mile intervals in the San Fernando Valley. Services along these three busways began in 1973, 1995, and 2005 respectively, with 2019 daily ridership at 16,700, 16,000, and 22,500 passengers. The El Monte Busway and the Harbor Transitway were initially designed for shared use with high-occupancy vehicles such as three-person carpools and were upgraded from high-occupancy vehicle lanes to high-occupancy toll lanes from 2012 onward (Niles and Jerram, 2010).

The busway system was instituted primarily to improve bus service quality by increasing bus operating speeds and reducing bus bunching, passenger waiting time, and passenger standing loads. A two-phase incremental deployment strategy was adopted by incorporating measures of frequent service, BSP, headway-based scheduling, simple route layout, less frequent stops, integrated with local bus service, level boarding and alighting, color-coded buses and stations, high-capacity buses, exclusive lanes, off-vehicle fare payment, and feeder network. In this case, the G Line evolving from the Orange Line becomes a showcase corridor that offers a premium, high-capacity rapid transit service. The Los Angeles Metro originally purchased the right-of-way of an abandoned railroad to build a rail line to provide rapid transit ride from San Fernando Valley to downtown Los Angeles but was unable to implement the plan due to political opposition. Building a BRT corridor became an alternative solution to offer rapid transit service at a lower cost than that of a light rail or subway line.

The 14-mile Orange Line including busway, stations, vehicle fleet, Intelligent Transportation Systems (ITS) installations, fare collection, and others was deployed within a budget of US$350 million with slightly more than 50% of the cost designated to the busway. For the BRT elements, the unique features of the corridor include dedicated at-grade bus running way with 36 intersections and five midblock pedestrian crossings, open design stations of ~1-mile spacing with protective shelter, seating, lighting, real-time traveler information displays, and other passenger amenities, 60 ft long high-capacity, low-floor articulated buses of with rail-like appearance referred to as "Metroliners," 4-minute headways at peak and 15- to 20-minute headways at off-peak, all-stop service only, off-coach fare collection, and synchronized signals at all intersections, coupled with strong marketing to identify premium "rail-like" rapid transit service. Compared with regular bus service prior to the BRT corridor deployment, the average BRT running speeds improved by 16% in the peak period, which are 16.5–16.8 mph for westbound and 20.5 mph for eastbound. The non-peak speeds increased by 18% (Niles and Jerram, 2010) (Figure 5.3).

5.1.5.3 Lagos BRT-Lite in Lagos, Nigeria

Lagos is the largest city, center of finance, commerce, and industry, and biggest port in Nigeria. The Lagos metropolitan area covers a land area of 2,700 km^2 with a population over 21.3 million in 2015, making it the second most populated city in Africa after Cairo, Egypt (Brinkhoff, 2015). The city's main commercial and government centers and largest business district are on Lagos Island, which has only five bridges connecting it to the mainland where most of the population resides. With a large population, unrestricted imports of secondhand automobiles, and subsidized gas prices, the auto ownership and usage by its residents are maintained at a relatively high level. Due to scarcity of land and lack of integrated land use and transportation planning, the roadway network in Lagos is inadequate in coverage, capacity, and condition. Also, it is one of a few megacities with a population of

Figure 5.3 Illustration of Los Angeles Metro Busways in California. (Source: Niles and Jerram (2010).)

over 10 million without a formally organized public transportation system. Transit service is largely provided by minibuses (*danfo* and *molue*), shared-ride taxis (*kabu-kabu*), motorcycle taxis (*okada*), and auto rickshaws (*keke*). The high travel demand and its intensities, coupled with insufficient transportation capacity, have led to severe recurrent traffic congestion. Typical travel times for commuters to Lagos Island from the main residential areas to the north and west of the city on the mainland are more than 2 hours and are even longer in presence of incidents caused by vehicle breakdowns, crashes, or flooding on the main roads leading to the few bridge crossings.

After years of struggling with a lack of an adequate roadway network and a reliable transit system, the Lagos Metropolitan Area Transport Authority (LAMATA), the World Bank, and transportation planning consultants, in conjunction with the support of the Road Transport Employers Association of Nigeria (RTEAN), representing interurban and large-bus sectors, and the National Union of Road Transport Workers (NURTW) representing small-bus sectors, began to review BRT possibilities. This led to Africa's first BRT system which began operations on March 17, 2008, in Lagos, referred to as BRT-Lite because of certain design compromises reflecting a limited budget and a politically motivated short implementation time. Specifically, its delivery cost was on average US$1.7 million per kilometer and the duration from conception to operation took only 15 months (Mobereola, 2009). This experience is relevant to the many cities throughout the world seeking to develop a BRT system.

The BRT-Lite corridor operates along a 22-km route of which 65% is physically segregated from the regular roadway and 20% is separated by road markings and signage. Curb-side stations, low platforms with branded shelters, and fencing were designed to facilitate ticket checking before boarding. Front-engine buses with a high floor, body-on truck chassis, and two narrow doors were used with the front door used for boarding and the back door used for alighting. The BRT-Lite system is operated by the 1st BRT Co-operative (FBC) established by NURTW under the private-public financing approach, which is also called Lagos NURTW (1st BRT) Cooperative Society Limited.

An evaluation of the BRT-Lite system completed by LAMATA in the fall of 2008, revealed daily ridership was about 200,000 passengers a day. Within its first 100 days, the system carried 9.7 million passengers, and within its first 6 months of operation, 29 million passengers. Passengers enjoy a reduction of 30% in average fares from 140 Naira in the past to

100 Naira and a reduction of 40% in journey time, cut average waiting time by 35%, and experience a welcomed absence of exposure to theft on public transit. This has been made possible by the introduction of discipline in operations (route franchising), the increase in average speed from <15 to 25 km per hour (Mobereola, 2009).

Critical success factors were defined as a significant and consistent political commitment, the presence and abilities of a strategic public transit authority in LAMATA, a scheme definition that concentrates on essential user needs, deliverability within a budget and program, work undertaken to engage key stakeholders and ensure that they benefit, and a community engagement program that has worked to assure the residents that the BRT-Lite system is a community project created, owned, and used by them. Given the scale of the undertaking, it was recognized that outsourcing vehicle maintenance function, financial management, and operational management of the BRT-Lite system would help sustain its overall performance in the long run (Figure 5.4).

5.1.5.4 Transit priority system in Los Angeles

In 2008, the city of Los Angeles implemented a transit priority system (TPS) in two corridors covering 211 signalized intersections. The TPS improves the on-time performance of Metro Rapid buses by adjusting signal timing plans of intersections when those buses are detected. Dynamic passenger information signs were installed on bus shelters to provide real-time bus arrival information based on bus travel time data updated within 1-minute time interval (USDOT, 2008). The TPS has led to total transit time savings by 25%, intersection-related bus delay time by 33%, and increases in the overall bus running speed along the two corridors by 23%–29% (Figure 5.5).

5.2 RIDESHARING MODES

5.2.1 General

The more traditional demand-responsive modes for public transit include carpool and vanpool aimed at commuters. These modes use vehicles such as automobiles and standard vans, or "body-on-chassis" minibuses to provide service. When integrated into public transit systems, they are often used in conjunction with financial incentives such as subsidies, tokens, and exemptions, or reductions in toll and parking charges. When these subsidies reduce or focus on reducing the SOV auto mode share, transportation planners call these strategies "demand leveling."

The routing, time scheduling, and frequency of carpool and vanpool services are highly demand-sensitive, which could be all fixed or all flexible. A factory-built standard van carries 12 or 15 passengers including the driver that might not be user friendly for elderly and disabled riders. An improved version of such a vehicle is the minivan between a car and a van that could load up to seven passengers with the driver counted with raised roof and drooped floor to help boarding and alighting for riders with constrained mobility (Furuhata et al., 2013; Zhong et al., 2020). Another type of vehicles is the "body-on-chassis" minibus that are wider and taller than standard vans with walk-in, front entry doors and a center aisle designed to accommodate 16, 20, 24, or 28 ambulatory passengers excluding the driver and could serve wheelchair riders by swapping four seats for addition of each wheelchair tiedown (Amirkiaee and Evangelopoulos, 2018).

In recent years, ridesharing services become increasingly popular. Service providers range from public transit authorities such as the Washington, D.C. Microtransit serves as an

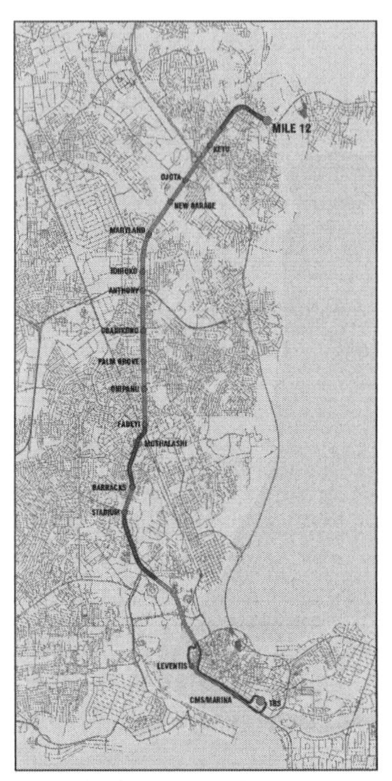

(a) BRT-Lite corridor

(b) Dedicated busway configuration

Figure 5.4 Lagos BRT-Lite system in Nigeria. (a) BRT-Lite corridor and (b) dedicated busway configuration. (Source: Mobereola (2009).)

Figure 5.5 Transit signal priority in Los Angeles, California. (Source: Frost (2019).)

extension of the existing public transit service, private enterprises including the traditional taxi companies, and newly emerging transportation network companies (TNCs) such as Uber and Lyft. The commercial rideshare industry emerged as a disruption of traditional taxi markets and has continued to evolve and innovate since then (Hensley et al., 2017). The defining characteristic of these services is the use of mobile applications to allow on-demand services that are hyper-flexible, simple, and convenient for passengers and provided by car drivers who also are given great flexibility by the service system to participate as providers. The services were pioneered in the United States, but their appeal to passengers and to potential drivers and car owners quickly spread them through the developed world and even into developing nations (Konishi and Mun, 2010; Asirin and Azhar, 2018; Eisenmeier, 2018; Paundra et al., 2020).

As they did with traditional carpool and vanpool options, now transit agencies are experimenting with working with ridesharing companies, and also with various forms of active transportation and microtransit discussed below, to provide more flexible and integrated mobility around the core transit network (Cmar, 2017; Van Wijnendaele, 2019; Hazan et al., 2019). In the United States and Canada, towns have even experimented with replacing traditional transit services with ridesharing (Hawkins, 2018; Cecco, 2019) (Figure 5.6).

5.2.2 Case in action: BART integrated carpool to transit access program

To promote carpooling ridership to Bay Area Rail Transit (BART) stations while ensuring control of demand intensities in peak periods, the Scoop app was developed for BART-based commuters to request or offer carpooling opportunities to their destinations at 9: 00 pm the day before their morning commuting or 3:00 pm in the afternoon or evening commuting. The app informs riders and drivers whether there is a carpooling match and provides map and navigation functions similar to those available from other apps such as Google navigation and Waze. The rider pays a nominal fee through the app to cover the driver's fee, plus US$1 premium to using the Scoop app (BART, 2018) (Figure 5.7).

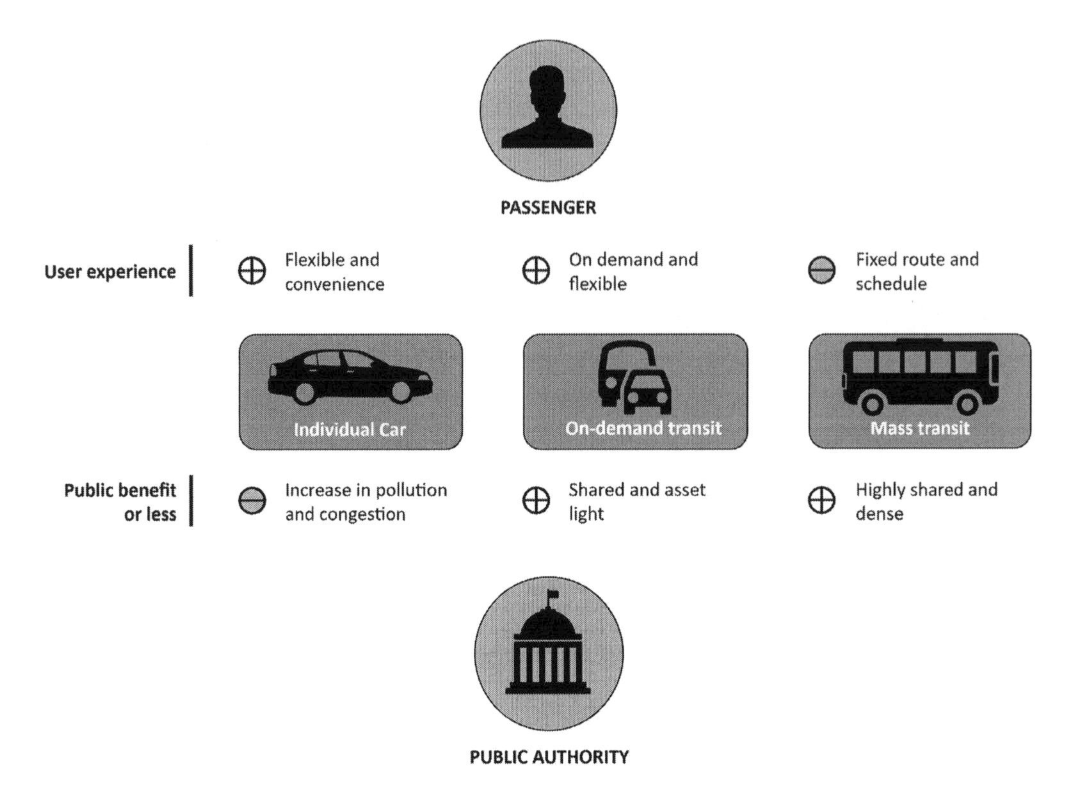

Figure 5.6 On demand transit versus personal vehicles/traditional transit.

5.2.3 Case in action: Seattle on demand microtransit

In 2019, the Seattle area started a pilot project of an on-demand service provided by Via, a private microtransit company (KCM, 2021). Riders in parts of Seattle can use a mobile application to request a ride from Via to a number of light rail or bust transit hubs. The goal of the project is to see if on-demand services to connect people to transit hubs will increase transit use and provide valuable mobility for riders. In its first year, the service averaged 4.2 rides per vehicle per hour, about 100,000 rides in 6 months, which exceeded initial goals (Hazan et al., 2019). Most trips are by people who live farther than 1 mile (1.6 km) from the transit hubs and opened up transit services to new passengers and cut down trip time and missed connections for many others (Figure 5.8).

5.3 ACTIVE TRANSPORTATION AND MICROMOBILITY MODES

Active transportation refers to human-powered, non-motorized modes of transportation such as walking, biking, and scooting. Expanding the capacity of active transportation can directly benefit the existing non-motorized travelers and those induced non-motorized travelers. Indirectly, it can bring additional travel cost savings, such as decreased automobile ownership cost and reduced traffic congestion. In addition, active transportation could lead to various types of equity benefits in terms of increased fairness in distribution of transportation impacts horizontally across individuals and groups, enhanced social justice in favor of lower-income groups to compensate for overall inequities, and mitigated barrier effect in

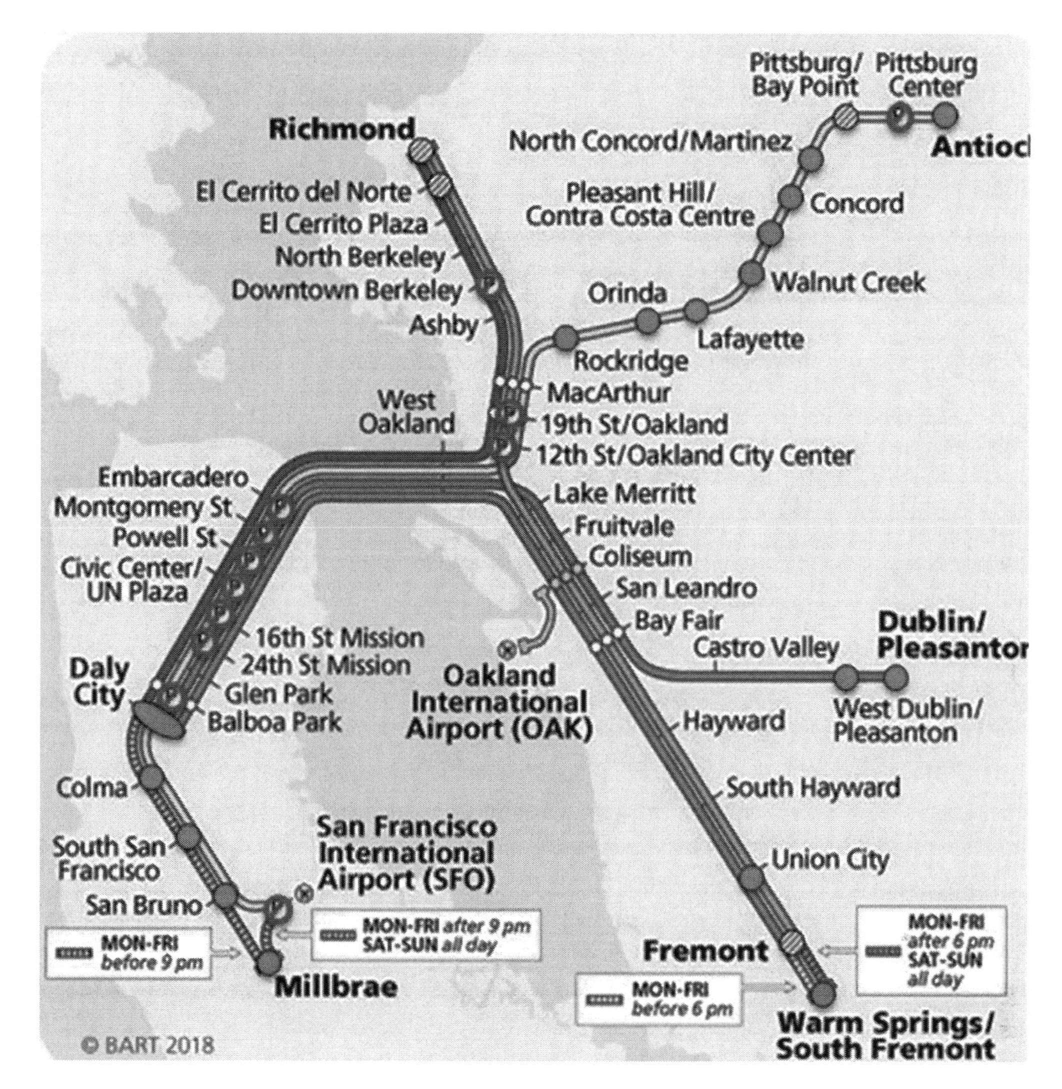

Figure 5.7 BART stations with Scoop app-supported ridesharing services in California. (Source: BART (2018).)

relation to travel delay in active transportation caused by the existing and induced vehicular traffic (Sallis et al., 2004; Litman, 2015).

Micromobility modes are personal mobility devices like electric-powered bikes and scooters, often offered as a shared mobility or on-demand system that uses docked stations where one can pick up and drop off a vehicle, or via a mobile application for dockless vehicles. But micromobility also tends to include active transportation modes such as human-powered bikes and scooters. Powered and unpowered becomes intermingled options in an ever more complex micromobility market and landscape.

A McKinsey analysis predicts that the number of passenger-kilometers traveled via micromobility will grow in the long run and will increase 5%–10% just by 2030. Their surveys find that currently over 80% of micromobility trips are for personal trips, not commuting, but also find wider willingness to use micromobility and greater attention among policymakers to policies to make micromobility more available and effective. Policy changes to

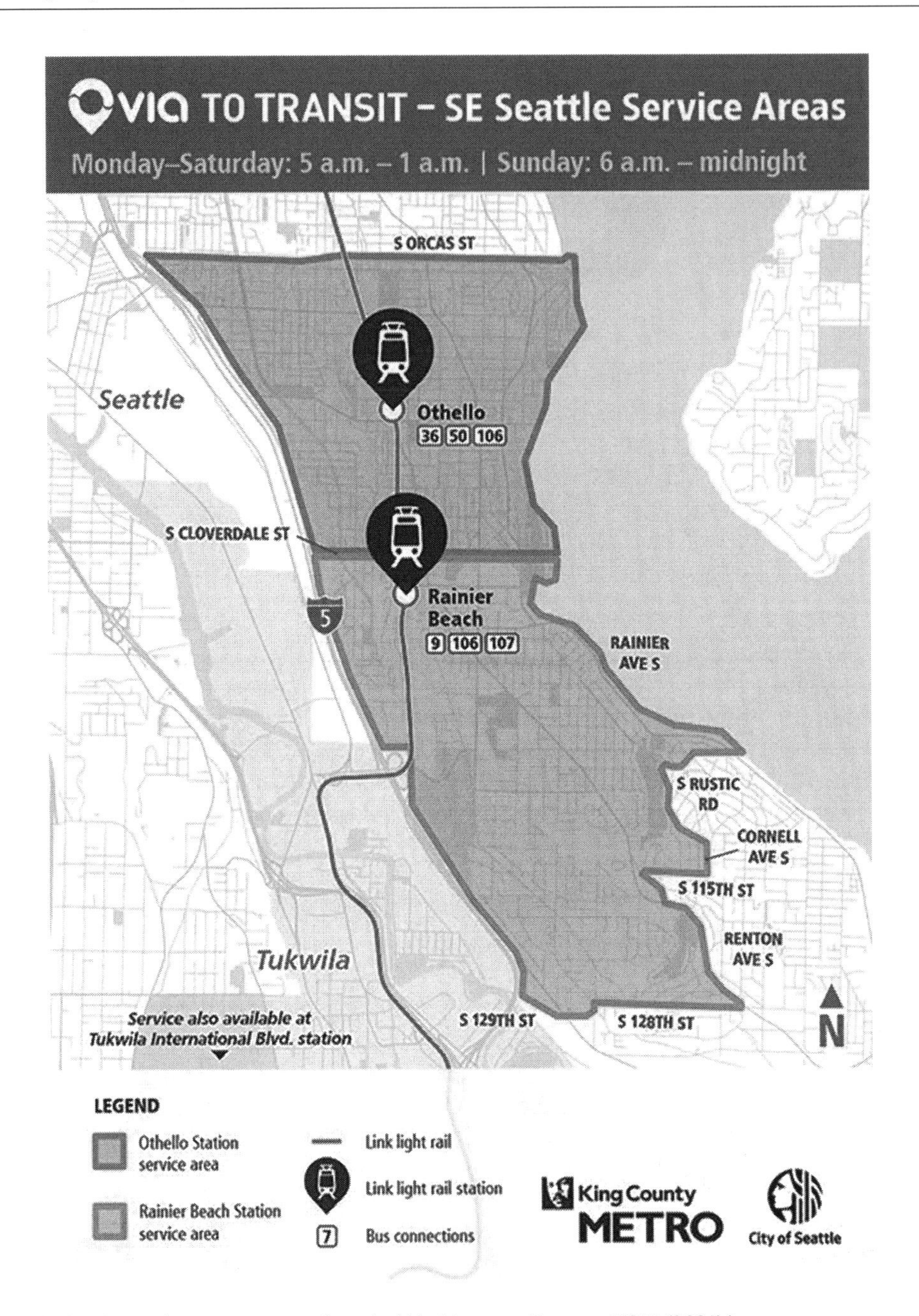

Figure 5.8 On demand microtransit in Seattle, Washington. (Source: KCM (2021).)

allow/encourage micromobility are crucial, ranging from removing barriers to private markets and innovations in this space, which are exploding, to helping create coordination mechanisms and conventions or rules where necessary to make use easy and convenient while managing externalities like sidewalk and curb misuse or safety issues (Chrysler et al., 2014; Hunkin and Krell, 2019).

More available, efficient, and popular active transportation and micromobility opportunities expand mobility and access over short distances and create better connectivity with traditional modes, especially improving travel via public transit by making it easier to connect to them. Many trips by transit become more feasible with the use of an active or micro-transit mode at either or both ends of the transit ride. A 2020 literature review of studies on integration of micromobility and public transit found a lot of research, most of which focused on understanding the potential of such integration, while actual measurement of impacts is less well developed in the literature (Oeschger et al., 2020). That early research does indicate that the potential for micromobility to improve transit access and utilization is promising.

At the same time, a richer micromobility landscape can also compete with public transit, providing new means of making short trips that might have been done before via transit buses or trains—one- or two-stop trips—which frees up transit capacity for longer trips. Entrepreneurs and others are creating business models around an "unbundling" of solo automobile trips into short trips via micromobility and longer trips via various transit modes, traditional or new (Dediu, 2017). An examination of micromobility in Brussels illustrates this point, exploring how microtransit fills niche traveler needs and helps meet a number of public policy goals for mobility and the environment (Figure 5.9).

5.3.1 Pedestrian walking facilities

Safe and efficient walking requires adequate pedestrian walkways, coupled with traffic control devices and safety hardware. Key components of pedestrian walkways include sidewalks, crosswalks, overpasses or underpasses, railing, fencing, curb ramps, and street furniture. Sidewalks are the most fundamental pedestrian facility separating pedestrian travel from motorized traffic. Crosswalks provide safer and easier crossings for pedestrians at complex locations. Pedestrian overpasses or underpasses help overcome impassable barriers and avoid conflicts with vehicle traffic from the designated right-of-way. Railing is often installed on the outside edge of a pedestrian overpass or underpass for the safety protection purpose. Pedestrian fencing provides positive separation between pedestrians, cyclists, and scooters from roadways to avoid conflicts with vehicular traffic. Curb ramps provide access for wheelchair users between the sidewalk and the street surface at intersections, bus stops, and midblock crossings. Street furniture, including benches, bus shelters, lighting, signage,

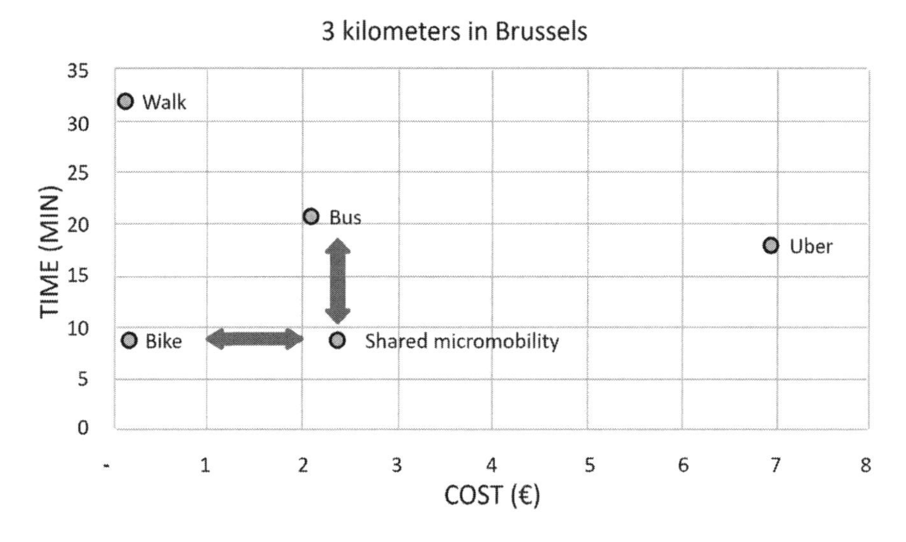

Figure 5.9 Short trips in Brussels, EU. (New drawing based on source: Van Wijnendaele (2019).)

trash receptacles, and water fountains, can enhance pedestrian's walking environment and walkability (Martincigh et al., 2010).

Traffic control devices and safety hardware that could enhance safety of pedestrians, cyclists, scootists, and vehicular traffic generally include signs, lighting, signals, and markings. Traffic signs provide critical information to help sustain safety and flow of vehicular travelers and nonmotorized travelers including pedestrians from improved information and alertness to hazards. Lighting enhances safety of all travelers, especially for nighttime travel, and enhances security of pedestrians from crime (Bushell et al., 2013). Traffic signals are designed for isolating conflict movements of vehicular and non-vehicular traffic entering an intersection meanwhile supporting efficient utilization of intersection capacity. The signal timing plan needs to ensure that the phase-based time duration is sufficient for pedestrians, cyclists, and scooter riders to safely cross intersection approaches. Some warning and alerting devices such as flashing beacons, pedestrian, bike, and scooter detection sensors, audible pedestrian cuckoos, pedestrian pushbuttons, and pedestrian countdown displays, as well as portable real-time speed trailers may be used to enhance pedestrian safety. Pavement markings showing pedestrian, bike, and scooter symbols, yield/stop signs, and island/curb/sidewalks markings will help pedestrians, cyclists, and scootists stay within the right-of-way and remind other travelers to stay off it to mitigate movement conflicts. In some roadway sections, bollards that are sturdy, short, vertical posts embedded in the ground can be installed to protect pedestrians, cyclists, and scooter riders by reducing vehicular speeds or limiting vehicle access to non-motorized travel areas.

5.3.2 Bike facilities

Bike facilities include bikeways, parking facilities, and traffic calming measures to reduce vehicle speeds and enhance safe travel of bicyclists. Bikeways include shared bicycle lanes, protected paths, and signed routes showing the safest way for O-D travel by bicycle. Bicycle parking facilities include bicycle racks, bicycle lockers, and bicycle stations to provide a closed space for safely storing, locking, and maintaining bicycles.

Traffic calming measures can be classified into measures of vertical deflections, horizontal shifts, and closures to ensure safe travel of cyclists. Vertical deflection measures include speed humps installed on pavements, long-speed humps or speed tables, and raised intersections. Horizontal shift measures largely contain curb extensions or chokers, a series of curb extensions called chicanes, and neighborhood traffic circles or roundabouts with raised islands placed in the centers of intersections to allow for reduced speeds of vehicles circulating on tight horizontal curvatures. Closure measures are islands built at intersections to prevent certain vehicular movements, including diagonal diverters, half closures, full-street closures, and median barriers.

5.3.3 Scooter facilities

As early as 1914, scooters were used as an active transportation mode, more affordable, easier to operate, and more convenient to park and store than automobiles. Starting in the late 1980s, the design of a modern two-wheel motorized scooter with step-through frame and integrated footrests for the rider, with a small-utility internal combustion engine or an electric motor or battery emerged. More recently smaller standing scooters like those in Figure 5.10 have become popular in cities across the globe.

Electric kick scooters have generally surpassed gas-engine scooters in popularity since 2000 (MyProScooter, 2020). The electric scooters in conjunction with bicycles, e-bikes, electric skateboards, shared bicycles, and electric pedal-assisted (pedelec) bicycles are

Figure 5.10 Scooter parking zone in San Diego, California. (Source: Shutterstock (2019) with permission.)

micromobility devices that are quiet, quick, and driven by users personally. Some of the electric scooters could technically reach a maximum speed of 40 mph (65 km/h) (MyProScooter, 2020). Owing to comparatively high speed, electric scooters are not suitable for use on sidewalks or pedestrian walkways. If a given roadway is bicycle compatible, scooter riders could travel within bike lanes and are allowed to make a left-hand turn by dismounting and crossing an intersection as pedestrians to ensure safely traversing through the intersection. Improvements in scooter technology and the addition of app-based shared and on-demand scooter services have led to a rapid growth in their use for urban mobility (VanderZanden, 2020) (Figure 5.10).

5.4 MULTIMODAL INTEGRATED PASSENGER TRAVEL

Perhaps more importantly than in highway capacity, mass transit interfacing with alternative passenger travel modes is crucial. Given the size of transit vehicles and the numbers of travelers carried by them, optimizing speed and delivering high-service quality is essential to transit performance. Fortunately, modern ITS technologies allow for these interfaces to be managed more efficiently and effectively. Moreover, given the regional footprint of megacities, new opportunities exist to deploy these technologies to travel modes extending from auto and transit modes to include ridesharing, active transportation, and microtransit modes on a larger scale and with a more comprehensive and holistic vision than in earlier decades.

5.4.1 General

Capacity expansion for passenger travel by means of multimodal integration essentially requires an integrated delivery of multimodal facilities primarily concerning auto and transit modes, transit sub-modes, shared mobility, and fixed guideway transit modes; and possibly a portion of freight shipment modes. At the same time, the growth of share ride services and share mobility devices with increasing ridership and mode shares in recent years need to be taken into consideration. Multimodal facilities bring together bus routes, fixed guideway lines, bus stops, rail transit stations, transfer hubs, terminals, parking and park-and-ride facilities, transit fleet maintenance and repair facilities, transit vehicle scheduling, vehicle monitoring, traffic control and access control facilities, ridership counting and yield management facilities, and traveler information provision facilities.

5.4.2 Case in action: multimodal passenger travel in Hong Kong, China

The uniqueness of multimodal passenger transportation in Hong Kong, China, is embodied in three aspects: a highly integrated multimodal passenger transportation network, preferential travel costs to different types of travelers especially those from aging and disabled groups, and provision and enforcement of laws for sustainable transportation development.

The primary non-auto travel modes in Hong Kong are bus, Mass Transit Railway (MTR), tram, taxi, and ferry modes. Collectively, these modes accommodate ~90% or 12 million daily person-trips generated by the residents in Hong Kong. For bus mode, over 300 bus routes are served by public light buses, franchised double-deck buses, and non-franchised resident minibuses, with complementary services provided by taxi cabs. Bus accommodate over 50% of all passenger travel. The vast majority of franchised buses are double-deckers with the highest utilization rate, leading to slightly higher than 30% of model share of all passenger travel. MTR consists of nine subway lines, one light rail line, and one airport express line with 152 stations in operation, which serves about 36% of all passenger travel. Trams mainly run along east-west routes carrying an average of about 230,000 passengers daily. In addition to special cruises to some important attractions, there are 18 licensed passenger ferry routes in Hong Kong providing services to and from outlying islands and inland ports, an important supplementary travel mode for residents and tourists. These diversified modes of passenger travel achieve accessibility and offer flexibility to choose preferred travel modes and routes.

To ensure efficient operations of the highly integrated multimodal passenger transportation system, preferential price measures have been implemented. For instance, the fare of MTR transit is set higher than that of bus transit. The MTR offers various forms of tickets, such as Badatong, one-way, and commuter tickets. For MTR subway lines, fares are divided into two types for general and concession travelers. Preferential fares are set for children under 12, seniors 65 and above, and full-time students aged 12–25. In addition, the MTR offers concessionary fares for Badatong users heavily traveling along the Airport express line. At the same time, the MTR subway riders can enjoy a 25% discount on travel fares during certain morning hours. For bus and tram services, concessionary fares are adopted for children under 12 and senior citizens 65 years or older, and eligible persons with disabilities. Concessionary fares are further expanded to cover MTR general lines, franchised buses, and ferries for eligible persons.

To effectively balance between urban travel needs and prohibitively expensive transportation facility deployments owing to land use scarcity, innovative transportation financing mechanisms such as transit-oriented development and transit joint development have been widely practiced, creating a win-win situation among travelers, transportation agencies, operators, and urban developers. The delivery of multimodal passenger travel facilities and operations management of public transportation is backed by the strict public transport law enforcement system in Hong Kong. For instance, increased penalties are imposed for misuse of one-way streets; illegal parking on streets and at terminals and transfer deports; unauthorized entry of bus travel lanes; and inadequate use of signalized intersections such as red-light running and conflicts with crossing pedestrian traffic; and travelers with disabilities. For the "black car" case involving violation of a private vehicle providing ridesharing services without a taxicab license, the first violation is a fine of 5,000 Hong Kong dollars and is punished by 3 months in prison. The second arrest is doubled, with 6 months in prison and a fine of 10,000 Hong Kong dollars. Further, the record of the offender will be used as a reference by the insurance company, making the driver renew vehicle insurance after the violation. On the "black car" rider side, insurance claim is disallowed in case of being involved with a vehicle crash incident. These strict regulations and their enforcement targeting both travelers and transportation providers flow from the high density of Hong Kong where bad actions have large consequences for travel flow and impact many others (Tiry, 2003; Lo et al., 2008).

5.5 MULTIMODAL FREIGHT TRANSPORTATION

5.5.1 Freight rails

Although freight rails are not typically included into the transportation system of a metropolitan area, the operational efficiency of freight rails directly affects timeliness of goods delivery that is vital to regional economic development. In some cases, commuter rails may need to rent rail tracks owned by freight rails to provide passenger services that are also affected by freight rail operations. All too often, excessive auto delays are expected at highway-rail grade crossings along major rail corridors. Improving freight rail facilities and operations could help handle passenger and freight traffic with reduced traffic delays, shorter commute times, better air quality, and increased public safety, which will lead to more employment and economic growth opportunities. Typical freight rail improvements include new roadway overpasses or underpasses at locations where auto, bus, pedestrian, bicycle, and scooter traffic crosses railroad tracks at grade levels; rail overpasses or underpasses to separate passenger and freight train tracks; viaduct improvements; track, switch, and signal system upgrades; highway-rail grade crossing safety enhancements; and integrated freight and passenger rail dispatch systems.

5.5.2 Cargo drones

Cargo drones range from plane-sized autonomous delivery vehicles capable of carrying hundreds of pounds for hundreds of miles. They are currently being tested to drop a single parcel near a designated location to smaller package delivery flying and sidewalk drones. Unlike goods movements by trucks or packages shipped by delivery trucks on prespecified schedules, flying cargo-carrying drones are designed to take off vertically and land on land or in the water with high cruising speeds (Lee, 2017). Sidewalk drones are emerging that can allow a delivery truck to park safely in one place and send out the drones to make deliveries, rather than parking in front of each delivery location. The nature of more flexible and faster goods delivery by this new generation of cargo carriers renders freight shipments more efficient with significantly reduced impacts.

5.5.3 Cases in action

5.5.3.1 The CREATE program in Chicago

For over 150 years, Chicago has been the U.S. rail hub due to its critical location at the nexus of the North American railroad network. Each day, Chicago handles close to one-third of the nation's freight rail shipments by 500 freight trains of 37,000 railcars heading to the region. In parallel, ~760 passenger trains serve more than 83 million passengers annually for regional commuting and inter-city travel. Currently, the average speeds of freight trains in the region typically run in the range of 5–12 mph. By 2045, the freight shipment quantity will double, and passenger travel by rail transit (with shared use of tracks of freight rail) is expected to increase significantly. The existing rail transportation system in the Chicago metropolitan area built more than a century ago was not configured for the volumes and types of goods being shipped today, not to mention future conditions.

Recognizing the growing urgency of the region's rail capacity needs, the Chicago Regional Environmental and Transportation Efficiency (CREATE) program was commissioned in 2003 via a partnership of the U.S. federal, state, and local governments, and major freight rails in North America passing through the region. The CREATE program will improve rail-based passenger travel and freight shipments in the region by deploying 7 passenger rail and 35 freight rail corridor improvement projects, and 25 safety enhancement projects through conversion of rail crossings to grade separations, totaling US$4.6 billion (Schneider et al., 2005) (Figure 5.11).

5.5.3.2 Rhaegal heavy-lifting drones

Developed by Sabrewing Aircraft in Camarillo, California, the Rhaegal heavy-lifting drone is a 60 ft long, 14 ft high, and 60 ft foldable wingspan unmanned aerial vehicle (UAV) powered by two gas turbines capable of reaching a cruising range of about 1,100 miles at 205 mph. The Rhaegal drone is remotely operated to perform semi-autonomous flight, which maintains hybrid capability of vertical takeoff and landing up to a maximum carrying weight of 5,400 lbs without runways or complex launch and recovery devices and performing conventional flight with a top payload weight of 10,000 lbs for long-range cruising at relatively high speeds in nearly any weather condition. It is well suited to carry lower-depth cargo containers to help ship raw materials to factories, conduct just-in-time delivery of parts from factories to clients, and shuttle cargo from logistics bases (De Reyes and MacNeill, 2020) (Figure 5.12).

5.5.3.3 Amazon's Prime Air

Prime Air is a drone delivery service currently developed by Amazon. The service uses a delivery drone to automatically deliver a single package to the customer within 30 minutes after ordering. In order to be eligible for a 30-minute delivery, the order must be less than 5 pounds (2.25 kg), small enough to hold the container that the cargo drone will carry, and the delivery location must be within 10-mile (16-km) radius of a participating Amazon order fulfillment center. The company was selected by the U.S. Federal Aviation Administration (FAA) together with seven companies including Zipline and Wingcopter to take part in the type certification program for delivery of drones (CBS, 2013; Lavars, 2015).

5.6 URBAN CURB SPACES AS MULTIMODAL PASSENGER/FREIGHT SHARED USE MOBILITY TERMINALS

For a typical weekday, the intensity and pattern of people travel and goods movements in an urban area vary greatly in nighttime (midnight to 6:00 am), morning (6:00 am to 11:00 am), midday (11:00 am to 4:00 pm), and evening (4:00 pm to midnight) periods. For the nighttime period, passenger movements are at a minimum through early hours of the morning and the predominant activities are from freight delivery services. For the morning period, delivering goods to stores continues before the AM peak of work-related trips and then gives

(a) Intermodal operations in Chicagoland

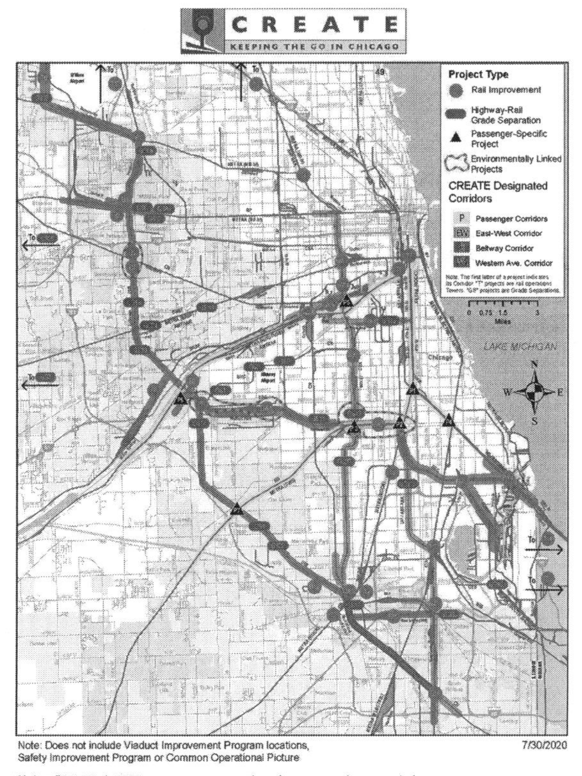

(b) CREATE program designated corridors

Figure 5.11 Illustration of the Chicago CREATE program for freight rail expansion in Chicago. (a) Intermodal operation in Chicagoland and (b) CREATE program designated corridors. (*Sources*: Chicago Department of Transportation (CDOT) (2020); Chicago Metropolitan Agency for Planning (CMAP) (2020).)

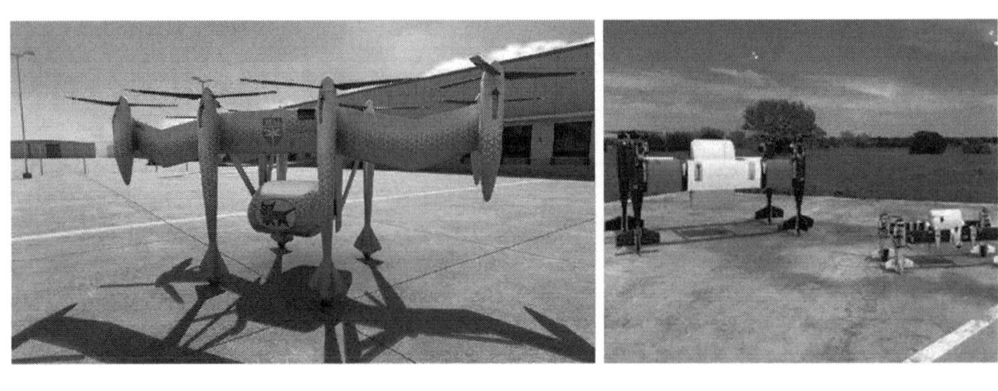

(a) Model I (b) Model II

Figure 5.12 Illustration of the Rhaegal heavy-lifting cargo drones in California. (a) Model I and (b) Model II. (Source: Choi (2020).)

way to passenger travel. For the midday period, freight delivery services resume back to handle late morning packages, mails, and so forth for local businesses and residents. Lunch activities begin at noon and last till 2:00 pm. Then, light delivery services follow through the start of PM peak with high intensities of passenger travel generated by employees returning back from work. For the evening period, passenger movements for activities like picking up kids, dinners, entertainment, and events continue into late evening. Afterward, freight delivery activities will gradually build up and proceed to the nighttime of the next day (Barth, 2019).

To promote multimodal integrated travel in consideration of passenger and freight movement dynamics, the traditionally envisioned usage of curbsides of urban streets for on-street parking, randomized passenger pickups and drop-offs, and freight loading has reached a time for deliberate reconsideration. Curbsides actually serve as origins, destinations, or transfer spots of passenger and freight movements that bring together many entities, including self-driving travelers, active transportation participants and providers, transit operators, ridesharing services such as the traditional taxi companies and TNCs like Uber and Lyft, delivery service firms, local businesses, enforcement, and emergency responders (Klein et al., 1997; Gonzalez–Feliu et al., 2014; Roe and Toocheck, 2017; Drees, 2020). With further advancements of automated vehicles and freight carriers such as cargo drones, they will likely boost the number of curbside passenger and parcel drop-offs and pickups while diminishing the need for and importance of on-street parking. This essentially shifts curbsides of urban streets to multimodal shared use mobility (MSUM) terminals.

With the shift of basic function of curb spaces from on-street parking, rider drop-offs, and package loading to MSUM terminals, the planning, design, construction, and management of curb spaces would greatly influence the mobility of urban core areas and passenger and freight trips connected with the core areas. This is because curb management is tied in with both travel demand management and multimodal integrated transportation with flexible capacity. On the travel demand side, each passenger trip from the origin to destination involves a sequential process of choices of destination location, arrival time, travel mode, departure time, travel path, and facility use. The availability, cost, and convenience of using the curbside either at the origin, destination, or transfer point are correlated with trip-specific travel time and out-of-pocket cost. Different curb management strategies could lead to changes in above choices. Moreover, curb management could even influence the origins of some trips for better utility of more travel options that offer tradeoffs.

On the flexible multimodal capacity side, the utilization efficiency of curb spaces could be significantly improved if the designated pay-to-use functions of curb spaces in urban core areas could be adjusted in response to dynamically changing passenger and freight movements in nighttime, morning, midday, and evening periods of weekdays and even weekends, as well as holidays and special event days. For an urban street network, permanent bus stops, passenger pickup and drop-off zones, ridesharing accessing and loading zones, freight loading and unloading zones, and metered parking spots may be deployed at a portion of curbsides of some urban streets. Conversely, the purposes and rates for shared use of the remaining curb spaces should be adjusted by location and time periods of the day (nighttime, morning, midday, and evening) in response to the intensity and pattern of passenger and freight movements. The delineation of fixed and flexible shared use of curb spaces as MSUM terminals in an urban area should be linked with land use purposes such as commercial, business, residential, mixed use, and industrial. In doing so, efficient management of curb spaces will not only reduce travel intensities but maximize multimodal shared capacity responsive to the passenger and freight demand with reduced intensities. The combined effect will lead to significant mobility improvements in the urban core areas and the metropolitan area that has passenger and freight movements connected with the core areas. Moreover, a greater extent of mobility gains could be achieved if efficient curb management is conducted in conjunction with effective traffic management.

As a result of fixed and flexible shared use of curb spaces in an urban area, sufficient multimodal integrated capacity is available to accommodate passenger and freight movements. This will substantially decrease driving and stopping violations, largely eliminate standing and parking violations and bus conflicts, and reduce safety concerns of cyclists, scooter riders, and pedestrians. The high rotation frequency of payable fixed and flexible shared use of curb spaces as MSUM terminals will help maximize curbside value capturing. The extensive revenue generated from paid use can be allocated to improve urban streets, traffic control devices, transit and active transportation systems, which will improve urban quality of life, health, and social equity.

5.7 CONCLUSION

This chapter has addressed primary dimensions of mass transit, its modes and sub-modes, and its applications to transportation networks in megacities. Readers should use this chapter as a jumping off point in rethinking the integrated role transit plays in today's growing megacities. A critical theme of this chapter in conjunction with Chapter 4 is that auto, mass transit, ridesharing and microtransit, active transportation and micromobility, and other travel modes must work in balance within a much larger system to ensure the continued growth and development of megacities. Overreliance on one mode risks forsaking the advantages provided by other modes. In a 3D framework with spiderweb interconnectedness, planners and decision makers can more thoughtfully and holistically consider more efficient and effective mobility solutions. Megacities, by virtue of their size and scale, do not lend themselves to siloed solutions. Multimodal integrated transportation with flexible capacity in the right place at the right time must be considered part of a comprehensive framework for promoting the mobility and accessibility necessary to ensure sustainable growth. The next chapter dives more deeply into how this capacity can be efficiently utilized to achieve and sustain high levels of mobility performance for megacities in the long run.

REFERENCES

AASHTO. 1978. *A Manual on User Benefit Analysis of Highway and Bus-Transit Improvements.* American Association of State Highway and Transportation Officials, Washington, DC.

Al Mamun, M., and Lownes, N.E. 2011. A composite index of public transit accessibility. *Journal of Public Transportation* 14(2), 4.

Amirkiaee, S.Y., and Evangelopoulos, N. 2018. Why do people rideshare? An experimental study. *Transportation Research Part F: Traffic Psychology and Behaviour* 55, 9–24.

Asirin, A., and Azhar, D. 2018. Ride-sharing business model for sustainability in developing country: Case Study Nebengers, Indonesia. *IOP Conference Series: Earth and Environmental Science* 158, 012053.

BART. 2018. Courtesy photo of Scoop to BART. San Francisco Bay Area Rapid Transit (BART) District, San Francisco, CA [Online]. Available: https://sharedusemobilitycenter.org/my-experience-taking-scoop-to-bart/ [Accessed on March 15, 2021].

Barth, B. 2019. Curb control. *Planning Magazine*. June 2019 Issue. American Planning Association, Chicago, IL [Online]. Available: https://www.planning.org/planning/2019/jun/curbcontrol/ [Accessed on February 10, 2021].

Brinkhoff, T. 2015. Nigeria: Metro Lagos local government areas. Oldenburg, Germany [Online]. Available: http://www.citypopulation.de/php/nigeria-metrolagos.php [Accessed on February 10, 2021].

Bushell, M.A., Poole, R.W., Zegeer, C.V., and Rodriguez, D.A. 2013. *Costs for Pedestrian and Bicyclist Infrastructure Improvements*. Highway Safety Research Center, University of North Carolina, Chapel Hill, NC.

CBS. 2013. Amazon unveils futuristic plan: Delivery by Drone. New York [Online]. Available: https://www.cbsnews.com/news/amazon-unveils-futuristic-plan-delivery-by-drone/ [Accessed on February 10, 2021].

CDOT. 2020. Courtesy image of Chicago Regional Environmental and Transportation Efficiency (CREATE) program designated corridors. Chicago Department of Transportation (CDOT), Chicago, IL.

Cecco, L. 2019. The Innisfil experiment: The town that replaced public transit with Uber. The Guardian, London, UK [Online]. Available: https://www.theguardian.com/cities/2019/jul/16/the-innisfil-experiment-the-town-that-replaced-public-transit-with-uber [Accessed on February 10, 2021].

Cervero, R. 2013. Bus Rapid Transit (BRT): An efficient and competitive mode of public transport. Institute of Urban and Regional Development, the University of California at Berkeley, Berkeley, CA [Online]. Available: https://escholarship.org/uc/item/4sn2f5wc [Accessed on February 14, 2021].

Choi, C. 2020. Cargo drones ease the way for urban air mobility. Inside Unmanned Systems, Red Bank, NJ [Online]. Available: https://insideunmannedsystems.com/cargo-drones-ease-the-way-for-urban-air-mobility/ [Accessed on March 3, 2021].

Chrysler, B., Honsberger, N., Irene Lambraki, I., and Hanning, R. 2014. Understanding healthy public policy processes: A multiple case study of the use of road modifications to improve active transportation: City of Hamilton case study. Propel Centre for Population Health Impact, The University of Waterloo, Waterloo, Ontario, Canada [Online]. Available: https://s22457.pcdn.co/wp-content/uploads/2017/11/active-transportation-policy-case-study-hamilton.pdf [Accessed on March 4, 2021].

CMAP. 2020. Courtesy image of Norfolk Southern intermodal operations in Chicagoland. Chicago Metropolitan Agency for Planning (CMAP), Chicago, IL.

Cmar, W. 2017. *How Cities are Integrating Rideshare and Public Transportation*. Harvard Ash Center, Harvard University, Boston, MA [Online]. Available: https://datasmart.ash.harvard.edu/news/article/how-cities-are-integrating-rideshare-and-public-transportation-978 [Accessed on March 3, 2021].

Cordahi, G., Shaheen, S., and Martin, E. 2018. MOD sandbox demonstrations independent evaluation Dallas Area Rapid Transit (DART): First and last mile solution evaluation plan. Report

No. FHWA-JPO-18-677. Federal Transit Administration, U.S. Department of Transportation, Washington, DC.
De Reyes, E., and MacNeill, J. 2020. Can cargo drones solve air freight's logjams? A drone startup says its big vertical-takeoff flier would be quick to land, load, and take off again. *IEEE Spectrum* 57(6), 30–35.
Dediu, H. 2017. Toward micromobility: The low end disruption of transportation. NERE, Boston, MA [Online]. Available: https://www.slideshare.net/asymco/toward-micromobility-the-low-end-disruption-of-transportation [Accessed on March 3, 2021].
DETR. 1997. Keeping buses moving, Local Transport Note 1/97. Department of the Environment, Transport and the Regions, London, UK.
DETR. 2000. Bus priority in SCOOT, Traffic Advisory Leaflet 8/00. Department of the Environment, Transport and the Regions, London, UK.
Drees, D. 2020. Managing the curb: Promoting accessibility, partnerships, and sustainable transportation modes. National Center for Mobility Management, Federal Transit Administration, Washington, DC [Online]. Available: https://nationalcenterformobilitymanagement.org/managing-the-curb-promoting-accessibility-partnerships-and-sustainable-transportation-modes/ [Accessed on March 15, 2021].
Eisenmeier, S. 2018. Ride-sharing platforms in developing countries: Effects and implications in Mexico City. Pathways for Prosperity Commission. Background Paper Series No. 3. The University of Oxford, Oxford, UK [Online]. Available: https://pathwayscommission.bsg.ox.ac.uk/sites/default/files/2018-10/eisenmeier_ride-sharing.pdf [Accessed on March 15, 2021].
Fielding, G.J. 1987. *Managing Public Transit Strategically: A Comprehensive Approach to Strengthening Service and Monitoring Performance*. Jossey-Bass, San Francisco, CA.
Fielding, G.J. 1992. Transit performance evaluation in the U.S.A. *Transpiration Research Part A* 26(6), 483–491.
Foth, N., Manaugh, K., and El-Geneidy, A.M. 2013. Towards equitable transit: Examining transit accessibility and social need in Toronto, Canada, 1996–2006. *Journal of Transport Geography* 29, 1–10.
Frost, A. 2019. Connected signals deploys its cloud-based V2I transit priority system in California. *Traffic Technology Today*, 2019-07-08 Issue, Mark Allen Group, London, UK [Online]. Available: https://www.traffictechnologytoday.com/news/connected-vehicles-infrastructure/connected-signals-deploys-its-cloud-based-v2i-transit-priority-system-in-california.html [Accessed on March 24, 2021].
FTA. 2020. Mobility Performance Metrics (MPM) for integrated mobility and beyond. Federal Transit Administration, U.S. Department of Transportation, Washington, DC [Online]. Available: https://www.transit.dot.gov/sites/fta.dot.gov/files/docs/research-innovation/147791/mobility-performance-metrics-integrated-mobility-and-beyond-fta-report-no-0152.pdf [Accessed on March 15, 2021].
Furuhata, M., Dessouky, M., Ordóñez, F., Brunet, M.E., Wang, X., and Koenig, S. 2013. Ridesharing: The state-of-the-art and future directions. *Transportation Research Part B: Methodological* 57, 28–46.
Gonzalez–Feliu, J., Semet, F., and Routhier, J.L. 2014. *Sustainable Urban Logistics: Concepts, Methods and Information Systems*. Springer, Berlin, Germany.
Hadas, Y., and Ceder, A. 2010. Public transit network connectivity: Spatial-based performance indicators. *TRB Journal of Transportation Research Record* 2143(1), 1–8.
Hawkins, A.J. 2018. Texas Town ditches its bus service for ride-sharing app via, The Verge, Vox Media, New York [Online]. Available: https://www.theverge.com/2018/3/12/17109708/via-arlington-texas-rideshare-app-replaces-bus [Accessed on March 15, 2021].
Hazan, J., Lang, N., Wegscheider, A.K., and Fassenot, B. 2019. On-demand transit can unlock urban mobility. Boston Consulting Group, Boston, MA [Online]. Available: https://orenvia-wpengine.netdna-ssl.com/wp-content/uploads/2020/04/On-Demand-Transit-Can-Unlock-Urban-Mobility_EN.pdf [Accessed on March 15, 2021].
Hensley, R., Padhi, A., and Salazar, J. 2017. Cracks in the ridesharing market–and how to fill them. *McKinsey Quarterly*, Article published online on July 17, 2017. McKinsey & Company, Chicago, IL [Online]. Available: https://www.mckinsey.

com/industries/automotive-and-assembly/our-insights/cracks-in-the-ridesharing-market-and-how-to-fill-them# [Accessed on March 21, 2021].

Hounsell, N.B., Farges, J.L., Fox, K., Friedrich, B., and Di-Taranto, C. 1996. Public transport priority at traffic signals: Results of a European collaborative study. *The 3rd World Congress on Intelligent Transport Systems*, Orlando, FL.

Hunkin, S., and Krell, K. 2019. Promoting active modes of transport: A policy brief. Interreg Europe Policy Learning Platform on Low-carbon Economy, Brussels, Belgium, EU [Online]. Available: https://www.interregeurope.eu/fileadmin/user_upload/plp_uploads/policy_briefs/TO4_PolicyBrief_Active_Modes.pdf [Accessed on March 21, 2021].

IHT. 1987. *Roads and Traffic in Urban Areas*. Institute of Highway and Transport, Her Majesty's Stationery Office, London, UK, 214–228.

IHT. 1997. *Transport in the Urban Environment*. Institute of Highway and Transport, Her Majesty's Stationery Office, London, UK, 329–348.

ITSA. 2004. An overview of transit signal priority. Advanced Traffic Management Systems Committee and Advanced Public Transportation Systems Committee, ITS America, Washington, DC.

KCM. 2021. King County via to metro. King County Metro, Seattle, WA [Online]. Available: https://www.kingcounty.gov/depts/transportation/metro/programs-projects/innovation-technology/innovative-mobility/on-demand/via-to-transit.aspx [Accessed on March 21, 2021].

Klein, D.B., Moore, A.T., and Reja, B. 1997. *Curb Rights: A Foundation for Free Enterprise in Urban Transit*. Brookings Institution Press, Washington, DC.

Konishi, H., and Mun, S.I. 2010. Carpooling and congestion pricing: HOV and HOT lanes. *Regional Science and Urban Economics* 40(4), 173–186.

Kumar, A., Zimmerman, S., and Agarwal, O.P. 2012. *The Soft Side of BRT: Lessons from Five Developing Cities*. The World Bank, Washington, DC [Online]. Available: https://www.ssatp.org/sites/ssatp/files/publications/BRT-Case-Studies.pdf [Accessed on March 21, 2021].

Lavars, N. 2015. Amazon to begin testing new delivery drones in the US [Online]. Available: https://newatlas.com/amazon-new-delivery-drones-us-faa-approval/36957/ [Accessed on March 21, 2021].

Lee, J. 2017. Optimization of a modular drone delivery system. In *2017 Annual IEEE International Systems Conference (SysCon)*, IEEE, Montreal, QC, Canada, 1–7.

Lee, Y.J., and Vuchic, V.R. 2005. Transit network design with variable demand. *ASCE Journal of Transportation Engineering* 131(1), 1–10.

Levinson, H.S., Zimmerman, S., Clinger, J., and Rutherford, C.S. 2002. Bus rapid transit: An overview. *Journal of Public Transportation* 5(2), 1–30.

Li, Z., Shen, J., and Budiman, J. 2008. Highway tunnels: An overview for planners and policymakers. In A Report for the Reason Foundation. Illinois Institute of Technology, Chicago, IL.

Litman, T. 2015. *Evaluating Active Transport Benefits and Costs*. Victoria Transport Policy Institute, Victoria, BC, Canada.

Lo, H.K., Tang, S., and Wang, D.Z.W. 2008. Managing the accessibility on mass public transit: The case of Hong Kong. *Journal of Transport and Land Use* 1(2), 23–49.

Martincigh, L., Tonelli, C., and Fuller, T. 2010. The walking environment design: Indicators and measures. *Pedestrians' Quality Needs* 358, 277–304.

Mobereola, D. 2009. Africa's first bus rapid transit scheme: The Lagos BRT-Lite system. Sub-Saharan Africa Transport Policy Program. The World Bank, Washington, DC.

Müller, J. 2020. Number of private hire cars in Singapore 2012–2019. Statista, Hamburg, Germany [Online]. Available: https://www.statista.com/statistics/953848/number-of-private-hire-cars-singapore/#:~:text=In%202019%2C%20there%20were%20around,ride%2Dhailing%20market%20in%20Singapore [Accessed on March 21, 2021].

MyProScooter. 2020. The history of the motorized electric scooter. *MyProScooter Magazion*, 2020-01-20 Issue. Middle Jutland, Denmark [Online]. Available: https://www.myproscooter.com/history-of-the-motorized-scooter/ [Accessed on May 14, 2021].

Niles, J., and Jerram, L.C. 2010. From buses to BRT: Case studies of incremental BRT projects in North America. MTI Report 09-13. Mineta Transportation Institute, San José State University, San José, CA.

Oeschger, G., Carroll, P., and Caulfield, B. 2020. Micromobility and public transport integration: The current state of knowledge. *Transportation Research Part D: Transport and Environment* 89. doi: 10.1016/j.trd.2020.102628.

Oliveira-Neto, F.M., Loureiro, C.F.G., and Han, L.D. 2009. Active and passive bus priority strategies in mixed traffic arterials controlled by SCOOT adaptive signal system: Assessment of performance in Fortaleza, Brazil. *Journal of the American Planning Association* 2128, 58–65.

Paundra, J., van Dalen, J., Rook, L., and Ketter, W. 2020. Ridesharing platform entry effects on ownership-based consumption in Indonesia. *Journal of Cleaner Production* 265. doi: 10.1016/j.jclepro.2020.121535.

Roe, M., and Toocheck, C. 2017. Curb appeal: Curbside management strategies for improving transit reliability. National Association of City Transportation Officials, New York [Online]. Available: https://nacto.org/wp-content/uploads/2017/11/NACTO-Curb-Appeal-Curbside-Management.pdf [Accessed on March 21, 2021].

Rondinelli, D., and Berry, M. 2000. Multimodal transportation, logistics, and the environment: Managing interactions in a global economy. *European Management Journal* 18(4), 398–410.

Sallis, J.F., Frank, L.D., Saelens, B.E., and Kraft, M.K. 2004. Active transportation and physical activity: Opportunities for collaboration on transportation and public health research. *Transportation Research Part A: Policy and Practice* 38(4), 249–268.

Schneider, P., Schwalbach, J., and Hamilton, L. 2005. Chicago Region Environmental and Transportation Efficiency (CREATE) program: Final Feasibility Plan. Chicago, IL.

Sen, L., Majumdar, S., Highsmith, M.K., Cherrington, L., and Weatherby, C. 2011. Performance measures for public transit mobility management. Texas Transportation Institute, Texas A&M University, College Station, TX [Online]. Available: https://static.tti.tamu.edu/tti.tamu.edu/documents/0-6633-1.pdf [Accessed on March 21, 2021].

Shutterstock, H.-S. 2019. The city has created dedicated parking spots for dockles bikes and scooters to relieve the hassles of blocked sidewalks in downtown San Diego, CA, USA on May 24, 2019. Shutterstock, New York [Online]. Available: https://www.shutterstock.com/image-photo/san-diego-ca-usa-may-24-1407263471 [Accessed on June 24, 2021].

SoundTransit. 2020. 2021 Financial plan and proposed budget. Central Puget Sound Regional Transit Authority and Union Station, Seattle, WA.

TCRP. 2002. Estimating the benefits and costs of public transit projects: A guidebook for practitioners. TCRP Report 78. Transit Cooperative Research Program, Transportation Research Board, National Academies Press, Washington, DC.

Tiry, C. 2003. Hong Kong, an urban future shaped by rail transport. *China Perspectives* 49. doi: 10.4000/chinaperspectives.647.

TunnelTalk. 2009. Seattle embraces new transit system. TunnelTalk [Online]. Available: https://www.tunneltalk.com/Sound-Transit-Jul09-Seattle-embraces-mass-transit.php [Accessed on March 21, 2021].

VanderZanden, T. 2020. 5 key innovations taking e-scooters to a half-billion rides in 2021. TechCrunch, Bay Area, CA [Online]. Available: https://techcrunch.com/2020/11/16/5-key-innovations-taking-e-scooters-to-a-half-billion-rides-in-2021/ [Accessed on March 21, 2021].

Van Wijnendaele, M. 2019. How cities deal with shared micro-mobility: Case study Brussels, Vox. [Online]. Available: https://www.linkedin.com/pulse/how-cities-deal-shared-micro-mobility-case-study-van-wijnendaele/ [Accessed on March 21, 2021].

Vuchic, V.R. 2005. *Urban Transit: Operations, Planning, and Economics*. John Wiley and Sons, Hoboken, NJ.

Wang, Y., Wang, Z., Li, Z., Staley, S.R., Moore, A.T., and Gao, Y. 2013. Study of modal shifts to bus rapid transit in Chinese cities. *ASCE Journal of Transportation Engineering* 139(5), 515–523.

World Bank. 2017. *Public-Private Partnerships Reference Guide, Version 3.0*. World Bank, Washington, DC.

Wunsch, P. 1996. Cost and productivity of major urban transit systems in Europe: An exploratory analysis. *Journal of Transport Economics and Policy* 30(2), 171–186.

Zhong, L., Zhang, K., Nie, Y.M., and Xu, J. 2020. Dynamic carpool in morning commute: Role of high-occupancy-vehicle (HOV) and high-occupancy-toll (HOT) lanes. *Transportation Research Part B: Methodological* 135, 98–119.

Chapter 6

Mobility management for efficient capacity utilization

6.1 GENERAL

6.1.1 Multiple, distinct goals in transportation system management

Transportation system management is naturally involved with multiple goals that reflect a broad range of transportation agency, user, and other stakeholder's perspectives of what the transportation system should achieve and what elements of system performance are important. This complexity is particularly acute for megacities, which are responsible for managing entire systems and networks of transportation services and infrastructure. The typical transportation system management goals include optimizing physical facility preservation, agency costs, user costs, accessibility and mobility, safety, and environmental impacts, as well as indirect management goals of economic development and quality of life promoted by the transportation system (Li and Sinha, 2009; Cambridge Systematics, 2000).

- *Physical facility preservation.* Facility preservation refers to the set of interventions implemented in a coordinated manner by type, timing, and frequency usually called strategies for maintaining a minimum acceptable level of physical conditions of transportation facilities.
- *Agency costs.* For preserving physical conditions of transportation facilities and sustaining system service levels, extensive investments are made in construction, maintenance, and repair, as well as improvements of transportation facilities in their service lifespans. System management strives for minimizing life-cycle agency costs of all types of transportation facilities by keeping the physical conditions of transportation facilities at or above the threshold levels.
- *User costs.* User costs are the costs incurred by transportation users, which may specifically refer to direct expenses of transportation users during their travel, such as vehicle operating costs and additional user charges, transit fares, and ridesharing service fees, and transportation costs of freight shipments. In the broader context, user costs also cover costs of losses of travel time, vehicle crashes, and vehicle air emissions associated with people and goods movements. To keep the consistency of the time span with estimation of agency costs, user costs are also calculated in transportation facility service life cycles.
- *Accessibility.* As an important function of transportation that supports people and goods movements, accessibility reflects the ease by which people move between residential, employment, recreational, or shopping locations; gain access to services; and use various travel modes; as well as goods and services between production and distribution points.

DOI: 10.1201/9780429345432-6

- *Mobility.* Ensuring mobility of people, goods and services, and factors of production and distribution by various travel modes is a basic role of any transportation system with efficiency considered as a policy goal.
- *Safety.* The provision of transportation is to ensure efficient movements of people and goods while ensuring safety of system users, operators, and other stakeholders.
- *Environmental impacts.* Most travel activities and transportation improvement actions, while yielding benefits of agency and user cost savings, accessibility and mobility improvements, and safety enhancements, tend to have adverse impacts on the quantity of resource consumption and quality of the environment. Such negative impacts need to be minimized simultaneously controlled as the minimum.
- *Economic development.* Most transportation improvements are geared toward enhancing operational efficiency, effectiveness, and equity. The end goal may explicitly or implicitly be to provide a top-class transportation system for an area to retain and expand business establishments that will lead to economic prosperity in the long run through enhanced mobility and access to markets.
- *Quality of life.* Besides sustaining economic development, the transportation system that could efficiently and effectively accommodate the needs of all groups of travelers and all types of goods shipments in an area will likely foster community spirit, promote social equity, help disadvantaged or disabled population, and promote the overall well-being that are essential to improve the quality of life.

Optimizing a transportation system or network over such a wide range of factors is challenging. Megacities face particular challenges due to their size, scale, and scope. Thus, implementing performance-based management systems is even more important to ensure sustainability and balanced growth.

6.1.2 The need for performance-based management

For each system management goal, a minimum acceptable level of performance or execution of a required function needs to be maintained by the transportation system in order to adequately accommodate people travel and goods movements. In the system management process, performance measures that are specific quantitative or qualitative indicators are adopted to analyze the current performance levels of individual system management goals, forecast future performance trends, and assess the extent to which a transportation stimulus realizes the system management goals. In this way, performance-based management is practiced. The adopted performance measures could help provide pertinent information on the operational efficiency and effectiveness of the transportation agency with which resources are transformed into services (outcomes), the quality of services (the extent to which the transportation user is satisfied with the services after resource allocation), and the performance outcomes (the extent to which the performance levels meet the prespecified performance targets after resource allocation). To support effective performance-based management, performance measures should be specific, measurable, achievable, relevant, and timely (SMART).

6.2 MOBILITY-CENTERED, PERFORMANCE-BASED TRANSPORTATION SYSTEM MANAGEMENT

For mobility-centered, performance-based management of a megacity's transportation system, the top priority is mobility (i.e., mobility first) to ensure that the target performance level of mobility for the entire system is always sustained. Meanwhile, other system management goals including facility preservation, agency costs, user costs, safety, and environmental impacts, as well as economic development are holistically addressed to strive for achieving their respective target performance levels for the system to the maximum extent possible pursuant to the available budget, technical capability, and management quality. Nonetheless, the bottom line of mobility-centered, performance-based transportation system management is to achieve maximized benefits of mobility improvements, without leading to a net loss among the positive and negative impacts on the performance of the remaining system management goals. In this respect, tradeoff analyses need to be conducted in budget allocation that would simultaneously lead to the highest level of performance gains to one or more system management goals other than mobility, such as safety and environmental impacts.

6.2.1 General

- *People/freight mobility versus vehicle mobility.* When dealing with multimodal travel, mobility management can be centered on either the mobility of people travel and goods shipments or mobility of vehicles serving as carriers for passenger and freight movements. Accordingly, the analysis can be conducted either for O-D-based person trips by private vehicle, transit, ridesharing and microtransit, biking, scooting, walking, and micromobility modes; and O-D-based quantities of goods shipped by truck, rail, cargo drone modes; or for O-D-based vehicle trips associated with passenger and freight movements (Geruschat et al., 1998; Li, 2018).
- *Network-level and project-level mobility management.* Network-level mobility management focuses on analyzing the current mobility performance levels, predicting the future mobility performance trends, identifying the mobility improvement needs, and estimating mobility impacts of alternative funding levels for the entire transportation system, as well as developing budget allocation strategies to achieve maximized network-wide mobility performance gains, without causing a net loss for other system management goals. Conversely, project-level mobility management generally involves identifying the most cost-effective strategy for a needed mobility improvement to ensure meeting the target mobility performance level with the lowest level of investment. Due to distinctive management focuses, project level analysis is typically more comprehensive than network-level analysis, leading to differences in performance measures, data, and methods adopted for respective analyses (Poister, 1997; Shaw, 2003; NCHRP, 2008).

6.2.2 Mobility performance measures

Table 6.1 lists typical performance measures that can be adopted for mobility management assessment.

6.3 MEASURES AND STRATEGIES FOR MOBILITY MANAGEMENT

For both passenger and freight travel, measures for mobility management could be derived from travel demand management and transportation capacity expansion perspectives. The combined use of demand management and capacity provision measures in a coordinated manner will become strategies for effective mobility management to help achieve the highest level of efficiency in utilizing the available capacity. The effectiveness of deployed mobility management strategies needs to be rigorously assessed before and after implementation and be periodically adjusted to be kept abreast of dynamically varying travel demand and transportation capacity provision.

Table 6.1 Typical performance measures for mobility management assessment

Category			*Performance measure*
Network-level	Amount of travel		Total person miles of travel (PMT) Total ton miles of travel (TMT) Total vehicle miles of travel (VMT), total truck VMT
	Travel time		Person-hours: $\sum\left(\left(\text{Length/actual travel speed}\right)\times\text{volume}\times\text{vehicle occupancy}\right)$
	Travel time index		$\dfrac{\sum\left(\left(\dfrac{\text{Freeway actual travel time}}{\text{Freeway free-flow travel time}}\right)\times\text{freeway peak VMT}\right)+\sum\left(\left(\dfrac{\text{Arterial actual travel time}}{\text{Arterial free-flow travel time}}\right)\times\text{arterial peak VMT}\right)}{+\sum\left(\left(\dfrac{\text{Arterial actual travel time}}{\text{Arterial free-flow travel time}}\right)\times\text{Arterial peak VMT}\right)}$
	Delays		Person-hours: $\sum((\text{Actual travel time} - \text{Free-flow travel} \times \text{volume}\times\text{vehicle occupancy}))$
	Congestion		Congested PMT: $\sum\left(\text{Volume}\times\text{vehicle occupancy}\times\text{congested segment length}\right)$
Project level	Passenger or freight	Amount of travel	PMT, TMT, VMT, truck VMT
		Travel time	Person-hours: $\left(\text{Length/actual travel speed}\right)\times\text{volume}\times\text{vehicle occupancy}$ Vehicle-hours: $\left(\text{Length/actual travel speed}\right)\times\text{volume}$
		Travel time index	Actual travel time/free-flow travel time
		Travel time buffer index	$\left(\text{95th percentile travel time} - \text{Average travel time}\right)/\text{Average travel time}$
		Average speed and reliability	Hourly, daily, weekly, monthly, and annual average travel speeds Average speed/free-flow speed Average speed/peak-hour speed Variations in daily, weekly, monthly, and annual average travel speeds
		Delays, congestion	Annual delays per traveler: $\left(\text{Actual travel time} - \text{free-flow travel time}\right)\times 250\ \text{weekdays/year}$ v/c ratio Level of service (LOS)
	Freight specific	Intermodal facilities	Average travel time and delays Average transfer time and delays Delay of trucks at facility per TMT

6.3.1 Managing multimodal travel demand

Central to demand management is to control physical travel and reduce travel intensity especially the intensity of passenger and freight trips in daily peak periods. The occurrence of each trip involves a sequential process of travel choices ranging from destination location, arrival time, travel mode, departure time, and trip path to facility use. Therefore, measures for travel demand management should be developed by identifying factors that could influence the above travel choices and possible ways to make factor-based adjustments to help control the total quantities of physical travel and reduce travel intensity.

Key factors governing travel choices include vehicle ownership and use; trip frequency and length; transit availability and service quality; ridesharing availability, and cycling, scooting, and walking-related non-motorized travel conditions; as well as flexible working arrangements. Vehicle ownership and use are affected by population and employment densities, transportation network connectivity, parking supply and management, and multimodal travel availability. Measures that could potentially limit conservation and use of private vehicles especially single-occupancy vehicles and concurrently stimulate the use of alternative travel modes are expected to be effective in travel demand management. Moreover, measures of urban/metropolitan land use that promote development of compact and mixed land use communities are likely to cut down auto ownership, trip frequency, and travel distance, as well as increasing mode shares of transit and non-motorized travel (Oppenheim, 1995; Rupert, 2020).

For work-related passenger trips, measures of flexible working arrangements could affect the trip frequency and choices of trip destination and arrival time and, in turn, be effective in travel demand management. Measures of compressed workweek, teleworking, and staggered working times could decrease physical travel, while measures of flexible working hours are likely to reduce demand intensity. Strategies developed based on flexible working arrangements as applicable could reduce both the work-related travel demand and intensity.

For travel mode choices, significant factors generally fall within the categories of travel time and out-of-pocket costs. Travel time mainly consists of out-of-vehicle travel time, waiting time, in-vehicle travel, and transfer time. Out-of-pocket costs occur in the forms of vehicle operating costs, tolls, parking fees, transit fares, ridesharing charges, and bike and scooter renting fees. If measures of park-and-ride service, carpooling, and vanpooling in conjunction with tolling and parking remissions and transit tokens could be actively promoted by transportation agencies, operators, and employers, they could help reduce the travel time of trips involving transit and non-motorized modes and comparatively raise out-of-pocket costs of trips by private vehicles. This will lead to modal shifts from private vehicles to transit and non-motorized travel modes and minimize trips and trip lengths by private vehicles (FHWA, 2004).

To control the daily private vehicle volumes and traffic flow intensity, high occupancy toll (HOT) lanes and providing pre-trip and en route traveler information could affect choices of departure times, routes, and travel lanes associated with private vehicle users to avoid time periods of the day and highway corridors that contribute to recurrent traffic congestion (Li, 2001; Dahlgren, 2002). Moreover, traffic incidents such as vehicle crashes, vehicle disablements, travel way debris, temporary workzones, and inclement weather conditions could trigger non-recurrent traffic congestion (Anbraoglu et al., 2014). Measures of dynamic travel lane control, active lane merging, and temporary use of paved shoulders both upstream and downstream of the incidents, coupled with overhead messages of speed control; displays of lane closure, merging, and diversion; and real-time monitoring of traffic condition and updating of traveler information could minimize the duration of each incident from its detection, response, and clearance to system recovery.

6.3.2 Multimodal integrated, expanded, and flexible transportation capacity

The efforts of transportation capacity expansion are multi-fold, which consist of expanding the capacity of multimodal transportation facilities, enhancing the integration of multimodal passenger travel, and promoting integration of intermodal and multimodal freight shipments. Typical measures of facility-related capacity expansion include adding traffic lanes and transit routes; building new roadways; introducing managed (high occupancy vehicle (HOV), HOT, exclusive-bus, and truck-only) lanes; adopting improved designs of unconventional intersections, overpasses/underpasses, and interchanges; and creating interconnected one-way streets for enhanced traffic circulation. Measures of multimodal integration strive for integration of private vehicle-related auto mode with transit and non-motorized modes, and integration of transit sub-modes for passenger travel, as well as integration of truck, freight rail, and cargo-drone modes for freight shipments. This will significantly broaden the conventional options of freeway/non-freeway travel by autos, bus/rail transit travel by transit riders, and freight shipments predominantly through freeways by trucks and freight rail by offering new passenger travel options of HOV and HOT lanes for private vehicles; truck-only lanes for hauling trucks; exclusive-bus lanes for regular, rapid, and express buses; bikeways and pedestrian walkways for bikes, scooters, and pedestrians; and ridesharing travel across private vehicle, transit, biking, and scooting modes; as well as expanded intermodal and multimodal freight shipment and cargo delivery services by trucks, freight rails, and cargo drones.

6.3.3 Efficient capacity utilization

Efficient utilization of the transportation capacity concentrates on matching the dynamically varying passenger and freight demand with the available multimodal transportation capacity to ensure the mobility of people/goods movements or traffic mobility by time of the day and by location is consistently maintained at or above the target performance level. Further, the performance levels of other system management goals concerning facility preservation, agency costs, user costs, safety, and environmental impacts, as well as economic development are holistically addressed and are sustained at the most desirable levels, respectively. In this respect, performance-based mobility management becomes a data-driven systematic decision-making process where data details of travel demand and available multimodal transportation capacity are utilized to assess the current and predict the future mobility performance levels using performance measures adopted for mobility management. By comparing the current or future mobility performance levels with the desirable performance targets, mobility performance deficits in presence or to be expected in the future could be identified. Mobility improvement strategies containing a list of combined measures for travel demand management and transportation capacity expansion could be proposed, evaluated, prioritized, and implemented to help mitigate the mobility issues.

6.4 FROM REACTIVE TO PROACTIVE MOBILITY MANAGEMENT

The process of developing mobility improvement measures and strategies and prioritizing a subset of most cost-effective strategies for mobility management is data driven. Benefited from technological advancements of data collection, transmission, processing, management, and analytics, real-time travel and traffic data for large network coverage and fine-grained time intervals has become increasingly available and is more widely utilized for mobility analysis and management decision-making. However, the effectiveness of mobility improvement strategies containing a list of combined and coordinated measures of travel demand management and transportation capacity expansion with analysis conducted using field data collected at a certain time window might not be as expected because the travel and traffic conditions at the time point of strategy implementation could be drastically different from the time point of data collection. Due mainly to the time lag effect, mobility management decisions made using field data collected even in real time will still be a reactive way of addressing mobility performance issues.

Alternatively, mobility improvement assessment and management decisions could be made using predictive travel and traffic data with high accuracy and precision that could be generated using high-fidelity large-scale simulation-based toolboxes for urban/metropolitan transportation planning, traffic operations, and evacuation planning/emergency management analyses. One exemplary toolbox that maintains the above analytical capability is the TRANSIMS (TRansportation ANalysis SIMulation System) or POLARIS toolbox developed in the United States nearly one decade ago (Li et al., 2012). It combines merits of the traditional regional travel demand forecasting models capable of conducting passenger and freight trip generation, trip distribution, mode split, and traffic assignment, and microsimulation models that could readily handle interactions of different types of vehicles and vehicles and non-motorized travelers in response to various traffic control measures. The development of such as high-fidelity large-scale analysis toolbox goes through a rigorous process of collection of data on comprehensive passenger and freight demand and multimodal transportation network configuration, calibration, and validation of factors affecting choices of destination locations, arrival times, travel modes, departure times, travel paths, and facility use for passenger and freight trips. The travel and traffic data collected in the field plays a key role in calibration and validation of analytical results in various steps of toolbox development.

The properly calibrated and validated toolbox could be employed to generate highly accurate and precise data on travel and traffic predictions. The predictive analysis capability of a high-fidelity large-scale toolbox will serve for two general purposes for mobility management. First, it will help generate predictive travel and traffic data for a future time point or time period to help identify mobility performance issues to be expected. Second, the toolbox could be employed to generate new sets of predictive travel and traffic data after implementing alternative mobility improvement strategies. The predictive data sets before and after implementing the mobility improvement strategies could be utilized to evaluate and prioritize the most cost-effective mobility improvement strategy. In contrast to mobility analysis using field data that is inherited with time-lagged limitations, mobility analysis relying on predictive data is fundamentally advantageous in identifying mobility performance issues to be expected, assessing the effectiveness of alternative mobility improvement strategies, and identifying and prioritizing the most-effective mobility improvement strategy. This leads to a paradigm shift from reactive to proactive performance-based mobility management.

6.5 MOBILITY MANAGEMENT SYSTEM

In the United States, annual person miles of travel, vehicle miles of travel, and ton miles of shipments by truck between 1980 and 2018 have increased by 109%, 112%, and 61%, respectively. However, the highway mileage for the same period has only extended by 8.2%. As a result, the annual person-hours of travel delays per auto commuter have increased from 36 person-hours to 83 person-hours, which is equivalent to an increment of 131%. More severe traffic congestion for longer time duration and larger geographic coverage has caused delays to commercial vehicles in delivering goods and providing services, leading to increases in transportation costs of the entire supply chain/logistics system. Congestion has also contributed to air-quality degradation (Shefer, 1994; Zhang and Batterman, 2013).

In early 1990s, Intelligent Transportation Systems (ITS) was initiated in the United States with the vision to enhance the utilization of highway capacity, safety, and environmental quality through efficient traffic management using emerging sensing, communications, electronics, and computer technologies, as well as management strategies in an integrated manner. Unlike the ITS-based approach that addresses traffic operation issues of the existing transportation system, the mobility management system specifically focuses on efficiently managing the mobility of people travel and goods movements or mobility of vehicles accommodating people and goods movements, namely, traffic mobility, in which mobility improvement strategies are sought from both travel demand management and transportation capacity expansion perspectives to ensure the mobility is consistently maintained at or above the target performance levels. Moreover, ITS-based management decisions are made based on information extracted from field-collected real-time traffic data. Because of the time lag effect between data collection and field implementation of traffic management strategies, the ITS-based management decisions are still reactive. In addition to using field collected data to help assess the current mobility performance level, mobility management systems also utilize predictive travel and traffic data to forecast future mobility performance trends, identify expected mobility issues, evaluate the effectiveness of proposed mobility improvement strategies, and prioritize the most cost-effective strategies. The shift from field data-driven reactive decisions to predictive data-driven proactive decisions will lead to more effective mobility performance outcomes.

6.5.1 Mobility management bundles and user services

A mobility management system involves integrated applications of information processing, communications, provision, and control technologies to multimodal transportation components: system users, travelers and freight items, vehicles, and physical facilities. Mobility management technologies are considered in terms of a collection of interrelated bundles and user services for applications to mobility performance issues currently in presence or to be expected. Table 6.2 presents typical mobility management bundles and user services. Individual user services are building blocks that may be combined for implementation in a variety of ways depending on local priorities and needs for mobility improvement. These services may require multiple technological elements or functions that are common for several services. For example, a single user service will usually require several technologies, such as advanced sensing, surveillance, communications, mapping, and visualization technologies, which may be shared with other services.

6.5.1.1 Travel demand and traffic management

With both travel demand and transportation capacity simultaneously considered, mobility management essentially handles mobility of people and goods movements or traffic mobility in an integrated manner. Therefore, the user services provided by a mobility management system gather and process travel demand and traffic data collected from the field and predicted by advanced analytical toolboxes, extract pertinent mobility performance information from the data, and provide commands to various travel demand management, transportation capacity provision, and traffic management tasks.

Table 6.2 Typical mobility management bundles and user services

Bundle	*User service*
Travel management	Travel demand management
Traffic management	Freeway corridor management Arterial management Intersection traffic control Highway-rail grade crossings
Traveler information	Pre-trip travel information En route driver information Route guidance Traveler service information
Advanced vehicle technologies	Longitudinal collision avoidance Lateral collision avoidance Intersection collision avoidance Vision enhancement for crash avoidance Safety readiness Precrash restraint deployment Automated vehicle operation
Transit management	Public transportation management En route transit information Personalized public transit Public travel security
Commercial vehicle management	Commercial vehicle electronic clearance Automated roadside safety inspection On-board safety monitoring Commercial vehicle administrative processes Hazardous material shipment management Commercial fleet management
Incident and emergency management	Emergency notification and personal security Emergency responder management No-notice emergency preparedness
Electronic payment	Electronic payment services Toll revenue management

Real-time travel and traffic information about travel demand intensities, alternative travel modes, ridesharing options, transit schedules and fare, travel routes, facilities, traffic conditions, toll rates, and parking availability and rates can be provided to travelers, goods suppliers and shippers, and drivers directly through personal apps, to in-vehicle infotainment systems, or through highway advisory radios (HARs), social media, and variable message signs (VMS). Other information may include incidents triggered by vehicle crashes, vehicle disablements, work zones, and inclement weather conditions, as well as special events or emergency cases. If pre-trip, this information can assist travelers and goods suppliers and shippers in planning destination locations, arrival times, travel modes, departure times, travel routes, and facilities of trips that will help reduce the total travel demand, demand intensities, and demand leveling in minimizing the trips made by single-occupancy private vehicles. If en route, this information can improve the efficiency of transportation capacity use. Eventually, it will ensure that the mobility of people travel and goods shipments or traffic mobility be consistently maintained at or above the target mobility performance level.

Additional traveler service information such as the location, operating hours, and availability of food, lodging, recreational places, public facilities, hospitals, and vehicle services may be provided. This information can be accessible at home, in office, or at other public locations, as well as en route.

6.5.1.2 Travel and traffic information dissemination

The effective dissemination of highly reliable travel and traffic information to travelers, goods suppliers and shippers, and drivers directly through personal apps, to in-vehicle infotainment systems, or through HAR, social media, and VMSs plays a vital role to the success of deploying a mobility management system. A higher level of responses to the deployed mobility improvement strategies by the transportation system users will increase the chance to fill in the gaps of mobility performance issues (Panichpapiboon and Pattara-Atikom, 2011). Information may be provided in real-time and static fashions. Real-time information comes from a variety of data sources including personal apps, drivers, intrusive and non-intrusive sensing equipment installed along transportation facilities, and travel and traffic prediction toolboxes. Static information comes from such sources as transit schedules and fare, planned work zones and travel way closures, and traveler service portals. Both the static and real-time information can be used for assisting travelers in the choices of destination locations, arrival times, travel modes, departure times, routes, and facility use for pre-trip planning and updating the choices of routes and facility use while en route.

Regardless of how travel and traffic information is provided, it must be timely and complete, accurate and precise, available on demand, and perceived by individual travelers, goods suppliers and shippers, and drivers as being relevant to their needs and providing a value when followed.

6.5.1.3 Advanced vehicle technologies

These services are primarily directed to improve vehicle safety through collision avoidance and warning systems, including vision enhancement, safety readiness, and precrash restraint deployment. In addition, the automated highway system (AHS) is a long-term goal to provide a fully automated, hands-off vehicle operating environment. In recent years, connected vehicle (CV) technologies with equipment, applications, or systems that use vehicle-to-everything (V2X) communications are rapidly introduced to address vehicle safety and mobility. The CV concept uses data from dedicated short-range communications (DSRC) broadcasts and peer-to-peer exchanges within ~300m to "sense" what other travelers (autos, buses, trucks, bicycles, scooters, pedestrians, wheelchairs, and others) are doing and identify potential hazards. Automated vehicles (AVs) are those in which at least some aspects of safety-critical control functions including steering, throttle, and braking occur without direct driver input. The CV technologies can be deployed in AV systems. CVs and AVs have the potential to dramatically improve vehicle safety and mobility and reduce vehicle air emissions (Mahmassani, 2016; Talebpour and Mahmassani, 2016). At present, CV technologies are being tested to improve safe and efficient truck movements along I-80 in Wyoming. Technologies of vehicle-to-vehicle (V2V) and intersection communications are being evaluated to improve vehicle flow and pedestrian safety in New York City. Similar pilot projects are underway for safety and mobility improvements along reversible freeway lanes in Tampa, Florida.

6.5.1.4 Transit mobility management

These services are the applications of technology-oriented travel and traffic management, information dissemination, and AV technologies to planning and management functions of transit systems to achieve the most desirable level of mobility performance. For example, in Kansas City, Missouri. 160 transponder-equipped signposts were installed along 38 bus routes to communicate with the transit management center and detect bus passes. A display appears on the bus indicating if it is running early or late. Additionally, buses are equipped with communications equipment that allow them to respond quickly to emergencies by immediately notifying the transit operator.

6.5.1.5 Commercial vehicle mobility management

These services are the applications of technology-oriented travel and traffic management, information dissemination, and AV technologies to commercial vehicle operations ensuring the highest possible level of mobility performance. Most of the applications are to provide (i) safety assurance through automated inspections and reviews; (ii) credential administration which allows filing credential and permit information through a single transaction; (iii) electronic screening which automates weight, safety, and credential screening at roadside weigh stations and facilities; (iv) seamless travel through state and international borders; and (v) carrier fleet management through AV location (AVL) and optimal route planning. An example of these services is the PREPASS program currently in use on the U.S. West Coast and in southwestern states, allowing truck preclearance in each state. A participating truck has a transponder on its dashboard that provides information electronically to weigh-in-motion (WIM) facilities. As a truck on the highway proceeds at a normal speed, a reader embedded in or above the roadway identifies the truck and a computer at the WIM station verifies the truck's credentials. Additionally, a WIM device embedded in the road checks the vehicle weight. If the truck passes inspection, it is given a green light to continue. If not, a light flashes to indicate that the truck must pull over at the upcoming weigh station. This system allows a truck to clear a WIM station without slowing, thus saving truck operator's time, keeping highway traffic moving smoothly, and reducing fuel consumption and vehicle air emissions.

6.5.1.6 Incident and emergency management for more resilient mobility

In addition to the normal traffic condition, mobility management also deals with random incident-affected and rare emergency event-affected traffic conditions, respectively. Sustaining more resilient travel and traffic operations in presence of incidents and under emergency circumstances is an inseparable mission of mobility management. Conventionally, resilience is quantified as a dimensionless quantity representing the rapidity of the system to revive from a damaged condition to the pre-damaged functionality level. It was initially introduced to measure the ability of transportation facilities to prepare for, absorb, recover from, and adapt to disturbances (Wan et al., 2018). For mobility management, resilience as a performance measure for transportation facility preservation viewed from the transportation supply side could be extended by taking travel demand and traffic components into consideration. Delivering and preserving more resilient transportation facilities, coupled with user services of effective travel and traffic management, information dissemination, advanced vehicle technologies, transit management, and commercial vehicle management, will make travelers, drivers, and truck operators more responsive and adaptive to mobility management strategies and more confident in planning for and managing personal travel and goods shipments in presence of incidents and in case of emergency events. This will help sustain more resilient mobility.

The incident management services use advanced sensing, data processing and analysis, and communications technologies to promptly identify, respond, and clear the incidents to minimize their effects on the normal travel and traffic operations (Owens et al., 2010).

The emergency management services rely heavily on proactive evacuation planning and emergency management strategies developed based on scenario analysis of emergency events with pre-warnings or no-notice emergency events that could be anticipated in an urban/metropolitan area. Whenever an emergency event occurs, a portion of the travel demand will be affected, and the transportation system might not be affected or might lose its partial serviceability, such as temporary suspension of a transit service owing to flooding or power outage. In both cases, the transportation system needs to help emergency responders, law enforcement, and rescuing forces to gain access to the evaluation area; remove the population of evacuees or affected goods from the affected area; and handle the background traffic. As an example, the Chicago regional transportation analysis model has been utilized to dynamically evaluate the travel demand change in response to roadway and transit line closures in some low elevation areas caused by yearly recurrent flooding problems in Spring and Fall seasons. For different precipitation levels anticipated, estimation has been made on the expected flooded area, severity, and duration; affected population; and roadway and transit closure. An emergency management plan that includes alternative trip plans, drivers' rerouting options, and bus-bridging scenarios in conjunction with emergency vehicle dispatching plans has been developed to minimize travel disruptions to the population of evacuees. The emergency management plan has been added to an electronic map to facilitate field deployment as needed.

6.5.1.7 Electronic payment

A common electronic payment medium is envisioned for all travel modes and functions, including tolls, transit fares, and parking fees, using "smart cards," apps, or other technologies. The flexibility offered by such a system can facilitate implementation of travel demand management strategies, road pricing schemes, and revenue management (Hensher, 1991; Turban and Brahm, 2000). For example, the Go Card used by the Washington D.C. Metropolitan Area Transit Authority in the United States enables a commuter to pay for bus rides, trips on metros, and parking fees. In New York City, the MetroCard similar to a credit card is used by the Metropolitan Transit Authority with a variety of choices, such as weekly or monthly Unlimited Ride, daily Fun Pass for unlimited subway and bus rides, and Pay-Per-Ride. Various toll bridges in New York City have incorporated E-ZPass, utilizing a sticker inside the windshield near the rearview mirror. As the vehicle enters the toll lane, the E-ZPass system recognizes the individual code on the sticker. The appropriate discounted toll is automatically deducted from the E-ZPass account and the gate goes up to pass the vehicle. Recognition activation requires about 2 seconds, which helps immensely with the traffic mobility.

6.5.2 Mobility management system architecture

Mobility management architecture defines the framework around which multiple design approaches can be developed, each one is specifically tailored to meet the needs of individual user services. The architecture defines the functions (for example, gathering travel and traffic data and extracting pertinent information) that must be performed to implement a given user service, the physical entities or subsystems where these functions reside (for example, the user, vehicle, facility), the interfaces or information flows between the physical subsystems, and the communicating requirements for the information flows (for example, wireline or wireless). The system architecture also identifies and delineates types of standards needed to accommodate national and regional interoperability, as well as product standards needed to support economic feasibility in deployment (Kim et al., 2012; Chandra et al., 2017).

6.5.2.1 Logical architecture

The logical architecture provides a model of mobility management functions that can be depicted using data flow diagrams. Logical architecture includes traveler, driver/operator, emergency, and electronic payment services. It also incorporates traffic and transit management, commercial vehicle operations, vehicle and monitoring control, and emergency planning and deployment, as shown in Figure 6.1.

6.5.2.2 Physical architecture

The physical architecture partitions the functions defined by the logical architecture into systems and subsystems as shown in Table 6.3. Traveler subsystems represent platforms for mobility management system functions of interest to users including travelers, goods suppliers, and shippers. Center subsystems refer to functions that can be performed at control centers. Roadside subsystems include functions that require convenient access to a roadside location for the deployment of sensors, signals, programmable signs, or other interfaces with users and vehicles. Vehicle subsystems represent functions (for example, navigation and tolls) that may be common across all types of vehicles.

6.5.2.3 Communications

Mobility management system architecture provides the framework that ties the transportation and telecommunication fields together. Four communication media types are identified to support the communications requirements between the subsystems: wireline (fixed-to-fixed), wide-area wireless (fixed-to-mobile), DSRC (fixed-to-mobile), and V2V (mobile-to-mobile). Center subsystems can be linked together over a wireline network. There are two distinct categories of wireless communications based on range and area of coverage: wide area and short range. Wide-area wireless communications are further differentiated based on whether they are one-way or two-way. Short-range wireless communications include DSRC and V2V communications (Dressler et al., 2014; Du and Dao, 2014).

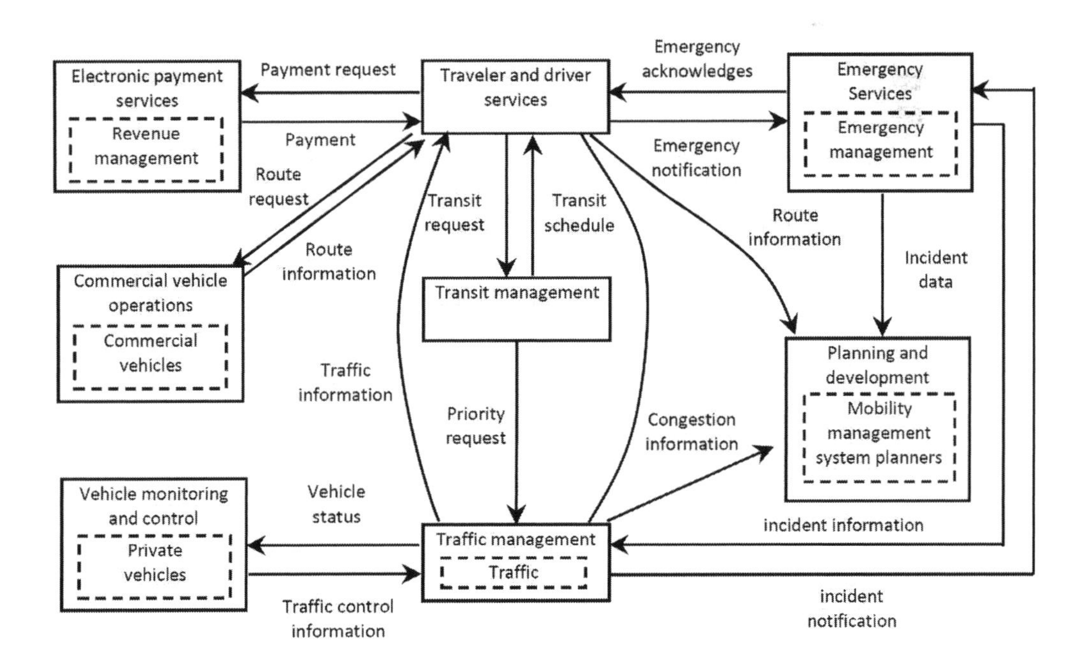

Figure 6.1 Simplified top-level data flow diagram.

Table 6.3 Physical architecture of systems and subsystems

System	*Subsystem*
User	Personal information access Goods supplier information access Freight shipper information access Remote user support
Center	Mobility management system planning Travel demand management Traffic management Transit management Commercial vehicle management Incident management Emergency management Emission management Traveler information services Toll administration
Roadside	Roadway Toll collection Parking management Commercial vehicle check
Vehicle	Private Transit Commercial Incident response Emergency management

6.5.3 Mobility management technologies

Mobility management technologies can be primarily grouped into surveillance, communications, user interface, control strategies, navigation/guidance, data processing, management, and analysis. Basic devices for surveillance are sensors, detectors, and video cameras. Sensors are sensitive to light, temperature, or radiation level and transmit a signal to a measuring or control instrument. Detectors are used to note the presence of an object. Video cameras are deployed to capture images of objects (Leduc, 2008). Communications technologies can be wireline or wireless. Common media include fiber optic wirelines, traditional telephone lines, commercial radios, cellular phones, satellites, video image processing, and infrared and microwave transmission. User interfaces may involve VMSs, touch screen, keypad/keyboard, voice recognition, voice output, visual display supported by on-board computers, and head-up display in vehicles. Control strategies can be realized by using ramp metering, HOV restriction, HOT enforcement, sign control, parking restriction, ramp or lane closures, road pricing, and reversible lanes. Navigation/guidance refers to pre-trip or en route advisory or instruction on trip choices (Anas and Lindsey, 2011). The advisory or instruction measures may include position display, map database, guidance display, and dead reckoning. One application for navigation purposes is the GPS technology. The GPS components for civilian use comprise a constellation of satellites, a network of base stations, radio communications, receivers, a database (for example, map data and application databases), and application software. Other mechanisms may include in-vehicle navigation systems (IVNS), VMSs, highway advisory radios, electronic mass media, and kiosks. Technologies used for data processing, management, and analysis include agent-based travel demand forecasting, driver, vehicle, transit, and freight shipment scheduling, and optimal resource allocation for mobility improvements (FHWA, 2001).

6.6 CASES IN ACTION

6.6.1 I-94 corridor ITS deployments in Minnesota

With the nature of high efficiency and flexibility, ITS deployments have been gaining momentum among transportation engineers and other stakeholders in Minnesota (MnDOT, 2020). Several deployments are worth noting. The SCorE-WZARD system deployed along the eastbound I-94 corridor improves traffic safety in inclement weather conditions related to snow and ice and in presence of construction, repair, and maintenance work zones that are prone to incident occurrences. The system provides advance and real-time notifications to raise the awareness of driving risks due to reduced vision and surface friction during snow. Truck Parking Information and Management System (TPIMS) aims to reduce the parking searching time for trucks. It helps truck drivers monitor the availability of parking sites and transmit the information to drivers via VMS, smart phones, and Websites. Signal priority system (SPS) is another important application. Aimed by the Connected Vehicle SPaT Deployment, the SPS contains a snowplow signal priority system (SSPS) which triggers a green signal or extended green phase for that allows the snowplows required green signal or extends current green phase before arriving at intersections via DSRC. It leads to savings of travel time and fuel consumption, and improved productivity for snowplows, and enhanced driving safety for the traveling public. The system expansion to support signal priorities for buses and emergency vehicles is underway (Figure 6.2).

6.6.2 I-90 Smart road in Schaumburg, Illinois

Interstate 90 (I-90) in Illinois, runs roughly northwest-to-southeast through the northern part of the state. From the Wisconsin state line at South Beloit, it heads south to Rockford before heading east-southeast to the Indiana state line at Chicago. The new 16-mile Jane Addams Memorial Tollway Smart Road of I-90 from the Kennedy Expressway to Barrington Road in Cook county is designed to deliver real-time information to drivers to provide safer, more efficient travel.

To fulfill mobility and safety goals, the Smart Road is equipped with a full complement of traffic monitoring and incident management equipment, traffic management measures, reliable communications technologies, and data analysis; decision-making and active traffic management (ATM) techniques. This suite of ITS technology is seamlessly integrated into the agency's traffic incident management system (TIMS). Up to October 2017, the Smart Road section is equipped with over 1,050 closed circuit television (CCTV) cameras, 45 VMS, 48 portable changeable message signs (PCMS), 250 microwave vehicle detection sensors (MVDS), 16 roadway weather information system (RWIS) sites, 5 WIM stations, and 3 ramp queue detection systems, among other ATM devices.

In addition to the CCTV cameras, equipment deployed for traffic monitoring include mainline detection, ramp detection, RWIS, and virtual WIM. The CCTV cameras provide visual coverage of the roadway for identifying the number and type of vehicles and the presence of drivers or passengers in the vehicles, reading placards of hazardous materials on commercial trucks, and monitoring incidents, congestion, stationary work zone, and partial system failure. Mainline detection is used to provide presence indication and measurements of traffic volume, lane occupancy, speed, and limited vehicle classification. Ramp detection devices are placed at ramps of both system and service interchanges and both exiting and entering ramps based on the functional needs, including vehicle counts, queue detection, and wrong-way detection. The RWIS is used in conjunction with VMS messages by tollway operators to notify motorists of expected and on-going weather that may affect roadway,

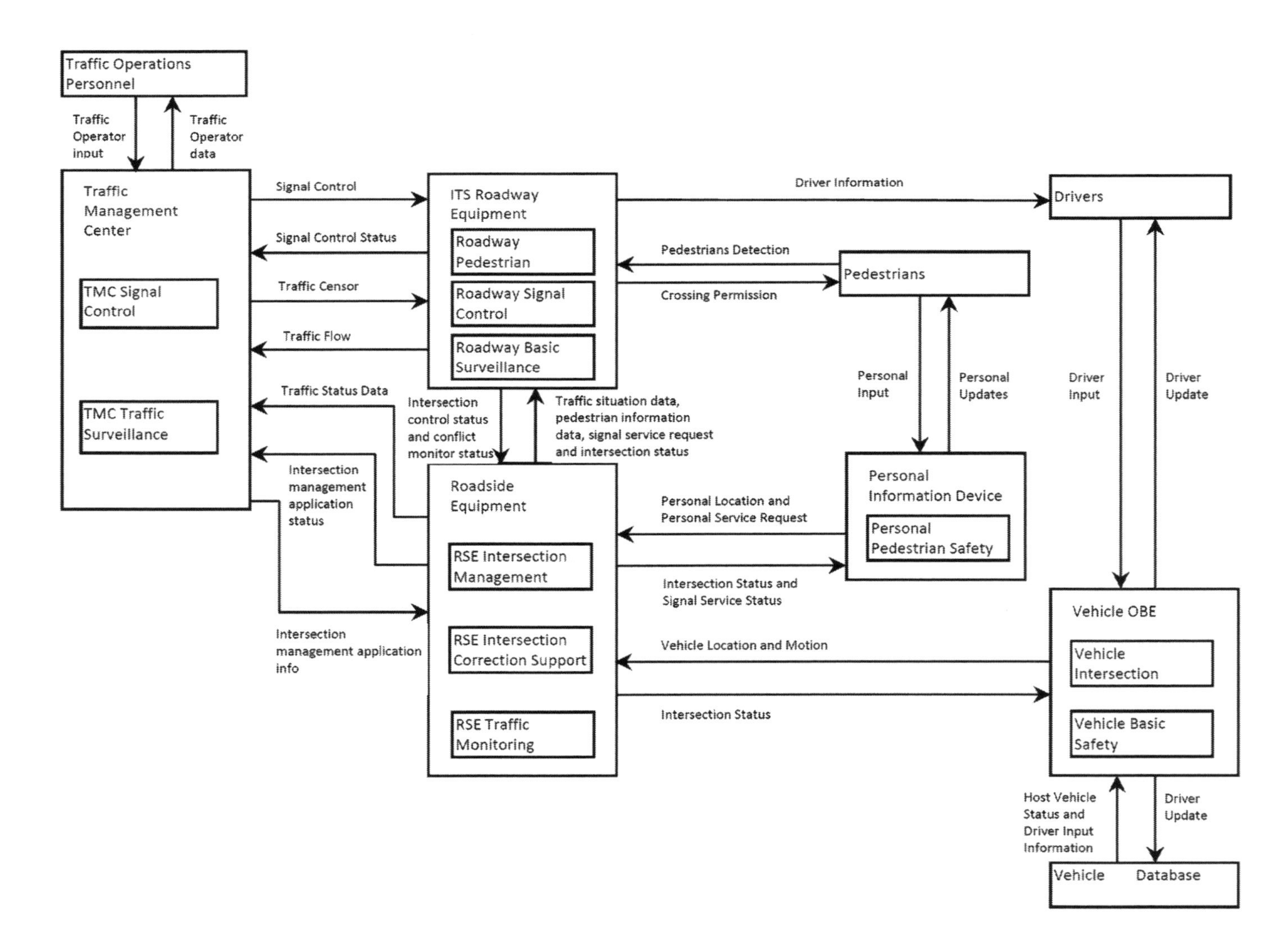

Figure 6.2 Signal, Phase, and Timing System (SPaT) in Minnesota.

sight, or driving conditions. The virtual WIM station measures dynamic tire forces of a moving vehicle and estimating the corresponding tire loads of the static vehicle to identify potentially overweight vehicles.

Equipment for incident management consists of lane control signals, VMS, and PCMS. The lane control signals are placed over the roadway travel lanes with green, yellow, and red arrows, respectively, indicating “Lane is open,” “Lane closure is imminent, merge left, right, or caution,” “Lane is closed,” or “Pace bus only” (referring to buses using a flex lane).

The dynamic message sign (DMS) is a large electronic sign deployed typically overhead at a fixed location to provide a wide range of timely and accurate information such as travel times to the downstream system interchange and major destination locations like downtown and airport to passing travelers that may influence users to choose alternate routes or even modify their destinations in normal but near-capacity traffic conditions. It is also widely used to inform travelers of congested locations, incidents, work zones, special events, icy or slippery pavement conditions, severe weather, and child abduction alerts. The PCMS is typically mounted on a trailer or a truck that can be deployed on short notice to a required location for a limited time as a modified traffic control measure in response to unusual changes in traffic conditions caused by road work, incidents, or special events.

The ATM measures encompass flex lane, dynamic lane use control, and dynamic speed limits. A flex lane is located on the left inside shoulder of the tollway mainline that is used for temporary parking only in the normal traffic condition and exclusively used by buses in presence of recurrent congestion. In times of incidents or emergency events, the flex lane can be used by emergency responders to route around congestion to safely reach the incident or emergency event sites and by general tollway users to safely navigate around incidents or emergency events. The availability of flex lanes is highly beneficial to avoid secondary vehicle crashes. Dynamic lane use control involves dynamically closing or opening individual traffic lanes with advance warning of the closure(s) through the use of lane control signals to mitigate mobility and safety impacts caused by traffic congestion and incidents. For vehicle speed control, dynamic speed limits are implemented according to real-time traffic, roadway, and/or weather conditions. The speed limits may vary by Smart Road section and by travel lane and are set as either regulatory subject to strict enforcement or advisory to ensure safe and efficient travel (Figure 6.3).

6.6.3 Spotlight ITS developments in Mainland China

The adoption of ITS concepts and field deployments in China could be traced back to the late 1990s. The early deployments mainly focused on traffic signal control using the off-the-shelf tools such as Split Cycle Offset Optimisation Technique (SCOOT) and Sydney Coordinated Adaptive Traffic System (SCATS). Accompanied by growing interests in applying information techniques to urban travel and traffic problems, the Chinese government developed the blueprint of its smart city initiative. This has provided an impetus to promote accelerated ITS deployments nationwide and in its major cities (Yan et al., 2012).

For urban traffic management, the ITS Command Center in Chengdu, a megacity of over 16 million people that is the capital of Sichuan Province in Southwest China, collects real-time traffic data covering the entire city, identifies traffic congestion based on information extracted from the data, and provides traffic information via more than 200 LED route guidance screens to mitigate travel delays. In 2018, an intelligent traffic light system was deployed in Wujiashan area, Wuhan, Hubei province. The system benefits from technologies of video image processing for data collection and advanced convolutional neural network, fuzzy logic, and genetic algorithms for data analysis to optimize signal timing plans for all intersections

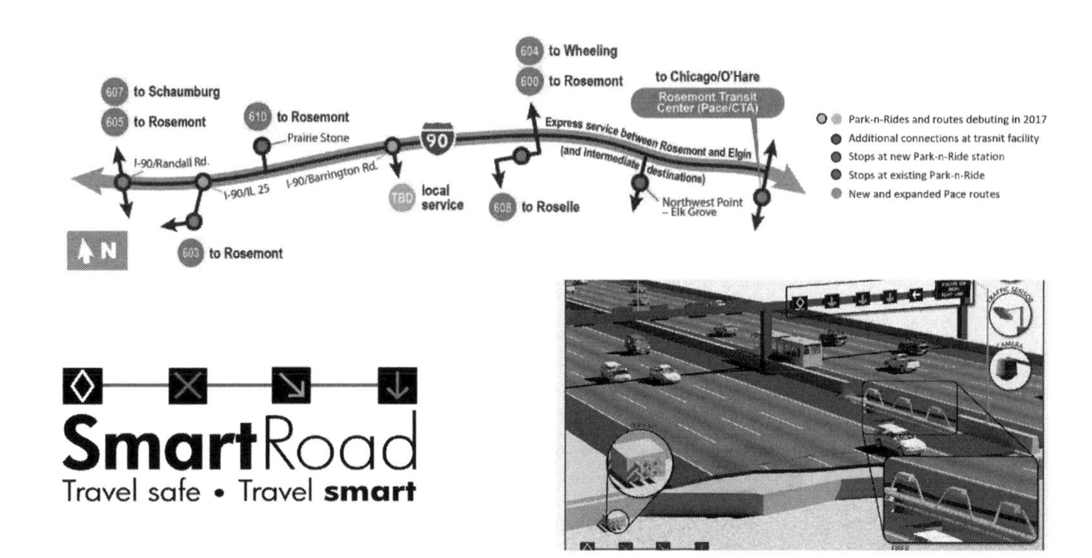

Figure 6.3 I-90 Smart road for multimodal mobility management in Schaumburg, Illinois.

within the area-wide street network aimed to achieve the lowest level of delays for travelers using the network. The delay time dropped by 24% after signal timing optimization.

Another major area of ITS applications is on transit management. For example, the Public Transport Command Center in Suzhou, Jiangsu province, performs real-time fleet scheduling, dispatching, and en route monitoring for 370 bus routes with 4,680 buses in operations at any given time point. Similarly, the Public Transport Command Center in Harbin, Heilongjiang province, conducts real-time bus location tracking for nearly 6,000 buses running on 208 bus routes (World Bank, 2018). In addition, ITS applications have been expanding to BRT operations that not only conduct bus position tracking in real-time but also incorporate functionalities of bus signal priorities that could drastically increase average bus operating speeds and reduce speed variations.

In recent years, electronic toll collection (ETC) for freeway travel has become increasingly popular across the country. As of 2019, over 40% of freeway drivers utilize ETC devices to pay the freeway tolls that could reduce the time delays at toll booths to manually complete toll payment transactions. With the 5G technology being rapidly implemented to the ETC system, its market concentration is expected to escalate soon (Huang et al., 2017) (Figure 6.4).

6.6.4 Intelligent traffic management system in Hong Kong, China

In terms of traffic management, Hong Kong's biggest feature is that it relies on information technology and intelligence in technology. The Hong Kong ITS Association, a public-private partnership, is the driving force promoting intelligent traffic management in Hong Kong, China. The intelligent traffic management system contains a regional traffic control system, passenger and traffic information system, automatic highway traffic payment system, and traffic monitoring and command coordination system.

There are 1,821 major signalized intersections in Hong Kong and 1,730 are controlled and operated by the regional traffic control system that collects and analyzes intersection traffic data in real time by 447 CCTV cameras, develops and deploys signal timing plans, and continuously refines then plans to achieve the lowest travel delays.

The passenger and traffic information system disseminates multimedia transportation information to major transportation hubs and terminals, such as airports, convention

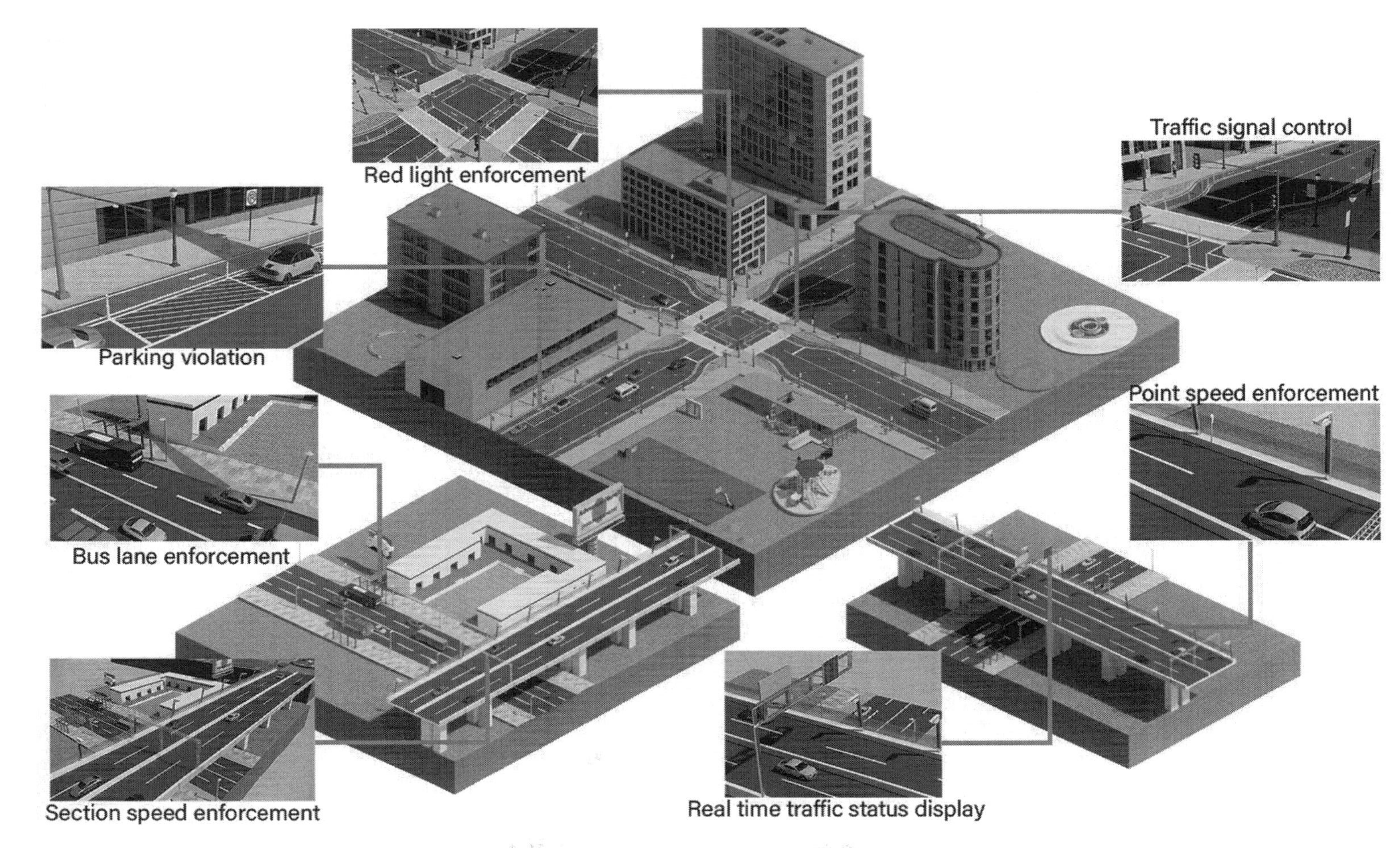

Figure 6.4 Intelligent traffic signal system in Wuhan, Hubei, China.

and exhibition centers, and transit transfer stations, broadcasts real-time traffic conditions on major roads and toll rates, and helps drivers find the best driving paths. Further, the system maintains driving speed maps capable of providing information on estimated driving speeds, cross-sea travel times, and snapshot images provided by the road surveillance cameras of the Hong Kong, Kowloon, and New Territories. The travel time displays installed on the cross-harbor passages provide estimated travel times for exits of the tunnels from Hong Kong Island to Kowloons which could help drivers in choosing the best crossline routes. Also, speed screens installed along arterial roads with real-time traffic and pedestrian flow information could help drivers pick the most arterial roads.

The automatic highway traffic payment system includes automatic payment system for non-stop road segments such as tunnels and bridges and an automatic payment system for Octopus smart-card electronic payment system. Some road segments maintain non-stop toll collection lanes. About one-half of vehicles passing through toll tunnels or toll roads use automatic toll collection systems, leading to significant reductions in vehicle delays, stop-and-start triggered fuel consumption, vehicle air emissions, and vehicle crashes. The Octopus electronic payment system is widely used by other modes of passenger travel, including rail transit, buses, minibuses, coaches, ferries, and parking meters and garages. In addition, small transactions in non-transport services such as supermarkets, convenience stores, restaurants, patisseries, self-service vending machines, home goods stores, telecommunications services, self-service photo stations, and cinemas accept payment by Octopus.

The traffic monitoring system mainly refers to the monitoring equipment installed on the main roads, such as CCTV, red light video recorder, emergency traffic coordination center in the speed monitor center, the New Territories regional traffic control system, Tsing Ma and Tsing Sha control district traffic monitoring system, as well as traffic control and monitoring system of Shenzhen Bay Highway Bridge, automatic vehicle detectors, lane control lights, and variable speed limit signs.

The command system consists of an emergency traffic coordination center, a regional traffic control system in the New Territories, a traffic monitoring system in Qingma and Qingsha control areas, and a traffic control and monitoring system for the Shenzhen Bay Highway Bridge. The system is currently being updated to an integrated transportation management center that brings together all transportation management departments, operating departments, and coordination centers in Hong Kong to support coordinated transportation management and service functions (Zegras, 2002). In the long run, the government intends to expand intelligent transportation management and service functions by developing and applying new technologies for vehicle detection and tracking, satellite image processing, cross-sector electronic tolling, and traveler information provision; and tapping on new management and service areas of travel time budgeting method, vehicle crash processing system, traffic data integration and public access, traffic information and navigation service, bus passenger information service, and online e-ticket taxi service.

REFERENCES

Anas, A., and Lindsey, R. 2011. Reducing urban road transportation externalities: Road pricing in theory and in practice. *Review of Environmental Economics and Policy* 5(1), 66–88.

Anbraoglu, B., Heydecker, B., and Cheng, T. 2014. Spatial-temporal clustering for non-recurrent traffic congestion detection on urban road networks. *Transportation Research Part C: Emerging Technologies* 48, 47–65.

Cambridge Systematics, Inc. 2000. *A Guidebook for Performance-Based Transportation Planning.* NCHRP Report 446. Transportation Research Board, National Academies Press, Washington, DC.

Chandra, Y.R.V.S., Shiva Harun, M., and Reshma, T. 2017. Intelligent transport system. *International Journal of Civil Engineering and Technology* 8(4), 2230–2237.

Dahlgren, J. 2002. High-occupancy/toll lanes: Where should they be implemented? *Transportation Research Part A: Policy and Practice* 36(3), 239–255.

Dressler, F., Hartenstein, H., Altinta, O., and Tonguz, O.K. 2014. Inter-vehicle communication: Quo vadis. *IEEE Communications Magazine* 52(6), 170–177.

Du, L., and Dao, H. 2014. Smart card-based electronic card payment systems in the transportation industry. *IEEE Transactions on Intelligent Transportation Systems* 16(1), 66–80.

FHWA. 2001. *Data Integration Primer.* Federal Highway Administration, U.S. Department of Transportation, Washington, DC.

FHWA. 2004. *Mitigating Traffic Congestion: The Role of Demand-Side Strategies.* Federal Highway Administration, U.S. Department of Transportation, Washington, DC.

Geruschat, D.R., Turano, K.A., and Stahl, J.W. 1998. Traditional measures of mobility performance and retinitis pigmentosa. *Optometry and Vision Science* 75(7), 525–537.

Hensher, D. 1991. Electronic toll collection. *Transportation Research Part A: General* 25(1), 9–16.

Huang, W., Wei, Y., Guo, J., and Cao, J. 2017. Next-generation innovation and development of intelligent transportation system in China. *Science China Information Sciences* 60(11), 110201.

Kim, W., Liu, Y., and Chang, G.L. 2012. Advanced traffic management system: Integrated multi-criterion system for assessing detour decisions during nonrecurrent freeway congestion. *TRB Journal of Transportation Research Record* 2324, 91–100.

Leduc, G. 2008. Road traffic data: Collection methods and applications. *Energy, Transport and Climate Change* 1(55), 1–55.

Li, J. 2001. Explaining high-occupancy-toll lane use. *Transportation Research Part D: Transport and Environment* 6(1), 61–74.

Li, Z. 2018. *Transportation Asset Management: Methodology and Applications.* CRC Press, Boca Raton, FL.

Li, Z., and Sinha, K.C. 2009. Methodology for the determination of relative weights of highway asset management system goals and of performance measures. *ASCE Journal of Infrastructure Systems* 15(2), 95–105.

Mahmassani, H.S. 2016. 50th anniversary invited article: Autonomous vehicles and connected vehicle systems: Flow and operations considerations. *Transportation Science* 50(4), 1140–1162.

MnDOT. 2020. Traffic engineering ITS projects (2016–2020). Minnesota Department of Transportation, Saint Paul, MN [Online]. Available: https://www.dot.state.mn.us/its/projects2016-2020.html [Accessed on March 24, 2021].

NCHRP. 2008. *Cost-Effective Performance Measures for Travel Time Delay, Variation, and Reliability.* National Cooperative Highway Research Program. National Academies Press, Washington, DC.

Oppenheim, N. 1995. *Urban Travel Demand Modeling: From Individual Choices to General Equilibrium.* Wiley–Interscience, Hoboken, NJ.

Owens, N., Armstrong, A., Sullivan, P., Mitchell, C., Newton, D., Brewster, R., and Trego, T. 2010. *Traffic Incident Management Handbook.* No. FHWA-HOP-10-013. Federal Highway Administration, U.S. Department of Transportation, Washington, DC.

Panichpapiboon, S., and Pattara-Atikom, W. 2011. A review of information dissemination protocols for vehicular ad hoc networks. *IEEE Communications Surveys & Tutorials* 14(3), 784–798.

Poister, T.H. 1997. *Performance Measurement in State Department of Transportation.* NCHRP Synthesis of Highway Practice 238. National Cooperative Highway Research Program, Transportation Research Broad, National Academies Press, Washington, DC.

Rupert, H.M. 2020. Mobility on Demand (MOD) sandbox demonstration: Fair value commuting. FTA Report No. 0167. Federal Highway Administration, U.S. Department of Transportation, Washington, DC.

Shaw, T. 2003. *Performance Measures of Operational Effectiveness for Highway Segments and Systems: A Synthesis of Highway Practice*. NCHRP Synthesis of Highway Practice 311. National Cooperative Highway Research Program, Transportation Research Board. National Academies Press, Washington, DC.

Shefer, D. 1994. Congestion, air pollution, and road fatalities in urban areas. *Accident Analysis & Prevention* 26(4), 501–509.

Talebpour, A., and Mahmassani, H.S. 2016. Influence of connected and autonomous vehicles on traffic flow stability and throughput. *Transportation Research Part C: Emerging Technologies* 71, 143–163.

Turban, E., and Brahm, J. 2000. Smart card-based electronic card payment systems in the transportation industry. *Journal of Organizational Computing and Electronic Commerce* 10(4), 281–293.

Wan, C., Yang, Z., Zhang, D., Yan, X., and Fan, S. 2018. Resilience in transportation systems: A systematic review and future directions. *Transport Review* 38(4), 479–498.

World Bank. 2018. Reducing traffic congestion and emission in Chinese cities. The World Bank, Washington, DC [Online]. Available: https://www.worldbank.org/en/news/feature/2018/11/16/reducing-traffic-congestion-and-emission-in-chinese-cities [Accessed on March 24, 2021].

Yan, X., Zhang, H., and Wu, C. 2012. Research and development of intelligent transportation systems. In *2012 11th International Symposium on Distributed Computing and Applications to Business, Engineering & Science*, IEEE, Guilin, China, 321–327.

Zegras, C. 2002. SUTP Module 1c— Private sector participation in urban transport infrastructure provision. In *Sustainable Transport: A Sourcebook for Policy-Makers in Developing Cities*. GTZ, Eschborn, Germany.

Zhang, K., and Batterman, S. 2013. Air pollution and health risks due to vehicle traffic. *Science of the Total Environment* 450, 307–316.

Chapter 7

Innovative transportation funding and financing

7.1 HISTORICAL REVENUE SOURCES

There is no simple and proven approach to funding megacity mobility. Global megacities currently receive funding from an array of sources, and all possible paths to funding mobility involve significant tradeoffs. A 2016 analysis of bridging global infrastructure gaps by McKinsey Global Institute points out long running and continued global underinvestment in transportation relative to its importance to the economy and personal mobility (Woetzel et al., 2016, 2017). Infrastructure investment overall fell as a share of gross domestic product (GDP) in the G20 economies following the 2007 global recession. Indeed, Organisation for Economic Co-operation and Development (OECD) data as seen in Table 7.1 on transportation spending from a sample of nations as many nations reduced transportation spending relative to GDP as did increase it since 2000. As more of the world's cities become megacities, keeping up with transportation funding needs becomes more imperative.

Nations and cities fund urban mobility with a wide variety of approaches. An OECD International Transport Forum (ITF) survey of ways urban public transportation systems are funded concludes "What is clear is that no single financing model emerges, reflecting the great diversity of local situations and needs. It also appears that there is no silver bullet for the funding of public Transport..." (ITF, 2013). This is even more true when looking at funding for all modes of urban transportation infrastructure and services and recognizing that the division of funding responsibility between national and subnational government jurisdictions varies greatly among global megacities.

One useful way to think of transportation funding is that revenues can come from users, beneficiaries, taxpayers, or any combination thereof. Users might pay via fuel taxes, tolls, or license fees. Fees levied on employers, developers, or builders can be viewed as beneficiary payments. And use of general revenues or consumption/sales taxes for transportation broadens the source to virtually all taxpayers.

Within those broad divisions lie many mechanisms of generating revenue to pay for urban mobility. It is worth looking at a few frequently used funding mechanisms.

7.1.1 Fuel taxes

Fuel taxes are levied in a great many nations, though most often as a source of general revenue (Sterner, 2012; OECD, 2018). In many such nations, general revenues are then used to pay for transportation infrastructure and services. In some, like Germany and the United Kingdom, fuel tax revenues exceed government road transportation expenditures, and the surplus is spent on other travel modes and on general expenditures (Price and Henebery, 2014). In others, like the United States, Brazil, and South Africa, at least some fuel tax revenues are dedicated to transportation spending (LOC, 2014).

DOI: 10.1201/9780429345432-7

Table 7.1 Total inland transportation infrastructure investment as a percentage of GDP (%)

Country	*2011*	*2012*	*2013*	*2014*	*2015*	*2016*	*2017*	*2018*	*Average*
Australia	0.8	0.6	0.4	0.9	0.9	1.0	–	–	0.8
Austria	1.8	1.9	1.6	1.4	1.2	1.3	1.5	1.7	1.6
Canada	1.2	1.1	1.0	0.4	0.6	0.6	0.6	0.5	0.8
China	4.1	4.4	4.6	5.0	5.3	5.5	5.8	5.6	5.0
France	1.0	1.0	1.1	0.9	0.9	0.8	0.8	0.8	0.9
Germany	0.6	0.6	0.6	0.6	0.6	0.6	0.6	0.7	0.6
India	0.8	0.8	1.0	1.1	1.2	1.1	–	–	1.0
Japan	1.0	1.0	1.1	1.1	0.9	1.0	–	–	1.0
Korea	1.6	1.7	1.7	1.5	1.6	–	–	–	1.6
Mexico	0.5	0.5	0.5	0.6	0.5	0.5	0.4	–	0.5
Russia	1.3	1.2	1.1	1.0	0.9	1.1	0.9	0.8	1.0
Turkey	1.1	0.9	1.2	1.1	1.3	1.2	1.0	–	1.1
United Kingdom	0.7	0.7	0.7	0.8	0.9	0.9	0.9	0.9	0.8
United States	0.6	0.6	0.6	0.6	0.6	0.6	0.5	0.5	0.6

Source: ITF (2021).

"–" means data unavailable.

At the subnational level, the United States and Canada allow "local option" fuel taxes to generate revenue for that jurisdiction, though few jurisdictions avail themselves of that option beyond cities in Florida in the United States and Vancouver and Montreal in Canada (Cooper and DePasquale, 1989; Ubbels and Nijkamp, 2002; Chen and Afonso, 2020).

Fuel taxes are an indirect user fee and charge consumers of fuel in rough proportion to their use of highways. They are a simple source of revenue with low collection costs but are very regressive and as technologies improve vehicles become more fuel efficient or even switch to electricity, fuel taxes become an ever less sustainable source of revenue and a less effective user fee.

7.1.2 Tolls or fares

Tolling of roads to pay for their costs, including sometimes congestion costs, is widely practiced globally. Several European countries, notably France, Spain, and Italy, pay for much of their intercity motorways with tolls levied on users, while Australia, China, Japan, and other Asian countries, Brazil, Chile, Mexico, and the United States also use tolls to pay for urban highways (Poole, 2018).

For urban areas, tolling is typically not used as a broad-based funding mechanism but is more often levied on specific facilities or the network of limited-access highways. Tolling has the virtue of most directly charging users to pay for the infrastructure and services they are utilizing, more tightly related to use than fuel taxes. They also can be, and often are, leveraged as the funding stream for a public-private partnership (PPP) to operate the tolled facility, which can speed construction and reduce operating costs (more on that later). Tolls are also adjustable for the cost of building and operating a facility and can easily be varied according to congestion, vehicle weight, or other factors that might justify differential charges.

However, tolls can be more expensive to collect than fuel taxes, though to what extent is hard to determine (Fleming, 2012). Tolls are often unpopular as road users seem to view direct charges less favorably than indirect ones. They can also create diversion of traffic from tolled highways to untolled ones, even if the latter are more congested and have less capacity for traffic (NSTIFC, 2009).

Fares for urban transit—ranging from rail to bus to paratransit—function much like tolls do for roadways. They charge users directly to help pay for the cost of the transit infrastructure and services. This has similar pros and cons to tolls as well. Fares help tie revenues to usage and provide useful signals to transit operators, and users are the primary, if not only, beneficiaries of the transit they use. Thus, asking direct beneficiaries to help pay the costs makes sense. But fares may, like tolls, discourage some travel and are regressive. In addition, in some higher-income countries, equity issues mitigate against full-cost pricing of transit services. Many users tend to be from lower-income households with limited access to alternative travel modes.

Most cities strike a middle ground and charge fares to transit users but not to cover the full costs the way that tolls typically do for highways. For transit, the balance is made up from other revenue streams, typically general revenues from local or higher levels of government and sometimes revenues from fees levied on road users (ITF, 2013). Many cities, particularly those in Asia, use land-value capture to monetize the benefits of improved transit access to fund transit services. Hong Kong in effect fully funds its transit system through this mechanism. Some cities, however, charge no fares at all, treating transit entirely as a public good.

Tokyo is an exception to all this (ITF, 2013):

> [W]here the largely private public transport operators are profitable and only require grants from the government in exceptional capital investment projects. Nonetheless, fare revenues fall far short of covering the costs of service provision and Tokyo region public transport operators derive the lion's share of their revenue from maximizing the return on their extensive real-estate asset base. In this business model, it is almost as if the public transport network's main purpose (from the operator's perspective) is to ensure an elevated and steady flow of customers to the operators' shops, hotels, health clubs, and other commercial undertakings.

With both toll roads and public transit, some jurisdictions utilize PPPs (discussed in more detail later in this chapter) to construct and/or operate facilities and services in ways that contain costs and allow private capital to finance projects based on existing revenue streams (Poole, 2018; Feigenbaum and Hillman, 2020). In the case of transit, it is also feasible to consider paying for the system entirely with fares while shifting subsidies from other revenue streams from the provider to users. So-called user-side subsidies would pay for the external benefits of transit use while allowing the full expression of benefits of a user fee approach that can easily be combined with private provision—an alternative path to something like the funding structure of the Tokyo transit system (Klein et al., 1997).

7.1.3 Weight-distance fees

Weight-distance fees charge a per-mile or per-kilometer rate varied by the weight of the vehicle, usually via vehicle type and often only required for heavy freight vehicles (VTPI, 2019). These fees are intended to charge for the additional wear and tear heavy vehicles impose on each mile of roadway they travel (Anderson and Auffhammer, 2014). Small et al. (1989) developed a framework to estimate the additional cost of pavement wear and modeled its impacts on highway financing, investment policy, and decision-making. This grew in the United States and other nations into forms of "cost allocation studies" designed to analyze and allocate the costs imposed on the highway system in order to help with infrastructure planning and enable better application of weight-distance fees and tolls or road pricing (FHWA, 1997; Garcia-Diaz and Lee, 2013).

Today, forms of weight-distance fees are applied to trucks in many nations (APIRG, 2009; Marcom, 2019). In Europe, Germany, Austria, Czech Republic, Slovak Republic, and Switzerland, all charge such fees on truck travel and their use is supported by EU policy (Robinson, 2008). For example, Austria taxes heavy vehicles on its roadways, as they cause excessive pavement damages from higher axle loads. The heavy vehicle pricing system was implemented in 2004, dividing all types of vehicles into two groups. Those vehicles having axle loads higher than 3.5 tons are taxed, and the others are exempt.

7.1.4 Value capture charges

Value capture refers to revenue mechanisms that impose charges or fees on development or landowners in a specific area for public improvements. In the case of transportation, these revenues are directed toward transportation facilities or services that impact the value of properties. The idea is that the improvements increase the value of a development or lands and value capture mechanisms are a means of having those beneficiaries help pay for the improvements (Lari et al., 2009).

Two dramatic examples of value capture funding for transportation networks exist in Tokyo and Hong Kong. These are particularly instructive examples because value capture works in both cities because of their large size and density, features of the many megacities of the future. In Tokyo, the firms running the metro system own properties around the rail stations and so capture the flow of value of benefits to improvements in either the properties or the transportation system (ITF, 2013). The Hong Kong Mass Transit Railway Corporation (MTRC) runs a massive rail transit system without government subsidy, and indeed makes a profit from an extensive real estate business (Verougstraete and Zeng, 2014). As in Tokyo, developing properties near MTRC station not only generates revenue for the system but also attracts residents and businesses, which adds to patronage of the rail system (Figure 7.1).

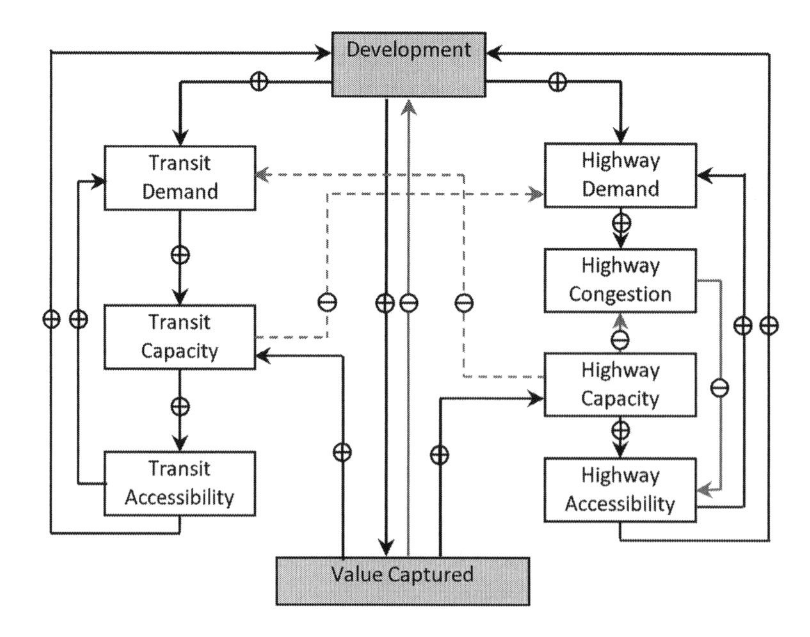

Figure 7.1 Value capture flow chart. (Source: Levinson (2014).)

More typically, value capture takes the form of property taxes, developer fees, or other mechanisms unique to each nation, such as tax-increment financing in the United States. Such revenue streams have been utilized in a number of U.S. and U.K. cities (APTA, 2015); in Germany, Japan, India, and Spain (Ubbels and Nijkamp, 2002); and in Ethiopia and Brazil (Mahendra et al., 2020).

Property tax and levy approaches come closest to the capturing of value inherent in this revenue concept. Especially if the taxes are tied to increased property values created by, for example, a public transit service. Development levies or impact fees are less value oriented. Instead, they are structured around the idea that a development creates the need or demand for more transportation infrastructure and the fees are intended to extract from the developer and new property owners some of the cost of transportation improvements made necessary by their land use.

Value capture, despite its theoretical benefits, has real challenges in practice. Mahendra et al. (2020) argue that value capture is often "ineffective due to lack of strong institutions and land-governance structures and subversion by political or private development interests." Verougstraete and Zeng (2014) point out that

> The importance of real estate activities could potentially distract MTRC from its core business of transport services. The question is then whether it would be appropriate to segregate the two types of business or whether a transport company is best equipped to manage real estate operations.

Calculating tax or fee amounts based on a determination of either benefits to a landowner, or impacts of a new development, is a process fraught with assumptions that must be made by a political or bureaucratic process. Disputes are frequent, with beneficiaries of, or advocates for, the revenues often feeling they are too modest, while those paying the taxes or fees often feel they are exorbitant. Opportunities for sweetheart deals or even outright corruption are rampant.

And there are genuine tradeoffs. Mahendra et al. (2020) find that economic and equity benefits are only sometimes achieved by value capture. In the United States, impact fees have supporters such as Nelson and Moody (2003) who argue they are beneficial revenue sources that fund needed projects and services. Others like Been (2018) and Millsap et al. (2019) find they restrict beneficial developments and have regressive effects.

7.1.5 General revenue

Many, if not most, urban transportation systems are funded in part by general revenues from local and national governments. Many jurisdictions have direct or indirect user fees such as fuel taxes but place those revenues in the general fund. Arguments for general revenue funding for transportation are that the benefits of transportation spending are broad and therefore justify a broad funding source. General revenues may protect transit system operating budgets and road maintenance budgets from the volatility of user fee revenues that tend to change as the economy expands or contracts. However, this protection is often not realized, as cuts in general fund expenditures following an economic contraction often include transportation spending. At the same time, general revenues for transportation can be cut when political decisions are made to prioritize other spending areas. So, a transportation system can, and often does, see rising demand and falling revenue. Relying on general revenue also removes links between those paying various fees and taxes and what those revenues are used for, and so general revenue appropriations for transportation are not particularly accountable to those who provide the funding or use the services which are funded.

7.1.6 Parking fees

Often, parking fees are a neglected revenue source for funding transportation-related facilities. Without these fees, travelers are shielded from recognizing the cost of providing facilities that are important components of the transportation network or system. Revenue from parking charges, or local fees on paid parking or non-residential parking spaces, are not likely to raise large amounts of revenue. But they are indirectly tied to use of the transportation system and influence incentives over mode choices (Istrate and Puentes, 2011; Kaufman et al., 2012; Shoup, 2017). Indeed, some analysts suggest that the incentive effects on mode choice may be more important than their revenue raising potential because they more fully reveal the "true" cost of parking to travelers (Fosgerau and de Palma (2013). They also tend to be an almost entirely municipally levied and controlled revenue source, serving as a local source of revenue and traffic management tool.

Many cities in the United States tax parking payments and in Australia and Canada some cities tax parking spaces (APIRG, 2009). Typically, the revenues go into the general fund, and the taxes are viewed as a means to reduce demand for parking and private vehicle use (Ubbels and Nijkamp, 2002). However, parking taxes in European cities are common and are more likely to be used to support transportation investments and services (Litman, 2013; Rye and Koglin, 2014).

7.1.7 Externality fees

Use of transportation systems can impose external costs that are not considered by standard taxes or fees, or even tolls and fares, charged to users. The leading examples are congestion and pollution. Vehicles using the roadway impose congestion costs and emit pollutants that negatively affect others.

In 1920, Arthur C. Pigou, an economist, argued that taxes should be levied on industries that cause negative externalities to account for their social costs since not all these costs were covered in private cost accounting for the business activity. Applying this theory to transportation was initially suggested by the RAND Corporation for freight transportation in 1949. This concept was formulated, developed, and championed in the United States and the United Kingdom during the 1950s. The prominent researchers involved in this process were James M. Buchanan, Alan Walters, and William Vickrey. They expanded on an idea by Pigou and proposed a flexible transit fare structure reflective of the ridership over time of a day for the New York subway system in 1952.

Today, some jurisdictions impose congestion charges in the form of fees to enter part of the city, most famously in Singapore, London, and Stockholm (VTPI, 2019). Meanwhile, 24 countries in Europe levy fees on vehicles directly related to their carbon emissions (ACEA, 2020). These latter charges are not practical to levy at the local level for local revenue, given current technology, but congestion charges can be an entirely local revenue source.

In current practice, the revenues are often not used for transportation. Rather they go into general funds. But if they were used to improve the transportation system and its efficiency, it might be possible to create significant incentives to reduce external costs without reducing mobility.

7.2 TRANSPORTATION FUNDING IS FRAUGHT WITH TRADEOFFS

7.2.1 General

The previous sections discussed but a sample of the many sources of revenue that can be, and are, used to fund transportation in cities. We discussed some of the tradeoffs involved but barely scratched the surface. Megacity transportation funding needs will likely be met by a combination

Table 7.2 Summary of some potential urban mobility funding sources

Strong	*Medium*	*Weak*	*Not evaluated/Flawed*
Vehicle mile/kilometer traveled fee (road usage charge)	Asset recycling (sales, leases, concessions)	Cordon pricing	Development and impact fees
Motor fuel tax	Vehicle sales tax	Passenger facility charges	Tourism taxes
Vehicle registration fee	General revenue	Bicycle tire tax	Tobacco, alcohol, and gambling taxes
Heavy vehicle use tax		Auto-related sales tax	Vehicle inspection and traffic citation surcharge
Sales tax on motor fuels		General sales tax	
Facility level tolling and pricing			

Source: NSTIFC (2009).

of sources differing among the many megacities across the globe. No one best mechanism can clearly do the job better than all the rest. Table 7.2 is derived from the analysis of the National Surface Transportation Infrastructure Financing Commission (NSTIFC) created by the U.S. Congress which used criteria that remain sensible today (NSTIFC, 2009). The break down of sources provide a broad sense of some merits to each source, but there is devil in the details. Not surprisingly, some such as Bornstein et al. (2018) point out the growing challenges to traditional means of funding mobility and argue that the future of transportation funding for cities will likely be a combination of usage-based charges, licenses and fees, and monetizing mobility data, combined with savvy use of PPPs and financing to leverage those revenue streams.

7.2.2 User fees

One clear theme emerging from analysis of the merits and potential drawbacks of various funding mechanisms is the virtues associated with user fees. The main beneficiary for some services is the consumer, and highway transportation is an exception. Although broader benefits via economic activity enabled by personal mobility accrue to the broader community, the primary benefits are to the individual traveler. Even a trip on public transit mainly benefits the traveler, although the scope and scale of broader social benefits may be larger than with individual vehicle travel if there are reduced externalities. Transportation services are well suited to user fees. Hence, the costs of providing infrastructure for roadway travel could be paid for entirely via user fees paid by travelers and goods shippers. The mobility challenges for those with low incomes can be met with the design of the fee structure or with the aforementioned user side subsidies. The same is nearly true for public transit, though the rebalancing between low-income and higher-income users would be much greater in cities that rely heavily on public transit compared to automobile use.

Having users pay directly for some or all of the cost of a municipal service has its tradeoffs, like all other funding mechanisms. Dewees (2002) puts it well:

> [F]or some services, user fees are not only feasible but economically desirable, because they help to allocate resources to maximize the satisfaction that we receive from those resources. User fees can constrain demand at a time when it is very expensive to expand supply. User fees can help to mediate situations where users are clamoring for more service and the agency does not have the resources to meet that increased demand. User fees can even help to deal with demand that varies greatly over time, through their tempering effect on such variations.

For many goods, user fees allow consumers to reduce costs by managing their consumption of the goods, which is why congestion pricing on roadways can be effective in managing traffic flows. At the same time, pricing public transit too high could induce people to use other means of travel that may have lower user costs but higher external costs or reduce mobility in ways that decrease economic growth and opportunity. This barely scratches the surface of how complex setting a user fee can be; setting them is like determining prices. History and experience show that no one person or organization can determine meaningful prices outside a well-functioning market. Prices provide information that reflects the multitude of choices, tradeoffs, and production and consumption decisions being made surrounding a given good by both consumers and providers (Weber, 2012). When the government provides a good and imposes a user fee, it cannot use a market process to determine the fee, but must make the best estimate possible, knowing the outcome will be imperfect.

Nevertheless, user fees have clear benefits over many other means of funding transportation. Broadly, these advantages include:

- *Fairness.* Those who benefit from the service help to pay for it. And what they pay and what they get for it is relatively transparent.
- *Choice.* User fees give those who pay them much more to the agency of what they pay for, and when and how often they pay it, than do more general taxes. They can make adjustments to lifestyle and location and other choices to improve their benefits from user fee-funded systems.
- *Flexibility.* User fees allow system operators the ability to adjust revenues and expenditures, as the economy, demand, and technology change.
- *Better incentives.* User fees create incentives for users to think seriously about the costs of the services they consume and make better decisions about how much to use them—and this is equally true for considering the benefits. At the same time, user fees give system owners and operators better information and incentives to strive for efficiency and quality that keep customers, and revenues, flowing in.
- *Constraint.* If users' costs don't change based on how much they use the system, they have no reason not to over consume it. "Free" roads are a classic example, and congestion, pollution, and lost time are the costs paid. User fees internalize those tradeoffs and avoid a "tragedy of the commons" created from broad funding sources that appear "free" to users when making discrete decisions to use the system at a given moment.

Specifically in transportation, Basso and Duvall (2013) point out that user fees improve the utilization, capacity, and operational efficiency of transportation facilities and services, are popular with users who experience the value transaction, and allow for the reality of variation in preferences while general revenue funding treats all users as identical. These practical realities have driven a rapid growth in the use of and exploration of broader applications for distance-based road user charges by many nations, which was the top outcome of the analysis behind Table 7.2.

7.3 PRINCIPLES OF SOUND TRANSPORTATION FUNDING

People often confuse the terms funding and financing and the concepts behind them, as explained in the following:

- *Funding.* It is a flow of revenue that can be used for necessary spending.
- *Financing.* It is various means of leveraging flows of funding to achieve more with the money.

In simple personal terms, a salary (or wages) funds a household, but a mortgage finances a home by redistributing future income to the present through annual payments paid out over time to the lender. So funding is the starting point. NSTIFC (2009) developed a set of principles for transportation funding that hold up well for many jurisdictions' interest in strategic and holistic approach to their systems and networks:

- The funding and financing framework must support the overall goal of enhancing mobility of all users of the transportation system. The range of mobility needs throughout the nation [megacity] requires a [3D multimodal] transportation network that ensures easy access, allows personal and business travel as well as goods movements without significant delays, and permits seamless transfers and choices among complementary transportation systems and services.
- The funding and financing framework must generate sufficient resources to meet national [megacity] investment needs on a sustainable basis, with the aim of closing a significant funding gap. The framework must enable the government to raise sufficient funds and also support the ability of other levels of government to raise sufficient funds and make appropriate investments.
- The funding and financing framework should cause users and direct beneficiaries to bear the full cost of using the transportation system to the greatest extent possible (including for impacts such as congestion, air pollution, facility damage, and other direct and indirect impacts) to promote more efficient use of the system. This will not be possible in all instances, and when it is not, any cross-subsidization must be intentional, fully transparent, and designed to meet network goals, equity goals, or other compelling purposes.
- The funding and financing framework should encourage efficient investment in the transportation system—recognizing the inherent differences between and within individual [megacities]—such that investments go toward projects with the greatest benefits relative to costs.
- The funding and financing framework should incorporate equity considerations—for example, with respect to generational equity, equity across income and disadvantaged groups, and geographic equity.
- The funding and financing framework should support the broad public policy objectives of energy independence and environmental protection. Revenue-raising mechanisms that impose the full cost of system use (including externalities such as carbon emissions) can support reduced petroleum consumption and improved environmental outcomes.

7.4 PRICE-BASED REVENUE GENERATION

To the extent user fees provide transportation revenues, they entail some form of pricing. Prices are simply the payment for consuming or using a given product or service. But they actually serve a much more important role. Economists have long understood that prices provide critical information to consumers and producers that coordinate resources (Decorla-Souza and Whitehead, 2003). For the consumer, prices provide a benchmark for determining how best to allocate their budget to achieve their goals and objectives given their preferred outcomes. For producers, prices set a benchmark for determining whether a product or service can be provided profitably. Prices serve the same function in transportation systems although the "public good" qualities of roads, transit, and other facilities and services complicate the supplier-driven component of service provision. Given the potentially broad economic and social benefits of transportation infrastructure, and the practical limits of pricing these facilities and services given current technology, policymakers and planners often must make decisions about investment without the benefit of effective market prices. Nevertheless, an objective of pricing as much of the transportation network as possible improves resource coordination, manages the demand for transportation facilities more effectively, and builds in sustainable revenue streams for financing current and future facilities.

As a starting point, pricing for transportation facility infrastructure entails charges to various types of users to recover costs in their service life cycles. Fees for operations and maintenance entail charges for some unit cost calculation of those costs, as well as possibly including costs of externalities such as congestion or pollution. Sustainable pricing would take into account costs to rehabilitate or rebuild infrastructure at the end of its life cycle. Life-cycle pricing would improve system performance by creating appropriate price signals for users in the short term while funding necessary maintenance. This can get quite dense; Figure 7.2 illustrates factors that might go into pricing a highway (Li, 2018).

Working through these costs and attempting to calculate these prices are worth doing even if the intent is not to charge the full price to users. These calculations will reveal much about the true resource consumption of mobility and can foster better assessments of costs and benefits for transportation investments, and thus establishing better framework for long-term decisions and policies. At the same time, these calculations should be used with humility. They involve many assumptions, especially about external costs, with little certainty of getting the right answer. Infrastructure costs are the easiest part to estimate, but these calculations are based on assumptions about travel trends, vehicle technologies, weather trends, and present versus future value of money. Considering the results as best guesses for decision-making purposes rather than scientific certainties is important.

7.4.1 Pricing to pay for highway infrastructure

Road infrastructure mainly covers pavements, bridges, tunnels, and parking lots and structures supporting individual vehicle travel and goods movements, bus transit, ridesharing, and some bicycle and microtransit. Transit-specific infrastructure includes bus and rail stations, park and ride lots, rail and tunnels, and facilities such as garages and repair shops for rolling stocks. One way to think about infrastructure costs is that a given facility has costs of construction, maintenance, rehabilitation, and replacement that should all roll up into a life-cycle cost calculation (Li, 2018). For simplicity, we will discuss the broad elements in each of those categories for a hypothetical roadway, though it is not difficult to see the parallels for a rail line.

- *Roadway construction costs* are the capital costs incurred in all phases of the design and construction such as feasibility studies, surveying, geometric and pavement design services, right-of-way (ROW) acquisition, and construction of pavements. The construction costs can be assumed as a linear function of pavement width. The portion that varies with width includes costs of preparing the roadbed, which do not depend on pavement thickness, and costs of the pavement itself, which are approximately proportional to the volume of pavement materials as the product of pavement width and thickness.
- *Pavement repair costs* are related to treatments periodically applied to pavements when they become worn. To a first approximation, the pavement is designed to withstand a certain number of passes of axles of a standard weight and configuration

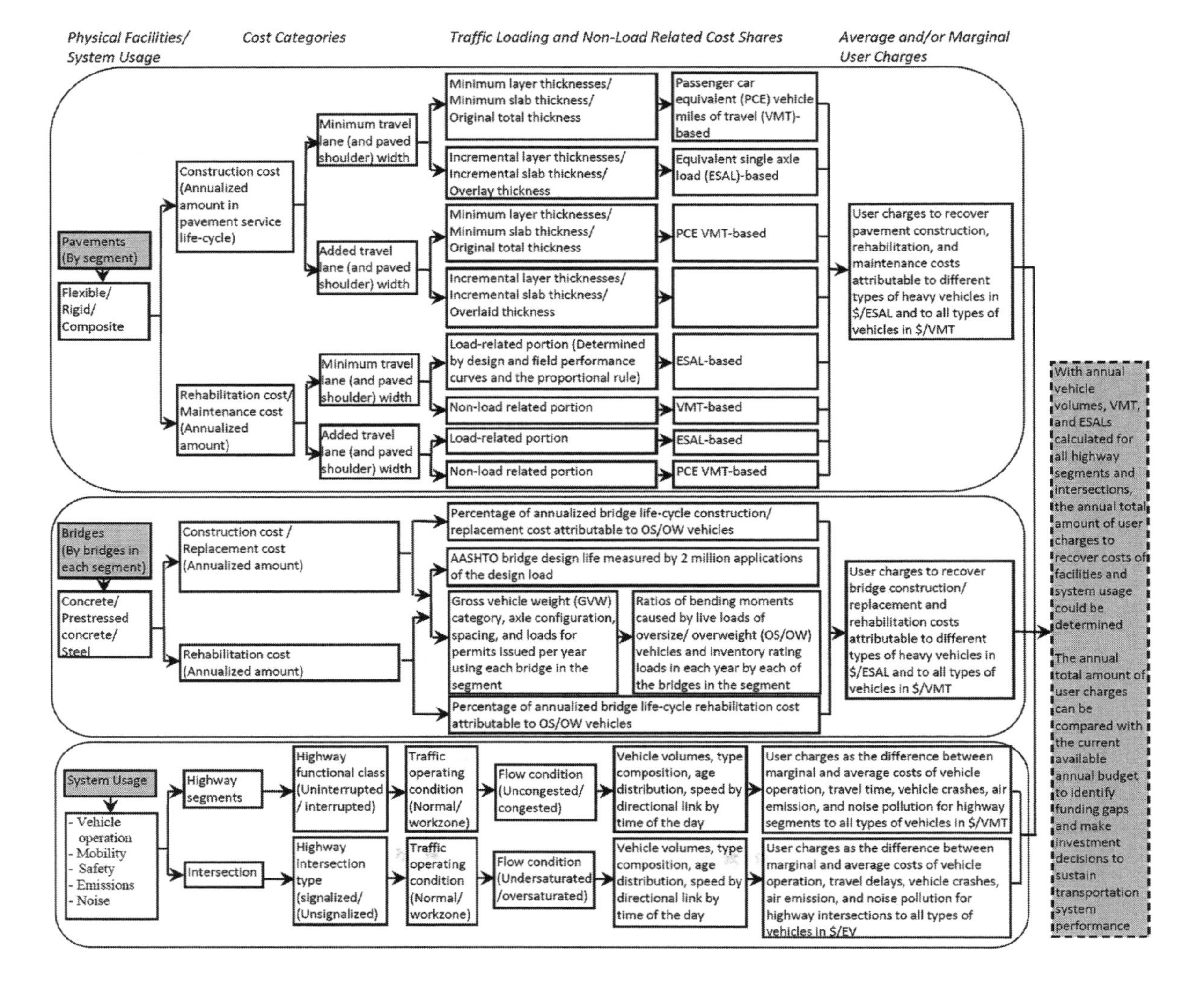

Figure 7.2 Cost allocation framework for pricing highways. (Source: Li (2018).)

before requiring repair treatments. The damages to the pavement are caused by the axle weight, not the total vehicle weight. The equivalent damages for an axle rise very steeply and roughly follow a fourth power of the ratio between the axle load and the corresponding standard load. For a single axle load, the standard load is 18,000 lbs. For a tandem or tridem axle, the standard load is 33,200 lbs for flexible pavements and 29,000 lbs for rigid pavements, respectively.

As axle loadings in the operating weight of an automobile are much lower than the corresponding standard loads, the pavement damages caused by automobiles are negligible and they are solely attributable to truck traffic. The equivalent damages of all axle loadings in the operating weight of a truck of a particular class is added together to arrive at the total equivalent damages per truck of a particular class. Provided information on the truck traffic volume in the traffic stream by truck class and the equivalent damages per truck of a certain class, the total damages caused by all trucks in the traffic stream can be determined. The conversion of axle loadings into equivalent damages to the pavement for various classes of trucks establishes pavement rehabilitation costs as a function of equivalent damages caused by truck traffic loadings and pavement durability as the cumulative damages designed to be withstood in the pavement useful service life cycle. The determination of which class of trucks share the rehabilitation costs depends upon a comparison of the equivalent damages caused by individual classes of trucks in the traffic stream.

- *Rehabilitation costs* include major repair and activities for bridge elements and require engineering analysis. These costs are allocated based on estimates of the fractions of these costs associated with different types of rehabilitation projects and costs broken down by roadway functional classification. Costs should be allocated on a per-mile/kilometer basis for each vehicle type, and the costs should be allocated only by weight category of truck traffic.
- *Replacement costs* represent the need to replace a facility at the end of its useful service life. These costs are allocated based on estimates of the percentage of these costs that are incurred per period and across users. Estimating the replacement costs for a transportation project is more complicated than might appear at first. Substantial uncertainties exist over the traffic growth for various types, the wear on materials used to build the facilities, or even what the proper discount rate should be for evaluating the costs and benefits for a facility that could last 50 years or more. Useful frameworks for estimating these costs over the life cycle under various certainty and risk scenarios have been examined in recent engineering literature (Li and Madanu, 2009; Li and Sinha, 2004).
- *Roadway maintenance costs* are incurred to preserve the capital investments made in the road infrastructure and to ensure that the road infrastructure provides a satisfactory level of service to the users. Maintenance treatments are implemented for either preventive or corrective purposes on a routine or periodic basis. Maintenance costs may be expressed in average unit accomplishment costs per project. To allocate roadway maintenance costs to road users, each of the different types of maintenance projects is designated to one of the following major categories: maintenance costs that relate only to VMT and are independent of weight; maintenance costs for flexible, rigid, and composite pavements based on equivalent damages; maintenance costs that are independent of weight but are dependent on axle configurations; maintenance costs for truck-only facilities such as weigh stations; maintenance costs for non-truck facilities; and maintenance costs for mixed-use facilities for both trucks and non-trucks. Then, the totals of all maintenance costs for the above categories are

summed up by highway functional classification. Finally, the cost share for each of the highway functional classification-specific allocators is established. The cost shares are further translated based on VMT by highway functional classification.

- *Operating costs* can include signals and signs, weather response such as snow plowing and salting, incident and emergency management, traffic management, and safety features and enforcement.

7.4.2 Pricing for specific facilities

A transit system might charge a dedicated price on certain parts of the system. A downtown trolley might have a special fare distinct from the fares for buses and trains in the rest of the system. A commuter rail line might have a special fare to pay for new, higher-quality rolling stock to attract more riders for those longer journeys. Underutilized parts of the system might have lower prices to attract people to try it, such as with shared use bikes and microtransit.

With roadways, lane pricing is electronically implemented via transmitters from drivers in added or converted highway lanes. Lane pricing includes charges to HOT lanes and fast and intertwined regular (FAIR) lanes. HOT lanes are HOV lanes that also allow non-HOV vehicles for a toll and are often proposed as a compromise between HOV lanes and road tolling as solutions to traffic congestion—often called express lanes or managed lanes (Fielding and Klein, 1993). HOV vehicles or low-emission vehicles may use the HOT lanes for free or for a reduced user charge rate. This allows excess HOV lane capacity to be used while maintaining an incentive for mode shifting. HOT lanes are able to finely tune the user charge rates for non-HOVs.

FAIR lanes are a concept to address potential equity concerns from higher-income users being able to use congestion-free, tolled lanes (FHWA, 2021). The FAIR lanes divide a road into fast lanes and regular lanes. Road users can opt for quicker trips by paying tolls and using the fast lanes. But with FAIR lanes, drivers who chose to remain in the regular lanes receive rewards in the form of transportation credits. These credits can be reserved for future use of the fast lanes.

7.4.3 Pricing for externalities

The two main externalities dealt with in transportation networks are congestion and emissions of harmful pollutants. Pricing has a potentially important role to play in addressing both of these types of externalities. In the first case, pricing helps manage flow on existing facilities. In the second case, pricing helps reduce and mitigate environmental impacts.

About half a century ago, Roth and Thomson (1963) developed the concept of road pricing that could potentially be used as a travel demand management strategy to mitigate congestion. Allais and Roth (1968) extended the concept by working out details of charges to highway users to achieve economic efficiency, as illustrated by Figure 7.3.

The short-run marginal travel time exceeds short-run average travel time along with traffic volume increases. This is because the marginal social travel time t_M not only includes the average travel time incurred by the traveler, t_A, but also the additional travel time that the traveler causes to all other travelers in the traffic stream, which is called marginal external travel time, t_{ME}. The efficient utilization of highway capacity is achieved when the benefit of each trip equals to marginal social travel time t_M represented by the demand and supply equilibrium point "1" based on the marginal travel time function. If the equilibrium point is at point "0" based on the average travel time function, user charge at a rate of $\alpha \cdot t_{ME}$ (α is the value of travel time) needs to be imposed to shift the equilibrium from point "0" to point "1."

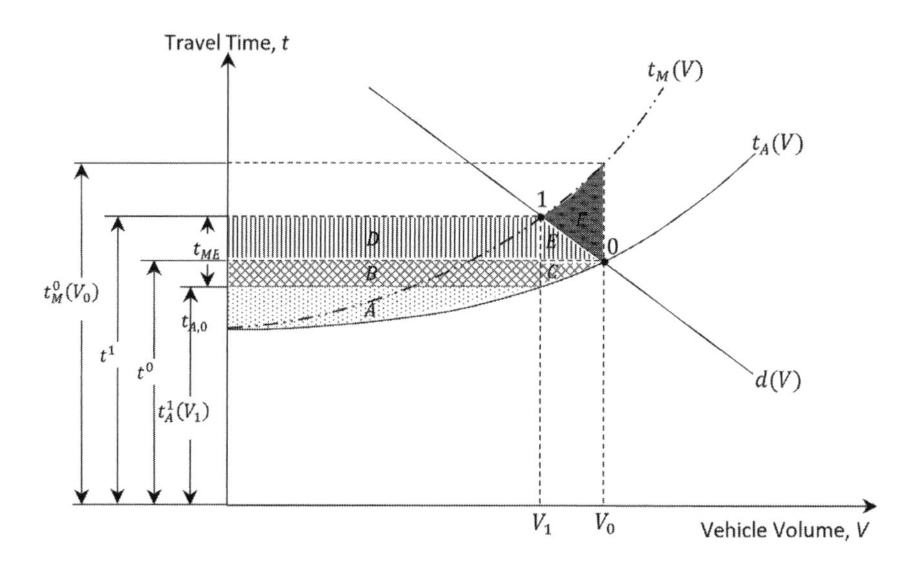

Figure 7.3 Short-run average, marginal social, and marginal external travel time. (Source: Li (2018).)

For the unpriced equilibrium point "0," the average travel time experienced by V_0 travelers is t^0. After shifting the equilibrium to point "1" by imposing the user charge, the average travel time experienced by V_1 travelers is reduced to $t_A^1(V_1)$. The generalized travel time for unpriced and priced equilibria are t^0 and $t^1 = t_A^1(V_1) + t_{ME}$, respectively. The quantity $(t^0 - t_{A,0}) \cdot V_0$ highlighted by shaded areas A+ B+ C can be interpreted as the extra total travel time experienced by travelers in the unpriced equilibrium compared with the travel using the base free flow speed. Specifically, area A represents efficient extra travel time and areas $B + C$ reflect inefficient extra travel time, which is eliminated by imposing the user charge. Further, the gain in social surplus is depicted by shaded areas C+ E+ F. The change in consumer surplus as user benefits is represented by shaded areas D+ E, which can be computed by $(t^1 - t^0 \cdot [(V_0 + V_1)]/2)$.

About the same period, Smeed et al. (1964) recommended congestion pricing for busy roads in the United Kingdom and made recommendations to consolidate various road costs, congestion costs, and social costs in a road pricing scheme in which the users pay not only for creating congestion on the road but also for using and causing various social impacts. The concept of road pricing conceptualized in the United Kingdom was eventually implemented in Durham and London as congestion mitigation strategies. Their economic analysis indicated that the largest economic benefits from road pricing were generating revenues. Demand management was a beneficial by-product (Ponkshe, 2019). The experiment showed the economic feasibility of the various infrastructure costs in today's times when budget is a major constraint to maintain and operate a highway system smoothly and efficiently (de Palma and Lindsey, 2011; Lindsey et al., 2012). In fact, road pricing has been used in a variety of contexts to fund new infrastructure facilities in lower-income countries experiencing rapid growth, including Mexico, Brazil, India, and China (Song, 2015).

The road pricing principle was implemented to address congestion problems in Singapore in 1975 by using electronic devices to charge the users (Development Asia, 2016). This was followed in Norway by implementing a cordon-based user charging system in the central business district (CBD) of Oslo in 1990, and then in London in 2003 (VTPI, 2019). The system in Norway was to tax the users to generate revenues for roadway maintenance and was not aimed at congestion mitigation. Similar systems are now in place in other cities.

Variable pricing is a rough form of congestion pricing that sets different tolls or fares, etc., for different times of day or different levels of demand. Prices are lower during hours when the road or transit system is not busy but higher during peak hours to help create incentives to encourage discretionary users to travel during less-congested periods, change routes, or to shift to another travel mode.

Cordon pricing is adopted in most practices of road pricing to control area-wide congestion in a city, and it may also be implemented for financing, efficiency, and/or environmental pricing purposes. In a typical cordon pricing system, each vehicle is charged a fixed toll when it passes through the specified cordon surrounding the central area of a city where traffic is most congested. The cordon pricing system is simple and easy to implement. However, it is not an easy task for a traffic control authority to rationally determine the cordon toll level and the cordon location.

For public transit congestion pricing in the form of variable pricing could improve operations and allow for price signals to share better information with transit operators (Cuddihey, 2019; Brown, 2018). It would also encourage discretionary users to travel during less-congested periods, change routes, or to shift to another travel mode like personal mobility or walking.

7.4.3.1 User charges for other externalities

User charges for other externalities have long been proposed as a means for having users pay for the marginal social costs of their actions (Maddison et al., 1996; Ramjerdi, 1996; Button, 2004; Ison, 2004). In practice, effective ways to directly measure external costs are unavailable or very complex, and pricing them is a difficult task laden with assumptions. Existing mechanisms tend to be crude fixed fees based on vehicle weight or fuel usage, not really a functional or dynamic price, nor are revenues dedicated to transportation. Vehicle license fees for carbon emissions in Europe illustrate that point (ACEA, 2020; Wappelhorst, 2018). The same is true of a new program in the Northeastern United States where a group of states are rolling out a sort of emissions-cap-and-trade system for large gasoline and diesel fuel suppliers where only some of the proceeds will go to transportation. Bigazzi and Rouleau (2017) claim that attempts to price externalities have very small effects. However, Cavallaro et al. (2018) are more optimistic about carbon charges.

7.4.4 Pricing controversies and challenges

That appropriate pricing of roads and transit can provide benefits to all users is well understood (ITF, 2018). Road pricing has reduced cross-subsidies among users, increased the horizontal equity, and potentially solved the funding gap between revenues and total highway costs (FHWA, 1998). Nevertheless, user acceptance is a major barrier. An example is Hong Kong's electronic road pricing system. This program had a successful run from 1983 to 1985 but was abandoned owing to public opposition (Litman, 2003). Charging vulnerable groups such as lower-income, elderly, and disabled users is a concern as well. But the degree of opposition depends on the percentage of lower-income users on the highway, the ride quality, and revenue utilization (Giuliano, 1994; Rajé, 2003; Parry, 2008; Burris et al., 2013). Kain (1994) indicated that road pricing could reduce the subsidies that users pay and increase their travel choices as well. Research studies also revealed that the issue of equity could be addressed by using a portion of the revenue generated from the road pricing to mitigate negative impacts created by the system (Levine and Garb, 2000; King et al., 2007). The resistance from all the other factions, such as the freight industry and government, is due to political motivations, distrust due to past mistakes, and inefficient usage of revenues (Samuel, 2000; Vassallo and Sánchez-Soliño, 2007; Bain, 2009).

7.5 TRANSPORTATION FINANCING

Financing allows efficient use of various revenue streams to help pay for mobility projects. Financing also allows transportation authorities and businesses to effectively borrow against future revenues to help build long-lived infrastructure or purchase long-lived rolling stock. Smart financing strategies can help achieve more with given revenue. As shown in Table 7.3, various public finance strategies and tools, including PPPs, can bring in private capital in the form of debt or equity to help finance a project and also private operational capability. All the options in the taxonomy are in use or have been used to finance transportation projects. Options further down the list lower the cost to the government but also increase the risk. Balancing the costs and benefits of cost-to-risk tradeoffs along with potential improvements in management and operations is at the heart of innovative financing and PPPs. Earning trust with good project design and transparency is important.

Table 7.3 Taxonomy of instruments and vehicles for infrastructure financing

Mode		*Infrastructure finance instrument*		*Market vehicle*
Asset category	Instrument	Infrastructure project	Corporate balance sheet/other entities	Capital pool
Fixed income	Bonds	Project bonds Municipal, sub-sovereign bonds	Corporate bonds, green bonds	Bond indices, bond funds, and exchange-traded funds (ETFs)
		Green bonds, sukuk	Subordinated bonds	
	Loans	Direct/coinvestment lending to infrastructure project, syndicated project loans	Direct/co-investment lending to infrastructure corporate	Debt funds via general partners (GPs)
			Syndicated loans, securitized loans via asset-backed security (ABS), and collateralized loan obligations (CLO)	Loan indices, loan funds
Mixed	Hybrid	Subordinated loans/bonds, mezzanine finance	Subordinated bonds, convertible bonds, preferred stock	Mezzanine debt funds (GPs) and hybrid debt funds
Equity	Listed	YieldCos	Listed infrastructure and utilities stocks, closed-end funds, real estate investment trusts (REITs), infrastructure investment trusts (IITs), and master limited partnerships (MLPs)	Listed infrastructure equity funds, indices, trusts, and ETFs
	Unlisted	Direct/coinvestment in infrastructure project equity, PPP	Direct/coinvestment in infrastructure corporate equity	Unlisted infrastructure funds

Source: OECD (2015).

7.5.1 Public financing of mobility

Governments have a number of means of public financing of mobility, including traditional debt, infrastructure banks, asset recycling, and PPPs. The latter merits particular attention in the next section.

7.5.1.1 Debt

The most conventional way in which governments finance transportation projects is through borrowing. Sovereign debt or national government debt is almost always used to build a road, rail, station, or other infrastructure that will be utilized for many years. This debt from borrowing or sales of financial securities such as bonds is paid off plus interest through future expenditures. If there are user fees, such as tolls or fares, among the available revenue streams, future revenues from those sources can be specifically borrowed against. The former debt is often called general obligation debt, while that latter is revenue debt. A steady and known source of user fees, or a reliable commitment of general revenue, combined with a good credit rating, can make the direct cost of government borrowing quite low.

In either case, debt finance involves many tradeoffs (OECD, 2015; UNECE, 2017). A government has practical, and often legal, limits on how much sovereign debt it can undertake, which must be allocated across many sectors. As such, national governments often do not have enough debt capacity to fund all desired transportation projects. Debt financing adds interest costs to a project which must start being paid immediately even if the project takes many years to be completed and start delivering benefits to those paying the debt. And general obligation debt leads to various forms of cross-subsidies from other sectors since revenues are not tied to specific sectors of government spending. Revenue debt can lead to cross subsidization, as when highway toll revenue is borrowed against to pay for transit projects. Layered on top of all of this is the reality that projecting future project needs, costs, and benefits has many uncertainties. Taking on large debt and building a project that either is not completed or winds up underutilized due to construction barriers (e.g., unexpected environmental impacts or political changes) or changing land use, travel patterns, or demographic shifts are always a risk.

7.5.1.2 Infrastructure banks

Infrastructure banks use initial capital provided by a government or group of governments or nations, through appropriations or low-cost debt and lend that money for selected projects and recycle the repayments into new rounds of loans. These "revolving loan" funds provide a continuing source of financing for projects. Some infrastructure bank concepts include allowing private capital to be added to the pool.

The most familiar infrastructure banks are mostly funded by wealthier countries to help infrastructure projects in developing nations. Examples include the World Bank, European Investment Bank, and Asian Infrastructure Investment Bank. However, many nations have set up national or subnational infrastructure banks to make loans to subnational projects (Uzsoki, 2018; GIH, 2019). Not all such institutions are equally successful. Having the right expertise to evaluate projects and loan requests and the right criteria for picking projects to which to lend is crucial. Many lessons have been learned about how to create a successful infrastructure bank program (Uzsoki, 2018; GIH, 2019).

7.5.1.3 Asset recycling

Asset recycling entails the sale or lease of government owned assets (toll road or bridge, power plant, building, etc.) and the use of proceeds to invest in new infrastructure projects.

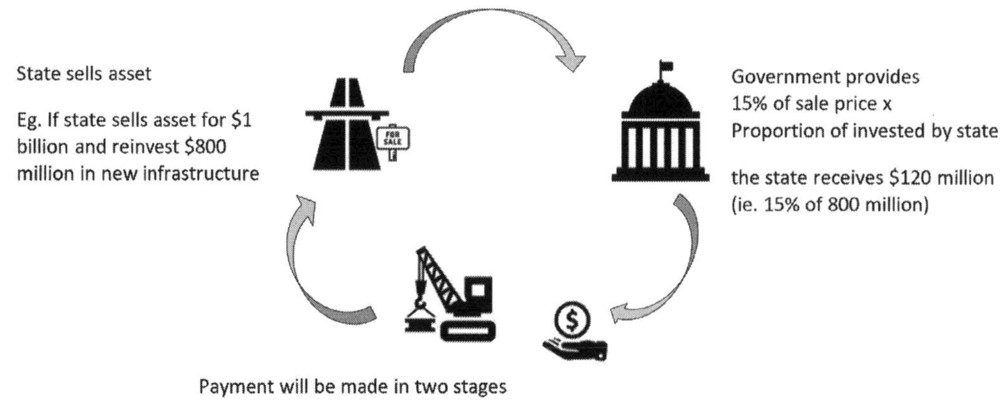

Figure 7.4 The asset recycling process. (Source: The Australian Government the Treasury (2019).)

Assets with a user fee revenue stream most readily lend themselves to this approach. This process not only unlocks the capital value of the assets, which otherwise goes unused by the government, but also creates incentives via ownership or lease terms for the private party taking over the asset to invest in upgrades and improvements or make up deferred maintenance (Figure 7.4).

Only one nation has used this tool in a systematic way—Australia—and it provides an instructive example. According to Chalmers et. al. (2018) and The Australian Government the Treasury (2019), a 3-year national program was commissioned to provide AU$2.3 billion in national incentive payments to subnational asset recycling projects that resulted in over AU$15 billion in additional infrastructure investments. One of the projects was a long-term lease of the electricity network by the state of New South Wales which used the proceeds to invest in expanding and improving the Sydney Metro, light rail transit, and road projects.

However, Poole (2018) points out there were a number of asset sales and leases in Australia before the national asset recycling program that specifically sought to unlock the capital value of government assets in order to make investments elsewhere, if not necessarily in infrastructure. He goes on to point out similar precursors of asset sales and leases in the United States, such as the Chicago leases of the Skyway toll road and city-owned parking garages, and The Ohio State University lease of parking garages. More importantly, he explains clear U.S. examples of transportation asset recycling, including the best example, the Indiana Toll Road (ITR) (see below).

- Case 1: *Indiana Toll Road*. The state of Indiana leased the 156-mile Indiana Toll Road (ITR) in 2006 to private investors in a 75-year deal that included a US$3.8 billion upfront lease payment and strict requirements for performance of operations of the facility and limitations on toll rate increases. Prior to the lease, the state was losing money on the toll road and it was accumulating deferred maintenance and obsolete technology.

  The asset was recycled to permit improvements in the facility itself, but the upfront payment allowed the state to invest in a broad range of new transportation projects including 500 miles of new roadways and 60 new or reconstructed interchanges.

They also created a fund for long-term maintenance of transportation infrastructure and a fund to pay for new road and bridge projects. Meanwhile, the state earned US$755.5 million in interest income from the upfront payment over the first 5 years of the lease. An asset that was running at a deficit every year was now bringing in substantial revenues even while the capital value was unlocked to pay for new projects and maintenance and at the same time improving the condition and performance of the asset.

This experience already sounds like an idealized example of asset recycling, but one last detail makes it even more so. After 2 years of massive investment by the private venture in technology upgrades and adding capacity to the ITR, the 2007 recession from the global financial crisis caused traffic and toll revenues to plummet. Soon the private venture filed for bankruptcy. While the private operators and financiers went through restructuring, services on the road were not interrupted, nor were payments to the state. The terms of the lease put the financial risks on the private venture, not the customers or the state. After the recession, the lease was bought out by a new private venture who continued to invest in improving the toll road and honor the terms with the state (Poole, 2018).

- Case 2: *Maryland's Seagirt Marine Terminal.* Maryland Port Administration (MPA) in 2009 leased Seagirt Marine Terminal to a private partner for 50 years. MPA received a US$140 million upfront payment which was recycled to transportation investments, specifically improvements to bridges and tunnels in the port area. In addition, the private partner invested several hundred million dollars in upgrading port facilities to accommodate more and larger container ships and increased shipping traffic (Poole, 2018).
- Case 3: *Puerto Rico Toll Road and Airport.* In 2011, Puerto Rico's Public-Private Partnership Authority leased two toll roads to private partners which entailed payments of US$1.136 billion over the 40-year lease and an investment of US$350 million worth of capital improvements to the toll roads. The capital value of the roads was recycled to pay for deferred maintenance and upgrades that the government could not otherwise afford. And in 2013, Puerto Rico leased the Luis Muñoz Marin International Airport for 40 years in return for an upfront payment of US$615 million and US$1.4 billion in capital investment in the airport and revenue sharing. The state used some of the upfront payment to improve regional airports and cruise ship terminals (Poole, 2018).

One way of looking at asset recycling is that it is a form of debt—the state receives money today that comes from future revenues from the asset. But as the ITR example shows, the structure of an asset recycling arrangement can deliver far more than any ordinary form of borrowing could hope to, including interest earnings on the capital value rather than interest payments to a lender, improvements to the asset, performance guarantees for the condition of the asset, and shift of financial risk away from the government.

It's no wonder then that asset recycling is proposed as a potential approach for nations struggling to meet infrastructure needs, including Asia and Africa (APRC, 2020; Ebobisse, 2020). However, while it has great potential as a tool, it is really a specialized form of PPP and so suffers from some resistance on that basis.

7.5.2 Public-private partnerships

PPPs are an array of tools for financing mobility projects, ranging from government purchase of, contracts for, or subsidies to private operations of transportation services, or to concession arrangements or outright privatization. The core concept that ties them together is some degree of decision-making about what risks and responsibilities will be assigned to the public or private sector or shared between them. Importantly, most PPPs include projects in which some private capital is invested in the project. While a PPP need not necessarily be based on user charges, user fees in the form of flat-fee structures or variable pricing initiatives make a PPP less risky and easier to structure. However, structures such as shadow tolling or availability payments have emerged to build PPPs around surrogates for user fees (AECOM, 2007).

In recent decades, use of PPPs for transportation and other infrastructure has become more widespread as familiarity with their structure and sophistication with their management has grown. Increasingly, public sector acceptance has found along with acceptance of mutually acceptable finance, construction, and management frameworks where public entities own transportation systems operated and managed by private entities, or where the government is the proxy for citizens as customers in contracting for services from and private owners and providers.

Today PPPs in transportation are practiced extensively on a global scale (PWF, 2011). As seen in Figure 7.5, the World Bank Private Participation in Infrastructure (PPI) database identifies 1,192 transportation PPPs between 1990 and 2020 in which a total of US$352.5 billion in private capital were invested. Looked at another way, PPPs account for an average of 3.1% of infrastructure investment globally, and 10%–15% in advanced economies (Woetzel et al., 2016).

For example, many European cities contract with private partners to operate and manage transit systems under a variety of PPP structures (Van de Velde et al., 2008b). The World Bank has studied a number of exemplary cases of PPPs in urban light rail transit (Mandri-Perrott, 2010). PPPs are even more extensive in highways, including urban roads. Poole (2018) describes their global growth over the last half-century:

> A wide array of countries, beginning with France and Italy in the 1960s, have brought private finance, development, operation, and maintenance into the highway sector. The French, Italian, Portuguese, and Spanish limited access highway systems were all developed and operated by what are now for-profit businesses, operating under long-term concession agreements. The toll concession model has also been the primary vehicle for major urban and intercity motorways in Australia, China, other Asian countries, and the major countries in Latin America. The reasons for this extensive use of tolling and private-sector investment and management vary from country to country, but one key factor has been the lack of a dedicated funding source for highways, such as our more-or-less dedicated federal and state fuel taxes. The types of projects done via concessions also vary, from developing intercity superhighways from scratch

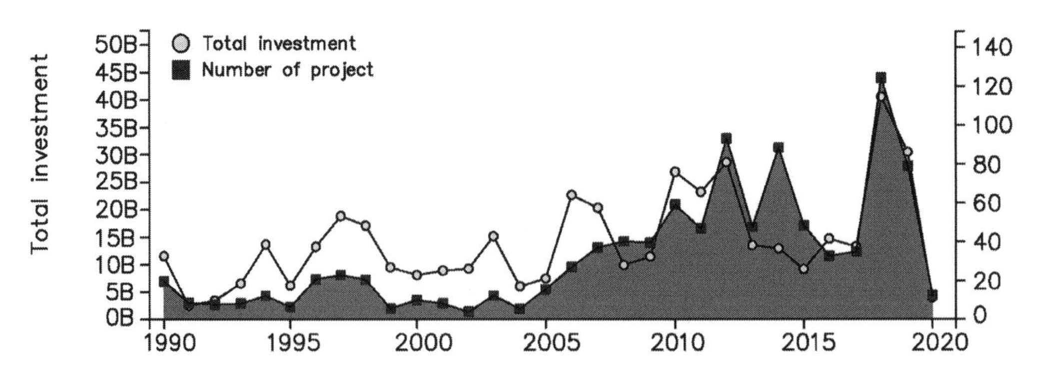

Figure 7.5 Global private investment in road infrastructure during 1990–2020. (New figure created using data in World Bank (2020a).)

in much of Europe, to upgrading two-lane highways into modern motorways in Latin America, and to developing modern urban expressways in Australia's largest cities, as well as in several major cities in Latin America.

As they have become more common, the structure of transportation PPPs has evolved. Traditionally, transportation projects follow the design-bid-build format, wherein the transportation agency designs the facility project by means of in-house staff or an external contractor and then bids out the project for construction. This traditional arrangement ensures that the design risk, along with the operation and maintenance, is retained by the agency while the construction is carried out by an external contractor. As PPPs matured, more innovative and sophisticated financing structures have emerged for facility project delivery, operations, and maintenance. These structures include design-build-finance-operate, build-operate-transfer, and long-term lease concessions, among others (Delmon, 2010).

The rise of private capital investment in PPPs with debt and/or equity opened a much wider range of possible partnership structures. Table 7.4 shows the wide array of sources of financing for PPP infrastructure projects in emerging markets and developing economies in 2020. Debt is still dominant, but there are many ways to pool funding sources in a PPP.

7.5.2.1 Strengths of PPPs

Like all policy choices, PPPs have tradeoffs. Many benefits and goals can be achieved through well-designed and managed PPPs. But there are risks as well, including the challenge of learning and executing well a new policy tool. A poorly constructed or managed PPP can carry the seeds of its own destruction.

The wide prevalence of PPPs around the globe testifies that governments must be seeing net benefits from them. PPPs are an expansion of or supplement to traditional funding and finance of transportation projects, not a replacement. Even entirely private transportation investments take place within a dense context of existing public infrastructure and services. PPPs may relieve the government of some responsibilities, but many remain, including managing externalities, rights of way, providing the public side of the PPP and ensuring performance first by not only effectively managing the process of creating PPPs but also enforcing standards and contracts during them. Clearly, data-driven thinking about project and service needs and growth and competent execution of infrastructure projects and transportation services is the context in which the most benefits

Table 7.4 Financing sources of infrastructure projects with private participation in EMDEs in 2020

Total investment	*Financing source*		
Equity (24%)	Public equity (0.2%)		
	Private equity (23.8%)		
Debt (76%)	International (51%)	Decentralized finance (DEFI) debt (24%)	Bilateral (14%)
			Multilateral (10%)
		Non-DEFI debt (27%)	Institutional (14%)
			Commercial (13%)
	Local (25%)		Public (16%)
			Commercial (9%)

Source: World Bank (2020b).

can be achieved from PPPs. Once PPPs are a piece of the overall transportation system of a megacity, then they should be prioritized for those projects where the maximum is to be gained from a PPP structure (Perkins, 2013).

PPPs have the virtue of narrow focus as the participants recognize success depends on mutual cooperation. Each has a stake in making a particular project or service succeed (Engel and Galetovic, 2014). A PPP to build, say, a new underground roadway to connect other parts of an urban road network is utterly focused on that project. In contrast, a government agency delivering the project in conventional fashion is simultaneously mired in all the other transportation priorities of the city and perhaps even competing policy priorities such as public safety, housing, arts, or recreation. Amid these competing interests and priorities, a single project may not get the management focus or budget priority it would get in a PPP structure. That said, putting the government in the role of oversight, rather than service provider, means PPPs tend to enjoy more scrutiny for performance and satisfaction, and even crucial matters like compliance with safety requirements—self-regulation is not as effective as oversight of one party by another.

The benefits of PPPs include cost savings, quicker and more certain project delivery, higher service quality due to competition, and a life cycle approach to managing spending and maintenance (Engel and Galetovic, 2014; Feigenbaum and Hillman, 2020; Kirk and Mallett, 2020). The World Bank, in its PPP Knowledge Bank, summarizes the value drivers of PPPs to include the following:

- *Whole-of-life costing.* Full integration, under the responsibility of one single party of upfront design and construction with ongoing service delivery, operation, maintenance, and refurbishment, can reduce project costs. Full integration incentivizes the responsible party to complete each project phase (design, build, operate, maintain) in a way that minimizes total costs and maximizes efficiency.
- *Risk transfer.* Risk retained by the government in owning and operating infrastructure typically carries substantial, and often, unvalued, hidden cost. Allocating some of the risk to a private party which can better manage it can reduce the project's overall cost to the government and minimize risk to the taxpayer.
- *Upfront commitment to maintenance and predictability and transparency of whole-of-life costs.* PPP requires an upfront commitment by the private operator to the whole-of-life cost of providing adequate maintenance for the asset over its lifetime. This commitment strengthens budgetary predictability over the life of the infrastructure and reduces the risks of funds not being available for maintenance after the project is constructed.
- *Focus on service delivery.* A contracting agency can enter into a long-term contract for specific services or facilities to be delivered when and as required. The PPP firm can then focus on delivery without having to consider other objectives or constraints typical in the public sector.
- *Innovation.* Specifying outcomes in a contract, rather than prescribing inputs, provides wider opportunity for innovation by the private partner. Competitive procurement of these contracts incentivizes bidders to develop innovative solutions for meeting these specifications.
- *Asset utilization.* Optimizing the utilization of assets for delivery of additional services leads to multiple revenue streams for the project, for example, the utilization of space in bus terminals for private vendors or unused space for advertisements.
- *Mobilization of additional funding.* Charging users for services can bring in more funding and can sometimes be done better or more easily by private operators than

Table 7.5 The challenges with infrastructure and how PPPs can help

Infrastructure challenge		*Essential role of the private sector*	*Complementary action/benefits*
Low coverage, quality, and reliability	Poor planning and project selection	Assisting in data-driven analysis and decision-making	Enhanced accountability, transparency, and governance of the public agency
	Inefficient or ineffective delivery	Providing incentives and participating in management in the facility delivery cycle	
	Inadequate preservation	Developing optimal investment strategies aimed to minimize agency and user costs in the facility service life cycle	
	Insufficient funding	Increasing the budget level from additional funding and financing sources	Improved conditions of facility preservation and performance of system usage

Source: World Bank (2017).

the public sector. Additionally, PPPs can provide alternative sources of financing for infrastructure where governments face financing constraints.

- *Accountability*. Government payments are conditional on the private party providing the specified outcomes at the agreed quality, quantity, and timeframe. If performance requirements are not met, service payments to the private sector party may be abated (Table 7.5).

Feigenbaum and Hillman (2020) point out that third-party evaluations of service quality for transit PPPs in the United States have typically found high quality in safety, reliability, and service quality. Transport for London contracts out all bus services in a performance-based structure and tracks metrics of quality, holding private partners accountable for meeting performance goals. As a result, ridership has grown with higher passenger satisfaction levels (Lotshaw et al., 2017).

The benefits of PPPs with respect to various risks merit a more detailed look. PPPs can mitigate budget risks. In a PPP structure, the payments the government must make are now and either fixed or set by formula or criteria. So, cash flows are more predictable than in traditional government projects where economic or budget shocks can cause interruptions in construction or services levels.

A well-designed PPP properly allocates risks to the public or private partner based on who can best control, manage, or cope with that risk—the idea is to let the risk incentivize the mitigation or insurance at the lowest cost. The public partner is typically best placed to manage risks of the right of way, environmental compliance, and political change. Risks best managed by private partners include legal liabilities, unexpected maintenance costs, financial market fluctuations, demand growth, and construction delays.

For transportation projects, construction cost overruns are a huge risk. Global research finds that traditional government infrastructure construction projects experience cost overruns five times those of PPPs (Flyvbjerg et al., 2003; Flyvbjerg et al., 2004; Duffield, 2008; Blanc-Brude et al., 2009; Blanc-Brude and Makovsek, 2013). Indeed, an analysis of European PPPs found that construction costs for transportation PPPs are modestly higher than those of traditional construction, but that premium is more than paid for by reduced construction cost overruns (Blanc-Brude et al., 2009; Tang et al., 2010).

The potential for PPPs to incentivize innovations is often overlooked. In a conventional sense much innovation is realized when private parties compete to be the private partner on a project—they have lots of incentive to come up with innovative ideas for how to build a facility or deliver a service. But PPPs can also build in opportunities to continue to seek innovative approaches. For example, Lotshaw et al. (2017) describes how Stockholm's bus contracts cover zones of the city, not routes. The private providers are incentivized to develop routes and customer outreach that combine basic service frequency and coverage with increasing ridership. Those that do better are advantaged in subsequent rounds of competition.

Approaches like this will become more important and the technology of urban mobility changes. Innovations in microtransit, ridesharing, and shuttle services open up many possibilities. While formal microtransit services are largely experimenting in the United States and Europe, they are beginning to be seen in places like Singapore and Jakarta, Indonesia. All of these services could potentially expand the network and convenience of traditional transit in ways that might appeal to more customers and lower the cost of mobility.

Feigenbaum and Hillman (2020) describe how Via, a private microbus transit company, has partnered with cities all over the globe including New York, Jakarta, London, and Tel Aviv to provide on-demand and other transit services that utilized apps and other technologies to better align transit services with demand. Other cities are working with Via to find ways to serve new sets of customers with new mobility solutions. This type of PPP can easily expand and grow transit services and might replace solo vehicle travel or provide mobility to people with few or no transit options more effectively than traditional transit services such as fixed guideways or conventional mass transit bus.

7.5.2.2 Risks and challenges of PPPs

PPPs may be a way to solve transportation needs in some cases, but in other cases may raise significant challenges and introduce new and unfamiliar risks (Rall et al., 2010). Many of those, in turn, may be managed by properly planning and execution of PPPs. What remains is precautionary tradeoffs.

For starters, one might think that if a government is having trouble building new infrastructure or providing a service how can it take on the entirely new challenge of putting together a successful PPP. If the original failing is due to severe conditions, lack of strong institutions, lack of competence in governance, or corruption, then bringing a PPP into the mix might make things worse, not better. A government with corruption problems might find a PPP exacerbates opportunities for corruption unless the process brings more transparency with it.

The World Bank (2017) PPP Knowledge Base states that:

> PPPs can achieve efficiency improvements in the delivery of infrastructure, as described above. However, creating the incentives to achieve efficiency gains, and ensuring the public and users reap the benefit, depends on the government effectively structuring, procuring, and managing the PPP project over its lifetime. This achieves competitive tension, real risk transfer, and ensures anticipated performance improvements materialize in practice. This can be difficult where low public sector capacity means that governments lack the resources and skill to structure and manage PPPs well.

Indeed, Marcelo and House (2016) point out that many nations struggle with one or more of these challenges to executing a successful PPP. But they also find that multilateral support reduces the chance a project will fail. "Multilateral funding often involves policy advice, capacity building, oversight and risk mitigation, project preparation assistance, assistance in mediation or renegotiation, and other forms of support that could create 'halo effects' over and above direct funding." In addition to multilateral assistance with capacity and governance building to manage PPPs, governments that use PPPs as a regular tool among their policy options develop the experience and governance skills to use them well. Experience matters.

Addressing those institutional and governance needs is crucial because PPPs require some new skills and knowledge not learned by doing traditional construction or service delivery, and governance failures are arguably the largest cause of PPP failures. Governments may need to spend time and resources on procurement processes more complex than the traditional budget and hire process within an agency; will need to provide performance measurement, oversight, and accountability not provided for in-house projects; and more often need skills and resources for dispute resolution and litigation.

Potential risks and pitfalls of PPPs are highlighted in many studies, and the most prevalent ones merit more explanation (Engel and Galetovic, 2014; Makovsek et al., 2015; Woetzel et al., 2016; World Bank, 2017):

- *Circumventing budget or accounting rules.* PPPs should be chosen to achieve efficiencies and solve real project or service delivery challenges, not simply budget constraints. PPPs driven primarily by fiscal constraints are often an off-balance sheet means of borrowing future revenues to spend today, meaning those future revenues will not be available for future uses. If the borrowed revenues are properly used to produce something that provides commensurate future benefits, as in completing long-lived and beneficial infrastructure, then there may be no problem. But if it is a means to fuel windfall spending by the current government, a PPP might beggar the future.
- *Lack of competition.* A one-off or poorly structured PPP may not engender sufficient competition among potential private partners or even give private parties too much bargaining power in the partnership to the detriment of citizens and taxpayers. One-off PPPs limit the incentive for private parties to invest in learning a government's needs, rules, ways of doing business, and willingness to spend a lot of money in the hope of winning the competition. When PPPs are a policy program and a tool to be used whenever appropriate, that pipeline of potential projects increases those incentives for private participation. At the same time, governments must have a well-structured PPP project and an appropriate tender process to get consistent and competitive bids. The World Bank PPP Knowledge Lab summarizes research from the United Kingdom, Latin America, and Portugal showing that inadequate preparation of a PPP project or flawed tendering process can lead to contract terms that give inappropriate financial windfalls to the private party.
- *Lack of transparency.* Many PPP failures can be attributable to lack of transparency. This starts with goals and expectations—the government, the private bidders, and the people/taxpayers/customers all must know and understand the goals of the PPP, the process that will be followed, and assurances that will be offered. Perception that the government or private parties have a hidden agenda can create political or popular pushback that can sink a project. The operation of the PPP must be equally transparent. Government partners need sufficient ability to observe and measure the activities and outcomes of the private partners. Private partners need to be confident in the flow of funding and government's management decisions. People/taxpayers/customers need the private service provider to be responsive to their needs and concerns, and the government to represent their interests. Lack of transparency breeds corruption or at least the perception of it.
- *Poor risk allocation.* If either the private or public party in a PPP is assigned risks that they are not best suited to manage, the odds of project failure go up

dramatically. Global experience with PPPs has made clear the proper allocation of most risks and PPP guides share similar guidelines. But the PPP process and negotiations must consider the relevant tradeoffs and make good decisions about risk allocations. At the same time, the partnership structure needs to have some flexibility or mechanism for making changes as the world changes— major shifts in the economy, development patterns, or transportation technologies—can change the fundamentals of a project. The ability to modify the project within a PPP to accommodate those changes is crucial.

- *Inadequate prior analysis or preparation.* The more thorough and accurate analyses of the existing conditions and needs and projections of future growth and needs are, the stronger will be the foundation of the PPP. This can be challenging in many cases, and renegotiation of PPP contracts is common when that is the case. Changes in the environment, new information, or errors in planning or analysis can all make a renegotiation sensible. This is not a problem if renegotiation is justified, and all parties gain from it. Obviously, the reasons for it and the new terms and why they benefit both parties must be transparent and understood by all.
- *Jurisdictional conflict and coordination.* Today's and future megacities may have multiple local jurisdictions, subnational higher jurisdictions, and national governments involved in large-scale transportation projects, or at least setting policies and helping to provide funding. Rare will be a major urban PPP with only one government partner. Indeed, Table 7.4 showed that today many PPPs involve multiple government partners just in the funding structure. Conflicts among jurisdictions are common. Therefore, the lines of decision-making, the intergovernmental agreements, and the policies and responsibilities need to be clear and spelled out in advance of building a PPP. Yet another place where flexibility and an ability to work together to deal with changes and challenges will be crucial, as is transparency so that all parties understand their responsibilities.
- *Rule of law and property rights.* If private debt or equity is involved in a PPP, private investors are likely putting money upfront in the project or in payments to the government and expecting future revenue in return. So, they must be confident that the risk of expropriation is exceptionally low, and the project faces minimal risk from political interference or reneging on the partnership. Institutional strength and governance ability on the part of government partners are crucial.
- *Governments lose talent.* The private parties in a PPP are often global companies or consortiums of local and international firms, all of whom may be experiencing growth and offer opportunities for talented individuals that the government partner cannot offer. So, the best government managers and employees in the partnership may be tempted to take jobs with a private partner, bleeding the government of some of its best talent. Governments will often need to ensure their employment is attractive and to plan for retention and developing new talent.

7.5.2.3 PPPs as a tool

Implicit in many of the strengths and pitfalls of PPPs are guidelines to how to successfully use PPP as a tool. These guidelines create the conditions for success and maximizing the potential for PPPs to help megacity's optimize mobility. How-to do-good PPPs is beyond the scope of this book. However, the OECD (2021) has provided a nice summary of principles and a list of key resources.

OECD RECOMMENDATIONS OF THE COUNCIL ON PRINCIPLES FOR PUBLIC GOVERNANCE OF PPPS

I. A clear, predictable, and legitimate institutional framework supported by competent authorities

- The political leadership should ensure public awareness of the relative costs, benefits, and risks of PPP and conventional procurement. Popular understanding of PPPs requires active consultation and engagement with stakeholders as well as involving end users in defining the project and subsequently in monitoring service quality.
- Key institutional roles and responsibilities should be maintained. This requires that procuring authorities, PPP units, the central budget authority, the supreme audit institution, and sector regulators are entrusted with clear mandates and sufficient resources to ensure a prudent procurement process and clear lines of accountability.
- Ensure that all significant regulations affecting the operation of PPPs are clear, transparent, and enforced. Red tape should be minimized, and new and existing regulations should be carefully evaluated.

II. Selection of PPPs grounded in value for money

- All investment projects should be prioritized at senior political level. As there are many competing investment priorities, it is the responsibility of the government to define and pursue strategic goals. The decision to invest should be based on a whole-of-government perspective and be separate from how to procure and finance the project. There should be no institutional, procedural, or accounting bias either in favor of or against PPPs.
- Carefully investigate which investment method is likely to yield most value for money. Key risk factors and characteristics of specific projects should be evaluated by conducting a procurement option pre-test. A procurement option pre-test should enable the government to decide on whether it is prudent to investigate a PPP option further.
- Transfer the risks to those that manage them best. Risk should be defined, identified, measured, and carried by the party for whom it costs the least to prevent the risk from realizing or for whom realized risk costs the least.
- The procuring authorities should be prepared for the operational phase of the PPPs. Securing value for money requires vigilance and effort of the same intensity as that necessary during the pre-operational phase. Particular care should be taken when switching to the operational phase of the PPP, as the actors on the public side are liable to change.
- Value for money should be maintained when renegotiating. Only if conditions change due to discretionary public policy actions should the government consider compensating the private sector. Any renegotiation should be made transparently and subject to the ordinary procedures of PPP approval. Clear, predictable, and transparent rules for dispute resolution should be in place.
- Government should ensure there is sufficient competition in the market by a competitive tender process and by possibly structuring the PPP program so that there is

an ongoing functional market. Where market operators are few, governments should ensure a level playing field in the tendering process so that non-incumbent operators can enter the market.

III. Transparent budgetary process to minimize fiscal risks and ensure integrity of procurement process

- In line with the government's fiscal policy, the Central Budget Authority should ensure that the project is affordable and the overall investment envelope is sustainable.
- The project should be treated transparently in the budget process. The budget documentation should disclose all costs and contingent liabilities. Special care should be taken to ensure that budget transparency of PPP covers the whole public sector.
- Government should guard against waste and corruption by ensuring the integrity of the procurement process. The necessary procurement skills and powers should be made available to the relevant authorities.

7.6 CASES IN ACTION

7.6.1 PPP for urban highways in Santiago, Chile

Engel and Galetovic (2014) provide an excellent example of a PPP that not only addressed needs for new urban highway capacity but also took and ground-breaking approach to creating competition in a sector typically considered to be a natural monopoly:

Between 2000 and 2008, a 140-mile (225-km) system of urban highways was built in Santiago, Chile's capital as shown in the blue lines in Figure 7.6. The system was divided among eight PPP concessions. Most of the funding to pay for the US$3 billion investment is coming from 20-30 year toll concessions. Tolls are charged by use and time of day through an electronic device attached to each car. Each month, users receive a bill and pay it like any other utility. Tolls vary by congestion and were fixed in each PPP contract.

How did Santiago build this system in less than a decade? The PPP program was planned and executed by a division in the Ministry of Public Works, which has authority over streets and highways across several municipalities. Urban PPPs were part of a broader national plan to upgrade Chile's urban highways through PPP concessions. The program began in the early 1990s, along with studies to build urban highways. A law was passed to regulate concessions in 1996, and PPPs were put to tender between 2000 and 2005.

Yet the system's origin dates to the late 1950s and early 1960s. In 1960, the Ministry of Public works issued its Santiago plan, PRIS (a Spanish acronym for intercommunal urban regulation plan). Planners anticipated that Santiago's rapid growth, which had begun in the 1940s, would eventually transform it into a polycentric city covering a substantial, ever-expanding area. It was thus crucial to plan and build streets connecting metropolitan subcenters and municipalities, avoiding trips passing through the city center. The plan anticipated the necessary transport investments, reserved strips of land for roads, and gradually executed the investments to put the plan to work. When PPPs came 40 years later, most roads had already been built, though they were in need of substantial upgrading.

7.6.2 Land value capture to fund urban metro in Hong Kong, China

Verougstraete and Zeng (2014) describe in some detail how Hong Kong funds its metro system with land value capture. The mechanisms follow the basic logic that enhanced accessibility to transportation systems adds value to land and real estate. As this value premium results from public investments, public authorities try to capture the surplus. The surplus captured can then be used to repay part of the cost of the transportation infrastructure.

The Hong Kong MTRC established in 1975 to provide metro services in the city and now carries over 4.5 million passengers each day. It includes a metro and light rail network with 104 stations. While most metro systems worldwide depend heavily on public financial support, MTRC operates without government subsidy and is highly profitable. This success is only possible because of the profits MTRC makes from its real estate business. Revenues are derived from profit sharing with private developers (mostly for residential projects) in real estate sales and from renting and managing MTRC-owned properties (in particular for commercial and office operations). Between 1998 and 2013, property-related operations generated almost twice the amount of money spent on railway line construction. The general property development process is as follows:

- When planning a new railway line, MTRC, in conjunction with the government, assesses the cost of construction and then prepares a master plan to identify property development sites along the railway.
- Having obtained all necessary approvals and having negotiated terms, MTRC purchases from the government the right for a period of 50 years to develop property above railway stations and depots, as well as on land adjacent to the railway (referred to as "development rights"). The "land premium" paid to the government for the rights does not take into account the increased value resulting from the transportation project (commonly referred to as the "before rail" land premium).
- MTRC then prepares a public tender for allocating these property development rights to private developers (development rights are usually divided into lots that are more manageable, in terms of cost, for developers).
- The private developers selected usually pay all development costs, including the land premium, for acquiring the exclusive development rights from MTRC. The private developers then have to bear the construction and commercialization risks and costs related to the residential and commercial properties.
- Profit-sharing mechanisms are included in the agreements with the private developers. For the residential units, MTRC will receive an agreed portion of the profit generated by the sales if the private partner manages to sell all the units before the contractual deadline. Otherwise, MTRC will obtain the unsold units and then determine whether to sell or lease in the open market. For shops and office units, MTRC generates profits by leasing directly with developers or by keeping part of the assets developed to generate long-term rental income.
- While MTRC is not in charge of the construction of the properties, it nevertheless supervises the work, carries out civil works, and enforces technical control standards and requirements for interfacing between its rail premises and the property development.

MRTC's model depends on grant of development rights from the government, while in other locations, land value capture might also be able to purchase new land and development rights as it grows.

Figure 7.6 The urban highway system in Santiago, Chile. (Source: Engel and Galetovic (2014).)

7.6.3 Congestion charges in London

The Centre for Public Impact (CPI, 2019) neatly summarizes the case of London's congestion charge zone:

- *The challenge.* London has limited road capacity but a high demand for road travel. At the turn of the millennium, "London suffered the worst congestion in the UK, with average traffic speeds slower than 8 miles per hour. The city lost between an estimated £2 million and £4 million every week in terms of time lost due to congestion." The congestion also affected air quality, with high levels of pollution.

  Studies assessing the viability of a congestion charge in London had been conducted since the mid-1960s, for example, the Smeed Report prepared by Reuben Smeed and his fellow committee members in 1964. This report advocated road pricing, but the government had resisted such recommendations as being too unpopular. London, though, was a special case.
- *The initiative.* The Greater London Authority Act 1999 defined the new Greater London Authority (GLA) and gave the mayor of London the power to provide guidance and directives to London's transport body, Transport for London (TfL). Schedule 23 of the Act included provision for road user charging schemes and Schedule 24 for workplace parking levies.

  In November 2000, the incoming mayor of London, Ken Livingstone, began preliminary consultation on his Draft Transport Strategy. This contained detailed information on how the proposed scheme might operate. The Technical Assessment Report on the Road Charging Options for London (ROCOL) was published in 2000, forming the basis for the Congestion Charging scheme of today.

  After the decision was taken to go ahead with congestion charging, an extensive public information campaign, to inform Londoners and visitors of the nature and details of the scheme, took place over the Fall and Winter of 2002–2003.

  The central London congestion charging scheme was successfully introduced on schedule on February 17, 2003. The charging area now extends from Kings Cross in the north to Elephant and Castle in the south, Hyde Park Corner in the west, and Old Street roundabout to the east. The charging period runs from 7:00 am to 6:00 pm and the standard charge is currently £11.50 per day. Charging zone residents receive a 90% discount while certain vehicles are exempt, such as buses, taxis, and electric vehicles and certain drivers, such as the disabled.

  The primary aim of the program was to reduce the number of private vehicles entering central London during the day and to facilitate greater use of transportation alternatives, produce environmental and safety improvements, and, in turn, raise substantial net revenues.
- *The public impact.* The scheme had significant impact from the outset:
  - There was a 37% increase in the number of passengers entering the congestion charging zone by bus during charging hours in the first year.
  - Greenhouse gas emissions were reduced by 16% from 2002 to 2003. NO_X and PM10 within the congestion charging zone by 2004 decreased by 18% and 22%, respectively.
  - The scheme generated £122 million net in 2005–2006.
  - By 2006, the congestion charging zone had reduced congestion in central London by 26% from its 2002 levels.
  - Vehicle crashes involving personal injuries have fallen between 40% and 70% within the zone.

- Based on the original £8 charge, the scheme was estimated to save £2.5 million per year as a result of a reduction in vehicle mile traveled, fuel consumption, and CO_2 emissions.
- The scheme achieved a cost efficiency of £78 million when all costs and benefits were considered.
- The congestion charge boosted sales of hybrid cars.

REFERENCES

ACEA. 2020. *ACEA Tax Guide*. European Automobile Manufacturers Association, Brussels, Belgium, EU.

ADB. 2008. *Public-Private Partnership Handbook*. Asian Development Bank Business Center, Manila, Philippines.

AECOM. 2007. *User Guidebook on Implementing Public–Private Partnerships for Transportation Infrastructure Projects in the United States*. Report work order 05-002. Federal Highway Administration, U.S. Department of Transportation, Washington, DC.

Allais, M., and Roth, G. 1968. *The Economics of Road User Charges*. World Bank.

Anderson, M.L., and Auffhammer, M. 2014. Pounds that kill: The external costs of vehicle weight. *Review of Economic Studies* 81(2), 535–571.

APIRG. 2009. Why and how to fund public transportation. Arizona PIRG Education Fund, Phoenix, AZ [Online]. Available: https://uspirgedfund.org/reports/usp/why-and-how-fund-public-transportation [Accessed on March 4, 2021].

APRC. 2018. Infrastructure asset recycling: Insight for government and investors. Asia Pacific Risk Center, Marsh and McLennan Companies, New York.

APTA. 2015. Value capture for public transportation projects: Examples. American Public Transportation Association, Washington, DC [Online]. Available: https://www.apta.com/wp-content/uploads/Resources/resources/reportsandpublications/Documents/APTA-Value-Capture-2015.pdf [Accessed on March 4, 2021].

The Australian Government the Treasury. 2019. Review of the national partnership agreement on asset recycling. The Australian Government the Treasury, Commonwealth of Australia, Canberra, Australia.

Bain, R. 2009. *Big Numbers Win Prizes: Twenty-One Ways to Inflate Toll Road Traffic and Revenue Forecasts*. Project Finance International, Leeds, UK.

Basso, J., and Duvall, T. 2013. Proposal 9: Funding transportation infrastructure with user fees. In *Book: 15 Ways to Rethink the Federal Budget*. Editors: Greenstone, M., Harris, M., Li, K., Looney, A., and Patashnik, J. Brookings Institution, Washington, DC, 50–56.

Been, V. 2018. City NIMBYs. *Journal of Land Use & Environmental Law* 33(2), 217–250.

Bigazzi, A.Y., and Rouleau, M. 2017. Can traffic management strategies improve urban air quality? A review of the evidence. *Journal of Transport & Health* 7, 111–124.

Blanc-Brude, F., and Makovsek, D. 2013. Construction risk in infrastructure project finance. EDHEC Business School Working Papers. EDHEC Business School, Roubaix, France.

Blanc-Brude, F., Goldsmith, H., and Välilä, T. 2009. A comparison of construction contract prices for traditionally procured roads and public–private partnerships. *Review of Industrial Organization* 35(1–2), 19.

Bornstein, J., Dixon, S., Flynn, M., and Pankratz, D. 2018. Funding the Future of Mobility. *Deloitte Review* 23, 14.

Brown, A. 2018. Fair fares? How flat and variable fares affect transit equity in Los Angeles. *Case Studies on Transport Policy* 6(4), 765–773.

Burger, P., and Hawkesworth, I. 2011. How to attain value for money: Comparing PPP and traditional infrastructure public procurement. *OECD Journal on Budgeting* 11(1), 91–146.

Burris, M., Lee, S., Geiselbrecht, T., Baker, R., and Weatherford, B. 2013. Equity evaluation of sustainable mileage-based user fee scenarios. Report 600451-00007-1. Texas Transportation Institute, Texas A&M University System, College Station, TX.

Button, K. 2004. The rationale for road pricing: standard theory and latest advances. *Research in Transportation Economics* 9(1), 3–25

Cavallaro, F., Giaretta, F., and Nocera, S. 2018. The potential of road pricing schemes to reduce carbon emissions. *Transport Policy* 67, 85–92.

Chalmers, B., Liu, Y., and Viet. P. 2018. Marsh & McLennan Companies Asia-Pacific Risk Center.

Chen, C., and Afonso, W.B. 2020. The adoption of local option fuel taxes: Evidence from Florida counties. *Public Budgeting & Finance* 105, 106–115.

Cooper, T.W., and DePasquale, J.A. 1989. Local option motor fuel taxes. *TRB Journal of Transportation Research Record* 1229, 127–136.

CPI. 2019. London's congestion charge. Centre for Public Impact, London, UK [Online]. Available: https://ssrn.com/abstract=3310243 or http://dx.doi.org/10.2139/ssrn.3310243f [Accessed on March 4, 2021].

Cuddihey, A. 2019. Dynamic Pricing Can Help Transport Charge Accurately for Services. *Metro.*

de Palma, A., and Lindsey, R. 2011. Traffic congestion pricing methodologies and technologies. *Transportation Research Part C: Emerging Technologies* 19(6), 1377–1399.

Decorla–Souza, P., and Whitehead, R.E. 2003. The value of pricing the use of roads. *Public Works Management & Policy* 7(4), 267–276.

Delmon, J. 2010. Understanding options for public–private partnerships in infrastructure. World Bank Working Paper 5173. World Bank, Washington, DC.

Dewees, D.N. 2002. Pricing municipal services: The economics of user fees. *Canadian Tax Journal/ Revue Fiscale Canadienne* 50(2), 586–599.

Development Asia. 2016. The Case for Electronic Road Pricing.

Duffield, C.F. 2008. Report on the performance of PPP projects in Australia when compared with a representative sample of traditionally procured infrastructure projects. National PPP Forum – Benchmarking Study, Phase II. The University of Melbourne, Melbourne, VC, Australia.

Ebobisse, A. 2020. This strategy can help Africa rapidly finance its economic recovery [Online]. Available: https://www.weforum.org/agenda/2020/11/africa-finance-economic-recovery-asset-recycling/ [Accessed on March 4, 2021].

Engel, E.M., and Galetovic, A. 2014. Urban transport: Can public-private partnerships work? World Bank Policy Research Working Paper 6873. World Bank, Washington, DC.

Feigenbaum, B., and Hillman, J. 2020. Contracting mass transit services. Reason Foundation, Washington, DC [Online]. Available: https://reason.org/policy-brief/contracting-mass-transit-services/ [Accessed on March 4, 2021].

FHWA. 1997. 1997 Federal highway cost allocation study. Federal Highway Administration, U.S. Department of Transportation, Washington, DC.

FHWA. 1998. 1997 Federal highway cost allocation study final report. Federal Highway Administration, U.S. Department of Transportation, Washington, DC.

FHWA. 2021. Pricing Information. [Online]. Available: https://www.fhwa.dot.gov/policy/otps/pricingkit.cfm#FAIR [Accessed on September 4, 2021].

Fielding, G.J., and Klein, D.B. 1993. High occupancy/toll lanes: Phasing in congestion pricing a lane at a time. Paper No. 179. University of California Transportation Center, Berkeley, CA.

Fleming, D.S. 2012. Dispelling the Myths: Toll and fuel tax collection costs in the 21st century. Policy Study No. 409. Reason Foundation, Washington, DC.

Flyvbjerg, B., Bruzelius, N., and Rothengatter, W. 2003. *Megaprojects and Risk: An Anatomy of Ambition*. Cambridge University Press, Washington, DC.

Flyvbjerg, B., Skamris Holm, M.K., and Buhl, S.L. 2004. What causes cost overrun in transport infrastructure projects? *Transport Reviews* 24(1), 3–18.

Fosgerau, M., and de Palma, A. 2013. The dynamics of urban traffic congestion and the price of parking. *Journal of Public Economics* 105, 106–115.

Garcia-Diaz, A., and Lee, D.T. 2013. Models for highway cost allocation. In *Game Theory Relaunched*. Editor: Hanappi, H. IntechOpen, Rijeka, Croatia, 135–156.

GIH. 2019. *Guidance Note on National Infrastructure Banks and Similar Financing Facilities*. Global Infrastructure Hub, Sydney, NSW, Australia.

Giuliano, G. 1994. *Curbing Gridlock: Equity and Fairness Considerations of Congestion Pricing*, Volume 2. Transportation Research Board, National Academies Press, Washington, DC.

Guasch, J.L. 2004. *Granting and Renegotiating Infrastructure Concessions: Doing It Right*. The World Bank, Washington, DC.

Ison, S. 2004. *Road User Charging: Issues and Policies*. Ashgate Publishing, Farnham, Surrey, UK.

Istrate, E., and Puentes, R. 2011. *Moving Forward on Public Private Partnerships: US and International Experience with PPP Units*. Brookings Institution, Washington, DC.

ITF. 2013. Funding urban public transport systems: Case study compendium. International Transportation Forum, Organization for Economic Co-operation and Development, Paris, France [Online]. Available: https://www.itf-oecd.org/sites/default/files/docs/13compendium.pdf [Accessed on March 24, 2021].

ITF, 2018. The social impacts of road pricing. International Transport Forum, Organization for Economic Co-operation and Development, Paris, France.

ITF. 2021. Total inland transport infrastructure investment as a percentage of GDP. In *Curbing Gridlock 2: ITF Transport Outlook 2021*. OECD iLibrary, International Transport Forum, Organization for Economic Co-operation and Development, Paris, France [Online]. Available: https://doi.org/10.1787/ea627644-en [Accessed on March 24, 2021].

Kain, J. 1994. Impacts of congestion pricing on transit and carpool demand and supply. In *Curbing Gridlock 2*. Editor: Ackerman, N. Transportation Research Board, Washington, DC, 502–553.

Kaufman, M., Formanack, M., Gray, J., and Weinberger, R. 2012. Contemporary approaches to parking pricing: A primer. Report No. FHWA-HOP-12-026. Federal Highway Administration, U.S. Department of Transportation, Washington, DC.

King, A.D., Manville, M., and Shoup, D. 2007. The political calculus of congestion pricing. *Transport Policy* 14(2), 11–123.

Kirk, R.S., and Mallett W.J. 2020. Reauthorizing highway and transit funding programs. CRS Report R45350. Congressional Research Service, Washington, DC.

Klein, D.B., Moore, A.T., and Reja, B. 1997. *Curb Rights: A Foundation for Free Enterprise in Urban Transit*. Brookings Institution, Washington, DC.

Lari, A., Levinson, D.M., Zhao, Z.J., Lacono, M., Aultman, S., Vardhan, D., Junge, J., Larson, K., and Scharenbroich M. 2009. Value capture for transportation finance: Technical research report. Final Report CTS 09-18. Minneapolis, MN.

Levine, J., and Garb, Y. 2000. *Evaluating the Promise and Hazards of Congestion Pricing Proposals: An Access Centered Approach*. Floersheimer Institute for Policy Studies, The Hebrew University, Jerusalem, Israel.

Levinson, D. 2014. Land value capture presentation. *Transportist*, 2014-09-24 Issue, Saint Paul, MN [Online]. Available: https://transportist.org/2014/09/24/land-value-capture-presentation/ [Accessed on March 24, 2021].

Li, Z., and Sinha, K.C. 2004. Methodology for Multicriteria Decision Making in Highway Asset Management. *TRB Journal of Transportation Research Record* 1885, 79–87.

Li, Z., and Madanu, S. 2009. Highway Project-Level Life-Cycle Benefit/Cost Analysis under Certainty, Risk, and Uncertainty: A Methodology with Case Study. *ASCE Journal of Transportation Engineering* 135(8), 516–526.

Li, Z. 2018. *Transportation Asset Management: Methodology and Applications*. CRC Press, Boca Raton, FL.

Lindsey, C., Van den Berg, V.A.C., and Verhoef, E.T. 2012. Step tolling with bottleneck queuing congestion. *Journal of Urban Economics* 72(1), 46–59.

Litman, T. 2003. *London Congestion Pricing: Implications for Other Cities*. Victoria Transport Policy Institute, Victoria, BC, Canada.

Litman, T. 2007. *Socially Optimal Transport Prices and Markets*. Victoria Transport Policy Institute, Victoria, BC, Canada.

Litman, T. 2011. *Pricing for Traffic Safety: How Efficient Transport Pricing Can Reduce Roadway Crash Risk*. Victoria Transport Policy Institute, Victoria, BC, Canada.

Litman, T. 2013. *Generated Traffic and Induced Travel Implications for Transportation Planning*. Victoria Transport Policy Institute, Victoria, BC, Canada.

LOC. 2014. National funding of road infrastructure. *Library of Congress*, Washington, DC.

Lotshaw, S., Lewis, P., Bragdon, D., and Accuardi, Z. 2017. *A Bid for Better Transit: Improving Service with Contracted Operations*. TransitCenter and Eno Center for Transportation, Washington, DC.

Maddison, D., Pearce, D., Johansson, O., Calthorp, E., Litman, T., and Verhoef, E. 1996. *Blueprint 5: The True Cost of Transport*. Earthscan, London, UK.

Mahendra, A., King, R., Gray, E., Hart, M., Azeredo, L., Betti, L., Prakash, S., Deb, A., Ashebir, E., and Ibrahim, A. 2020. *Urban Land Value Capture in São Paulo, Addis Ababa, and Hyderabad: Differing Interpretations, Equity Impacts, and Enabling Conditions*. Lincoln Institute of Land Policy, Cambridge, MA.

Makovsek, D., Hasselgren, B., and Perkins, S. 2015. Public private partnerships for transport infrastructure: Renegotiations, how to approach them and economic outcomes. International Transport Forum Discussion Paper No. 2014-25. International Transport Forum, Organisation for Economic Co-operation and Development, Paris, France.

Mandri-Perrott, C. 2010. *Private Sector Participation in Light Rail-Light Metro Transit Initiatives*. World Bank Publications, Washington, DC.

Marcom, H. 2019. Four states with weight distance taxes for truckers. Apex Capital, Fort Worth, TX [Online]. Available: https://www.apexcapitalcorp.com/blog/four-states-with-additional-registration-requirements-for-truckers/ [Accessed on March 14, 2021].

Millsap, A., Staley, S.R., and Nastasi, V. 2019. Assessing the effects of local impact fees and land-use regulations on workforce housing in Florida. James Madison Institute, Florida State University, Tallahassee, FL [Online]. Available: https://www.apexcapitalcorp.com/blog/four-states-with-additional-registration-requirements-for-truckers/ [Accessed on March 14, 2021].

Nelson, A.C., and and Moody, M. 2003. *Paying for Prosperity: Impact Fees and Job Growth*. Brookings Institution Center on Urban and Metropolitan Policy, Washington, DC.

NSTIFC. 2009. Paying our way: A new framework for transportation finance. National Surface Transportation Infrastructure Financing Commission, Washington, DC.

OECD. 2015. Infrastructure financing instruments and incentives. OECD iLibrary, Organization for Economic Co-operation and Development, Paris, France [Online]. Available: https://www.oecd.org/finance/private-pensions/Infrastructure-Financing-Instruments-and-Incentives.pdf [Accessed on March 24, 2021].

OECD. 2018. Consumption tax trends 2018. OECD iLibrary, Organization for Economic Co-operation and Development, Paris, France [Online]. Available: https://doi.org/10.1787/ctt-2018-en [Accessed on March 24, 2021].

OECD. 2021. Recommendations of the council on principles for public governance of public private partnerships. OECD/LEGAL/0392, Organization for Economic Co-operation and Development, Paris, France [Online]. Available: https://legalinstruments.oecd.org/public/doc/275/275.en.pdf [Accessed on March 24, 2021].

Parry, I. 2008. Pricing urban congestion. RFF Discussion Paper 08-35. Resources for the Future, Washington, DC.

Perkins, S. 2013. Better regulation of public-private partnership for transport infrastructure: Summary and conclusions. International Transport Forum Discussion Paper No. 2013-6. International Transport Forum. Organization for Economic Co-operation and Development, Paris, France.

Ponkshe, A. 2019. Congestion pricing economics. Observer Research Foundation, New Delhi, India [Online]. Available: https://www.orfonline.org/expert-speak/congestion-pricing-economics-55360/ [Accessed on March 14, 2021].

Poole, R.W. 2018. *Rethinking America's Highways: A 21st- Century Vision for Better Infrastructure*. The University of Chicago Press, Chicago, IL.

Price, K., and Henebery, A. 2014. *The Life and Death of the Highway Trust Fund*. ENO Center for Transportation, Washington, DC.

PWF. 2011. 2011 Survey of public private partnerships worldwide. Public Works Financing, Sacramento, CA [Online]. Available: https://www.pwfinance.net/document/October_2011_vNov202011.pdf [Accessed on March 14, 2021].

Rajé, F. 2003. *Impacts of Road User Charging/Workplace Parking Levy on Social Inclusion/Exclusion: Gender, Ethnicity and Lifecycle Issues.* Transport Studies Unit, University of Oxford, Oxford, UK.

Rall, J., Reed, J.B., and Farber, N.J. 2010. Public-private partnerships for transportation: A toolkit for legislators. National Conference of State Legislatures, Washington, DC [Online]. Available: https://rosap.ntl.bts.gov/view/dot/18441 [Accessed on March 24, 2021].

Ramjerdi, F. 1996. *Road Pricing and Toll Financing, with Examples from Oslo and Stockholm.* Institute of Transport Economics, Oslo, Norway.

Robinson, F. 2008. Heavy vehicle tolling in Germany: Performance, outcomes and lessons learned for future pricing efforts in Minnesota and the U.S. The University of Minnesota, Minneapolis, MN.

Roth, G.J., and Thomson, J.M. 1963. The relief of traffic congestion by parking restrictions. *The Town Planning Review* 34(3), 185–198.

Rye, T., and Koglin, T. 2014. Parking management. In *Parking Issues and Policies-Parking Issues and Policies,* Volume 5. Emerald Group Publishing Limited, Bingley, UK, 157–184.

Samuel, P. 2000. *Putting Customers in the Driver's Seat: The Case for Tolls.* Reason Public Policy Institute, Reason Foundation, Washington, DC.

Shoup, D. 2017. *The High Cost of Free Parking*, Routledge, New York.

Small, K., Winston, C., and Evans, C. 1989. *Road Work, a New Highway Pricing and Investment Policy.* Brookings Institution, Washington, DC.

Smeed, R.J., Roth, G., Beesley, M.E., and Thompson, J.M. 1964. *Road Pricing: The Economic and Technical Possibilities.* Ministry of Transport, London, UK.

Song, S. 2015. Should China implement congestion pricing? *Chinese Economy* 48(1), 57–67.

Sterner, T. (Ed.). 2012. *Fuel Taxes and the Poor: The Distributional Effects of Gasoline Taxation and Their Implications for Climate Policy.* RFF Press, New York.

Tang, L.Y., Shen, Q., and Cheng, E.W.L. 2010. A review of studies on public-private partnership projects in the construction industry. *International Journal of Project Management* 28(7), 683–694.

Ubbels, B., and Nijkamp, P. 2002. Unconventional funding of urban public transport. *Transportation Research Part D: Transport and Environment* 7(5), 317–329.

UNECE. 2017. Innovative ways for financing transport infrastructure. United Nations Economic Commission for Europe, New York.

Uzsoki, D. 2018. Infrastructure banks: Solutions and best practices. IISD Discussion Paper. 7 International Institute for Sustainable Development, Manitoba, Canada.

Van de Velde, D., Beck, A., Van Elburg, J., and Terschüren, K.H. 2008a. *Contracting in Urban Public Transport.* European Commission, Amsterdam, Netherlands.

Van de Velde, D., Veeneman, W., and Schipholt, L.L. 2008b. Competitive tendering in The Netherlands: Central planning vs. functional specifications. *Transportation Research Part A: Policy and Practice* 42(9), 1152–1162.

Vassallo, J., and Sánchez-Soliño, A. 2007. A subordinated public participation loans for financing toll highway concessions in Spain. *TRB Journal of Transportation Research Record* 2297, 1–8.

Verougstraete, M., and Zeng, H. 2014. A brief introduction to China's PPP application in transport and logistics sectors. Economic Commission for Europe, United Nations, Geneva, Switzerland [Online]. Available: http://www.unescap.org/sites/default/files/Case%204-20Land%20Value%20-%20Hong-Kong%20MTR.pdf [Accessed on March 24, 2021].

VTPI. 2019. Distance-based pricing mileage-based insurance, registration and taxes. Victoria Transport Policy Institute, Victoria, BC, Canada [Online]. Available: https://www.vtpi.org/tdm/tdm10.htm [Accessed on March 24, 2021].

Wappelhorst, S., Mock, P., and Yang Z. 2018. Using Vehicle Taxation Policy to Lower Transport Emissions. International Council on Clean Transportation.

Weber, T. 2012. Price theory in economics. In *Oxford Handbook of Pricing Management.* Editors: Özer, Ö., and Phillips, R. Oxford University Press, Oxford, UK. 281–339.

Woetzel, J., Garemo, N., Mischke, J., Hjerpe, M., and Palter, R. 2016. Distance-based pricing mileage–based insurance, registration and taxes. McKinsey Global Institute, Chicago, IL [Online]. Available: https://www.mckinsey.com/business-functions/operations/our-insights/bridging-global-infrastructure-gaps [Accessed on March 24, 2021].

Woetzel, J., Garemo, N., Mischke, J., Kamra, P., and Palter, R. 2017. Bridging infrastructure gaps: Has the world made progress? Discussion Paper. McKinsey Global Institute, New York.

World Bank. 2017. Public-private partnerships reference guide, version 3.0. World Bank, Washington, DC.

World Bank. 2020a. Private Participation in Infrastructure (PPI) database. The World Bank, Washington, DC [Online]. Available: https://ppi.worldbank.org/en/ppi [Accessed on March 24, 2021].

World Bank. 2020b. Private Participation in Infrastructure (PPI): 2020 annual report. The World Bank, Washington, DC [Online]. Available: https://ppi.worldbank.org/content/dam/PPI/documents/PPI_2020_AnnualReport.pdf [Accessed on March 24, 2021].

Chapter 8

Performance-based, mobility-centered transportation budget allocation

8.1 GENERAL

Central to transportation budget allocation is to holistically consider economic, social, and environmental impacts of various investment options to achieve transportation policy goals of efficiency, effectiveness, and equity. To facilitate strategic and day-to-day management, the policy goals are defined more explicitly as strategic system management goals, which typically include facility preservation, agency costs, user costs, mobility, safety, environmental impacts, and economic development supported by the transportation system. Performance measures under various system management goals are used to assess the impacts of an intervention or investment on the performance of the transportation system. The overall impacts of an investment on the transportation system could be established by aggregating the positive and negative impacts assessed by various performance measures. Similarly, the total impacts on the transportation system caused by implementing a collection of investment options could be quantified. Following transportation asset management principles, performance-based transportation budget allocation aims to allocate the available budget to a subset of investment options that would lead to the maximized total benefits, namely, the largest net gain of positive and negative impacts to the transportation system. In presence of innovative financing practices such as one or more forms of private-public partnerships, the overall budget allocation decisions for a transportation system are made by the transportation agency or the agency in conjunction with private partners to ensure global optimality of the decision outcomes.

For mobility-focused management, transportation budget allocation strives for prioritizing a subset of investment options that would achieve the maximized level of mobility improvements for the transportation system according to the available budget, without leading to a net loss among the positive and negative impacts on the performance of the remaining system management goals. In this respect, tradeoff analyses need to be conducted in budget allocation to simultaneously help achieve maximized performance gains to one or more system management goals other than mobility, such as safety and environmental impacts.

8.2 MOBILITY-CENTERED, PERFORMANCE-BASED BUDGET ALLOCATION PROCESS

Figure 8.1 shows key components of a mobility-centered, performance-based budget allocation process using transportation asset management principles:

- Establishing transportation system management goals with top priority assigned to the mobility performance goal

DOI: 10.1201/9780429345432-8

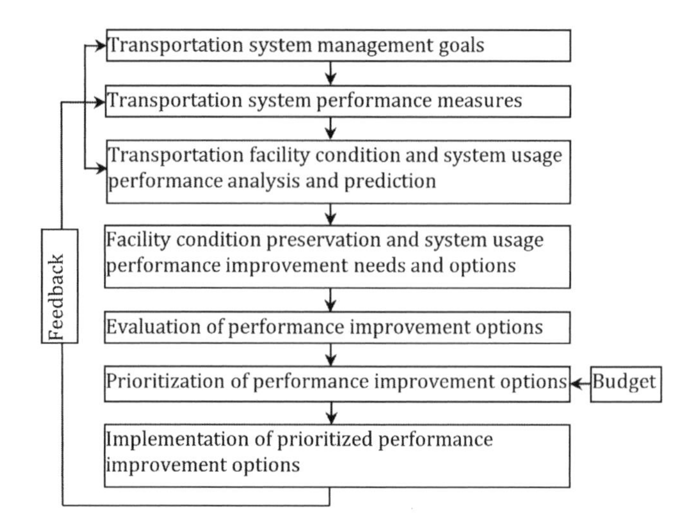

Figure 8.1 Performance-based transportation budget allocation process.

- Determining transportation system performance measures (Table 6.1 lists typical mobility performance measures)
- Conducting current performance analysis and future performance trend prediction
- Identifying current and future performance improvement needs
- Proposing investment options for performance improvements
- Evaluating the efficiency, effectiveness, and equity of performance improvement options
- Allocating available budget to a subset of improvement options to achieve maximized mobility-centered performance improvements
- Implementing the prioritized performance improvement options
- Providing feedback of the actual effectiveness of performance improvements.

8.3 MOBILITY MANAGEMENT DATA NEEDS AND DATABASE MANAGEMENT

8.3.1 General requirements

For mobility-centered transportation development and management, performance measures play the central role from data collection and analysis to budget allocation decision-making. Data type, time fragmentation, geographic coverage, sampling, and updating frequency essential to data collection are all driven by the performance measures adopted for mobility management (Bauer et al., 2016). The impacts of performance measures are further extended to the sequential steps of data analyses for determining the current performance levels, predicting future performance trends, identifying performance improvement needs, proposing and evaluating performance improvement options, and conducting the most cost-effective budget allocation.

Since performance measures could be adopted for mobility analyses at both the network level and project level, the collected data should help network-level planning, programming, and policy decisions. In parallel, it should support project-level decisions about the best budget allocation strategy to sustain or improve mobility performance of the transportation system and meanwhile help achieve other system management goals including facility condition preservation, agency costs, user costs, safety, and environmental impacts, as well as economic prosperity supported by the transportation system.

One of the key features of performance-based mobility management is the shift from reactive management caused by the time-lag effect between field data collection and deployment of mobility management strategies to proactive management that utilizes predicted data to identify the expected mobility improvement needs, develop a list of proactive mobility improvement strategies prior to the arrival of mobility issues, deploy the proactive strategies to prevent the occurrence of mobility issues and mitigate the mobility impacts should they occur, and refine the list of strategies over time to keep abreast of changing mobility issues and improvement needs for the entire transportation system.

8.3.2 Field collected data

At a minimum, the following types of field data are required:

- *Demographic/socioeconomic data.* Population, household numbers and sizes, auto ownerships, and income distributions of the megacity region served by the multimodal transportation network to help synthesize regional travel demand by trip purpose and its diurnal distribution by time and location.
- *Transportation system inventory data.* Locations, geometric design standards, material types, and ages of highway, transit, active transportation, and freight transportation networks.
- *Physical facility condition data.* Pavement, bridge, tunnel, traffic control and safety hardware, transit, active transportation, and intermodal facility conditions for a relatively long period of time.
- *Agency cost data.* Records of reconstruction/replacement, rehabilitation, and maintenance treatments for different categories of physical facilities over a long period of time.
- *Traffic flow data.* Vehicle volume, composition, speed, and travel time by travel lane for highway segments; and through and turning movement counts and delays by approach of intersections.
- *Vehicle trajectory data.* The related data provides the exact time and space distribution of all roadway users throughout a day. This is the most fundamental data to construct traffic details. Besides, it provides the traffic demand between each O-D pair and each link, which can be used to develop effective means to conduct travel demand and supply management, and further dynamically improve the balance between demand and supply.
- *Transit data.* The traffic demand in passenger trips can be estimated based on traffic volume data and transit data since the total passenger travel demand will be split into different modes, mainly into auto and transit. The transit data including route, ridership and transfer, frequency of usage, schedule, mode of access, and transit vehicle capacity can help the transportation decision maker gain a better understanding of travel patterns and calibration of the total travel demand.
- *Non-motorized traffic data.* Bike, pedestrian, and scooter traffic counts along bikeways, pedestrian walkways, and scooter pathways by time of the day.
- *Freight data.* This category of data mainly contains details of multiple commodity flows, truck, freight rail, and cargo drone movements. Truck traffic could potentially be a major component of the total vehicular traffic on some highways.
- *Traffic safety data.* This category of data covers auto, truck, transit, bike, scooter, pedestrian, and freight rail crash and incident records on locations, severities, types, and causes.

8.3.3 Predictive traffic data

In addition to field-collected travel demand and traffic flow data, predictive traffic data is required to support proactive mobility management. This set of data is a part of the input to predict temporally and spatially varying mobility of people travel and goods movements or vehicle mobility, which is essential to support proactive mobility improvement decision-making. Two practical approaches can be employed to generate detailed predictive traffic data. The first approach applies statistical methods and artificial intelligence (AI) techniques to historical data for traffic forecasts. Statistical methods have long been used and AI techniques such as neural networks and machine learning have increasingly become popular in recent years. However, the predictability of calibrated statistical or trained AI models could decrease drastically when extensive traffic redistribution occurs due to changes in regional population, land use patterns, or transportation capacity. The prediction power of those models may also be compromised due to the lack of highly reliable historical data.

The second approach for deriving detailed predictive traffic data is by using the travel demand forecasting models typically built based on high-fidelity large-scale simulation-based platforms. Typical outputs of travel demand forecasting include predicted traffic volumes, speeds, and travel times in fine-grained time intervals for travelers and goods using the transportation network (Wilson, 1974; Ben-Akiva and Lerman, 1985; Oppenheim, 1995; Meyer and Miller, 2000). The ability to derive highly accurate and precise predictive traffic data is essential to assess the effectiveness of mobility improvements where field traffic data for the after mobility-improvement period at the decision-making stage is not available. More importantly, this approach could predict traffic details as a result of simultaneously implementing multiple mobility improvement projects that would lead to multiple rounds of traffic redistributions in the transportation network. Hence, it could help capture the interdependency of mobility impacts caused by multiple improvements.

8.3.4 Data sampling methods

Performance-based mobility management is data intensive and the budget allocation decision-making for mobility improvements is data driven. Although collection of data on multimodal travel demand, transportation capacity provision, and dynamically varying traffic flow is guided by the performance measures adopted for mobility-centered management, issues of where the data comes from, how it is gathered, how to ensure the representability, accuracy, and precision of the collected data need to be considered in the first place. Practically, data details of the transportation system inventory need to be obtained. The total person-trips and goods shipment flows and trips are estimated based on sampled data largely collected using survey instruments. Due to the plenitude of travel demand, transportation supply, and traffic data, as well as and the cost involved with data collection efforts, only sampled data is collected from the field and derived using predictive methods and models. It is therefore imperative to utilize adequate data sampling methods to ensure data accuracy, precision, coverage, timeliness, and cost-effectiveness of data collection. This way, the timely available sample datasets could be used to extract highly reliable information on the current mobility performance levels and future mobility performance trends, identify mobility performance improvement needs, propose countermeasures, evaluate the effectiveness of countermeasures, and prioritize the available budget to achieve the highest level of mobility improvements without compromising performance levels of other transportation system management goals including facility preservation, user costs, safety, environmental impacts, as well as economic development.

The data sampling methods can be classified into probability and non-probability categories. The sampling methods well suited for mobility-focused data collection include simple random sampling, systematic random sampling, stratified random sampling, and cluster sampling. Simple random sampling selects every data sample in the target population with an equal and known chance. Systematic random sampling selects every Nth element from the randomly ranked population according to the required sample size. Stratified random sampling splits the target population into multiple data groups, determines data proportions across the data groups, and then conducts random sampling within each group. Cluster sampling divides target population into clusters and performs simple random sampling within each cluster (Li, 2018).

8.3.5 Data collection techniques

Over the last several decades, a wide range of techniques have been developed for data collection centered on multimodal travel demand, transportation inventory and facility condition, and traffic flow details. The following highlights some representative techniques for data collection:

- *Passenger demand data.* Traditionally, survey instruments are the primary means of data collection, which typically include (i) household travel surveys to record an individual's travel activity in a travel diary or through a recall interview for a certain time; (ii) roadside station surveys supplementary to household surveys to obtain travelers' travel characteristics; (iii) employer and special generator travel surveys to collect data on establishments such as airports, train stations, and tourist interest points with great trip attractions; (iv) on-board transit surveys to collect data on transit riders' characteristics and travel patterns; (v) hotel and visitor surveys on collect data absent from household surveys; and (vi) parking surveys to obtain details of people entering and exiting parking facilities (Wilson, 1974; Ben-Akiva and Lerman, 1985; Oppenheim, 1995). With technological advancements in recent decades, license plate matching, the Global Positioning System (GPS), and Bluetooth technologies have been adopted to collect O-D trip data directly. These methods have become increasingly popular because of low costs and being non-intrusive to travelers (Meyer and Miller, 2000).
- *Truck demand data.* Truck demand data could be collected from different perspectives, such as driver/carrier, producer/recipient, vehicle/trip, and consignment. Key freight data collection methods come from this typology, which include (i) commercial vehicle travel surveys or trip diaries filled in by truck driver to obtain O-D data for trucks and other commercial vehicles; (ii) roadside intercepts to truck drivers for short interviews or questionnaire surveys to record the nature of truck trips and commodities being shipped; (iii) automated counts/identification via license plate recognition and RFID to collect data on the number of commercial vehicles passing a certain intercept point during a certain time interval; (iv) enterprise surveys consisting of brief interviews of businesses, accompanied by aggregate data collected from other sources, such as published annual reports and historical data, and combined with statistical inference, to obtain information on commodity flows to help find freight trip characteristics; and (v) administrative by-product with data sources of consignment notes, vehicle log books, positioning information on freight, and vehicle trajectory record (Turner et al., 1998).
- *Transportation system inventory data.* Transportation inventory data is concerned with locations, geometric design standards, material types, and ages of highway, transit, active transportation, and freight transportation facilities. Typically, the design and construction details are stored in the geographic information system (GIS) database.

- *Transportation facility condition data.* The conditions of various types of transportation facilities such as pavements, bridges, tunnels, traffic control and safety hardware including traffic signs, signals, lighting, pavement markings, and guardrails, transit installations, bikeways, pedestrian walkways, and intermodal facilities affect safe and efficient utilization of the transportation system.
  - *Pavements* are the most common type of transportation facilities with which the conditions could be evaluated from surface and structural condition perspectives. For a homogeneous pavement segment classified according to characteristics of pavement type, material type, layer thickness, subgrade type, highway classification, and traffic loading, data on pavement conditions can be collected based on surface-related roughness, rut depth, skid resistance, and distresses, as well as structural characteristics such as deflection. For pavement roughness and rut depth measurements, notable techniques include dipstick profiler, profilograph, response-type road roughness meters (RTRRMs), and profiling devices. For pavement skid resistance measurements, the commonly used data collection equipment are the locked wheel tester and portable field devices such as keystone tester, California skid tester, and pendulum skid tester for skid resistance measurements. Apart from collecting pavement surface condition data, data on pavement structural conditions is essential to help select and design specific repair strategies. Pavement structural condition evaluation can be carried out using destructive and nondestructive testing. Destructive testing involves coring and removing surface, base, subbase, and subsoil samples for laboratory testing to determine the strength of materials and types of damages present in each layer, the cause of structural failure, and load-carrying capacity of the pavement. Nondestructive testing provides measurements of the overall pavement response to an external force or load without disturbing or destroying the pavement components by using such techniques as Benkelman beam, dynaflect, road rater, falling weight deflectometer, rolling deflectometer, and ground-penetrating radar (AASHTO, 1990; FHWA, 1991; Haas et al., 1994; Li et al., 2001, 2002).
  - *Bridge condition assessments* are carried out separately for substructure, superstructure, and deck components using destructive methods and nondestructive methods. Destructive methods including coring and chipping, and nondestructive methods are those that test the properties of a material, component, or system without causing any damage, including visual inspection, hammer sounding, or chain dragging. The half-cell potential, infrared, ground-penetrating radar, ultrasonic, impact echo, magnetic flux leakage, and Eddy current methods are commonly used nondestructive techniques for detecting corrosions, cracks, and welding flaws of reinforced bridge components (FHWA, 1987).
  - *Traffic control devices and safety hardware* play a significant role in ensuring safe and efficient utilization of the transportation system. For traffic signs and pavement markings, retroreflectivity specified in candelas per lux per square meter (cd/lux/m^2) is an important characteristic to assess the nighttime visibility. Notable instruments for collecting accurate retroreflectivity data include RetroView, digital imaging camera, RetroChecker, Line-inspector, RoadVista, StripeMaster II, Retroreflectometer, and GPS-Compatible Sign Management and Retroreflectivity Tracking System. For lighting system, the quantity of light or the light output and light levels are measured in lumens, lux, and foot-candles. Initial lumens/foot-candles reflect the amount of light produced by a lamp when it is installed. Supply voltage variations, the lamp's interaction with the ballast, and dirt build-up reduce the produced amount of light. Technologies applicable to collect data on the levels of illuminations emitted from a light source include HISLAT, spectroradiometer, and luminance photometer (Vonderohe et al., 1993; FHWA, 1985, 2001, 2005a,b, 2007, 2009).

- *Travel time data.* Field data on travel time is key to mobility performance analysis. In recent years, fine-grained real-time travel time data has increasingly been collected to assess the dynamic interactions between traffic flow derived from travel demand and transportation supply with flexible capacity. Probe vehicle and non-intrusive detection techniques are the two categories of techniques for travel time data collection in real time. Probe vehicle techniques generally include automatic vehicle identification, ground-based radio navigation, cellular geo-location, and GPS-equipped probe vehicle technique. Non-intrusive detection techniques use devices that cause minimal disruption to normal traffic operations and can be deployed more safely than conventional detection methods. The non-intrusive detection techniques mainly include infrared detection, magnetic detection, Doppler microwave devices, acoustic detection, ultrasonic detection, and video detection (FHWA, 2003, 2004a,b, 2005c).
- *Transit operations data.* Many of the transit performance data, especial supply data, could be found from transit planning and operation records, such as transit right-of-way, location, and equipment on bus stops, schedule. Data on transit operations including passenger boarding and alighting counts and speed-and-delay data must be obtained through field observations (Vuchic, 2005; Fielding, 2013). The detailed passenger boarding and alighting counts by the technique of automatic passenger counters provide the number of boarding and alighting passengers at each station along a transit route, which could help derive passenger volume and load counts, passenger miles of travel, and trip production and attraction of each station during a certain time interval. Speed-and-delay data that records the time a transit unit spent on running, accelerating, decelerating, waiting at stops, and other types of delay can be collected by automated vehicle positioning systems and further used to investigate the time profile over the service of individual transit lines. This measured time profile is important in operation speed computation and service reliability evaluation.

8.3.6 Data collection frequency

The data on multimodal travel demand, transportation supply, and dynamic traffic details for performance-based mobility analysis needs to be replenished at regular intervals. Data collection is cost and time intensive and consumes many resources. By considering tradeoffs of demand versus supply, passenger versus freight, and geographic coverage, the different categories of sampled data can be collected annually, quarterly, monthly, weekly, daily, hourly, in 15-min, 5-min, or 1-min intervals, or in real time.

8.3.7 Data quality assurance

Quality assurance for various sets of data collected using different techniques and equipment is an ongoing process, which could be achieved using system design, statistical methods, training, and auditing to maximize various attributes of data. Several important attributes are essential to data quality assurance (Fekpe et al., 2003; FHWA, 2013):

- *Accuracy.* Accuracy is a measure of rightness, indicating whether the estimation achieves the correct value. It refers to how close a sample statistic is to a population parameter.
- *Precision.* Precision is the measure of exactness that refers to how close estimates from different samples are to each other. For instance, each set of data contains two sources of errors attributable to the precision level of the equipment used for data collection and the operator conducting data measurement. The standard error of multiple data values is a measure of precision. Smaller standard error indicates higher precision and vice versa.

- *Coverage.* The extent of data coverage associated with a multimodal transportation system is a key design decision and a key limitation on its usefulness. For instance, the focus of data collection on freeways is of little relevance to identify mobility improvement needs for arterial roads and urban streets.
- *Details.* Maintaining an appropriate level of details ensures that the information extracted from the data could support effective mobility management.
- *Timeliness.* Timeliness refers to the age of data at the time it is used. Timeliness must balance competing requirements of time point at which to conduct data collection, the need to process the data, and the need for the extracting pertinent information from data in assessing the current mobility performance levels, predicting mobility performance trends, identifying mobility improvement needs, proposing and evaluating countermeasures, and allocating budget to achieve maximized mobility performance gains without compromising performance levels of other transportation system goals.
- *Accessibility.* This attribute of data quality refers to the ease with which the data can be put to use for some specific purposes in support of mobility improvements.
- *Assumptions and definitions.* Data sources may have definitional differences or inherent assumptions that make them more or less useful for different types of mobility management applications.

8.3.8 Data integration and database management

Integration and sharing of data on people travel and goods movements, transportation network capacities, and traffic flow details will enhance the ability of the transportation agency (and private partners as applicable) to carry out data-driven, fact-based mobility performance analysis, prediction, needs assessment, and resource allocation to achieve maximized mobility gains without compromising performance levels of other system management goals. Practically, the continuingly collected data needs to be stored in an interoperable database. In the database planning phase, the following issues need to be thought through: identification of data in the needs assessment, inclusion of data in the data model, creation of metadata, collection and entry into the database, updating and maintenance, and data retention according to the appropriate record retention schedule. In the database design phase, efforts are needed for identifying functions of hardware and software such as using the GIS-compatible technique, estimates of usage, and scoping the size of the database.

Database management faces challenges of data storage capacity, heterogeneous data, and noisy and missing data, which should be adequately addressed to meet the new data collection and analysis requirements. As a result of advancements in technology, the data collected could be more detailed and fine-grained that could enable data analysis in diverse ways and from different points of views. However, the exponentially increasing need for storage of replenished data can push the hardware and software to the limits and the cost could potentially exceed the benefit of data integration. Moreover, data is collected in the field by different equipment or generated by different analytical tools over time and is stored in different formats via various media. In consideration of data format standardization in representing, accessing, manipulating, transferring, and reporting information, the effort spent on data homogenization before standardizing the data format without compromising the quality of data ought to be a major task in data integration. Finally, data integration should respect the fact that data quality in terms of accuracy and precision changes over time. The early data collection techniques are not able to provide clean data, which brings impurities in the legacy data. Device failure will cause the issue of missing data. Procedures need to be developed for regularly auditing the quality of data and cleaning the data prior to conversion, integration, and subsequent use for analysis and decision-making.

8.4 MOBILITY PERFORMANCE ANALYSIS AND PREDICTIONS

The focus of mobility management could be put on mobility of people travel and goods movements or vehicle mobility. For both cases, traveler time and travel time reliability assessed by travel time index and travel time buffer index of trips for each O-D pair are important measurements of mobility performance.

8.4.1 O-D path travel time estimation

For an auto- or transit-accommodated passenger trip, the O-D path travel time consists of travel times of individual roadway segments, urban streets, and time delays at intersections along the travel path, as well as additional time for parking, or travel times of waiting, in-vehicle, and transfers along the transit route.

For estimating the vehicle travel time spent on a roadway segment, both static and dynamic models can be utilized:

- *Static models.* The static models generally conduct macroscopic traffic stream analyses, which can be classified into instantaneous speed-density-flow, space-average, and time-average models. Some representative speed-density-flow models include Greenshields model (1935), Pipes model (1953, 1967), Greenberg model (1959), Underwood model (1961), Drake–Schofer–May model (1967), Boardman–Lave model (1977), Inman model (1978), and Van Aerde model (1995). Notable space-average models encompass Smeed model (1968), Keeler–Small model (1977), Ardekani–Herman model (1987), and TRB (2010) model. Typical time-average models comprise Bureau of Public Roads (BPR) function (1964), Small model (1982), Akcelik model (1991), and Small–Verhoef model (2007). Both the speed-density-flow and space-average models could help determine the space-mean speed and indirectly derive the average travel time on a roadway segment. However, the space-average traffic stream relationships could not handle oversaturated traffic conditions in presence of congestion. In such situations, speed depends not only on contemporaneous flow but also past flows. In this respect, the time-average models could be employed to directly estimate average travel time on a highway segment.
- *Dynamic models.* Compared with the static models, dynamic models including shockwave, queueing, single-vehicle motion, longitudinal vehicle motion, and lateral vehicle movement models are advantageous in dealing with extreme time-varying congested traffic conditions (Rakha, 2016). Shockwave and queueing models belong to macroscopic dynamic models which simultaneously deal with multiple vehicles in dynamic traffic flow analyses, whereas single-vehicle motion, longitudinal vehicle motion, and lateral vehicle movement models could readily track the trajectories of individual vehicles. In this context, some models could update the location of each vehicle on a second-by-second basis, which are called mesoscopic dynamic models; and other models could even reduce the time interval of updating each vehicle's location to a sub-second unit of time, which are regarded as microscopic dynamic models.

  In particular, the shockwave model such as the Lighthill–Whitham–Richards (LWR) model introduced by Lighthill and Whitham (1955) and Richards (1956) could readily determine the direction and speed of a shockwave created by a stationary bottleneck that will help estimate the average vehicle travel time on the highway segment that encounters traffic disruptions. Shockwave models dealing with moving bottlenecks such as the Newell model began to appear in the 1990s (Newell, 1998). Queueing models can also be applied to estimate average travel time to traverse through a bottleneck (Newell, 1971).

Typical single-vehicle motion models focus on modeling of vehicle accelerations, decelerations, powertrains, engine performance, and gear selections. Vehicle accelerations can be characterized by a vehicle kinematic model, vehicle dynamics model, or gear-shifting model. Exemplary models include the point mass vehicle dynamics model specifying vehicle accelerations by Rakha et al. (2001), vehicle engine performance model by Ni and Henclewood (2008) applicable in conjunction with the engine torque model by Guzzella and Sciarretta (2007), and gear selection model by Wong (2001).

Longitudinal vehicle motion models that consider vehicle interactions are generally called car-following models. For instance, Gazis–Herman–Rothery model, also known as the General Motors (GM) model (Gazis et al., 1961), Gipps model (Gipps, 1981), Intelligent Driver model (Trieber and Kesting, 2013), and Rakha–Pasumarthy–Adjerid model (Rakha et al., 2004) allow for continuous space and time traffic dynamics where vehicles are treated as discrete entities, and the acceleration and deceleration rates of consecutive vehicles in the same travel lane that form a car-following pair are constrained by vehicle clearance, speed difference, and driver' varying reaction time (May, 1990; Jiao et al., 2020). This will help determine the travel time of each vehicle covering a certain distance.

Lateral vehicle movement models can be grouped into lane-changing and gap acceptance models that generally model discretionary and mandatory lane changes of vehicles in dynamical traffic conditions (Carter et al., 1999; Laval and Daganzo, 2006).

In particular, Wiedemann model (1974) and Wiedemann and Reiter model (1992), Fritzsche model (1994), and Van Aerde model (1995) and Van Aerde and Rakha model (1995) have been adopted by VISSIM, PARAMICS, and INTEGRATION microsimulation software, respectively, to model individual drivers' car-following and lane-change behaviors.

Apart from travel times spent on highway segments, urban streets, transit routes, or multimodal routes, travel delays could be experienced at intersections because of traffic movement conflicts. Both static and dynamic models can be employed to estimate vehicle and pedestrian delays at intersections. The U.S. Highway Capacity Manual (HCM) contains one of the most widely used static models for intersection delay estimation where the total delay per vehicle at a signalized intersection consists of uniform delay, incremental delay to cope with different vehicle arrival patterns, and initial queue delay caused by residual vehicles from the previous signal cycle of intersection (TRB, 2010). The HCM also covers methods for estimating pedestrian delays at an isolated intersection or multiple intersections along one side of an urban street.

In many urban areas, the insufficient supply of parking facilities renders drivers cruising in seeking available parking spots. Further, vehicles engaged in entering, exiting, or looking for parking spaces slow down other vehicles. Owing to inefficient curbside utilization that is typically for on-street parking only, traffic disruptions will occur from double parking of delivery trucks for lack of curbside loading zones or vehicle blockage at intersection corners by ridesharing services for lack of designated curbside pick-up and drop-off locations. In all those cases, added travel time resulting from parking activities need to be included into the total O-D travel time of individual travelers.

8.4.2 Travel time index and travel time buffer index

The travel time index is the ratio between the traffic volume and free flow volume of all highway segments, urban streets, transit routes, or multimodal routes of an O-D path weighted by vehicle miles of travel associated with the links in the peak period, which provides useful information on the travel quality associated with an O-D travel path comprised of different classes of highway segments, urban streets, transit routes, or multimodal routes. Further, travel time buffer index as the ratio of the extra travel time experienced by 90th or 95th percentile travel time over the average travel time with the average travel time can be calculated to incorporate reliability considerations for a route, a corridor, or an O-D travel path. The two indices can be used independently or collectively to identify one or more highway segments, urban streets, or multimodal routes comprised of an O-D path, or multiple O-D paths with peak period traffic mobility issues in need for improvements as long as one or both index values exceed the pre-defined threshold values.

8.5 MOBILITY IMPROVEMENT NEEDS ASSESSMENT

Using data on current or future peak-period traffic flow details, the travel time, travel time index, and travel time buffer index of one or more highway segments, urban streets, transit routes, or multimodal routes comprised of an O-D path, or multiple O-D paths can be computed, and be largely correlated with levels of service (LOS) of highway segments, urban streets, transit routes, or multimodal routes to assess mobility improvement needs.

8.5.1 Mobility improvement needs assessment by travel mode

For highway mobility assessment, highway segments could be classified by highway function class, such as rural and urban freeways, multilane highways with divided or undivided medians, and two-lane roads with class I and class II designations. For a freeway and multilane highway segment, LOS is determined by the density of the traffic stream directly correlated with space-mean speed. For a two-lane class I road segment, LOS for class I two-lane roads is calculated according to percent-time-spent-following and average travel speed, whereas in a two-lane class II road segment, the LOS is computed in accordance with percent-time-spent-following. For intersection-related mobility assessment, intersections are split into signalized and unsignalized intersections. The control delay per vehicle is estimated to correspond to different LOS levels. When LOS is employed for establishing threshold mobility performance levels, an LOS D is typically considered to be the minimum acceptable mobility performance level. If any highway segment or intersection in the highway network experiences traffic mobility worse than LOS D in the daily peak period, it is a candidate for mobility improvement treatment. The analysis can be expended to identify mobility improvement needs for the entire highway network (Table 8.1).

Transit mobility performance is viewed from availability and quality perspectives. The availability is concerned with spatial and temporal availability of transit service. With transit service available, the quality of service can be assessed in terms of comfort and convenience experienced by transit riders. For both the availability and quality aspects, the assessment can be carried out separately by transit element, including transit stops/stations, route segments, and transit system with multiple routes providing service to a specific area (Tables 8.2 and 8.3).

Likewise, the LOS analysis can also be utilized for shared use bicycle or scooter mobility assessment (Patten et al., 2006). Practically, LOS C is used as the threshold level for shared use bicycle or scooter path mobility improvements (Table 8.4).

Table 8.1 Traffic mobility assessment using LOS analysis

Network component		*Performance measure*		*LOS*	*Description*
Highway segment	Freeway	Density, D (pcpmpl)	D≤11	A	Free flow
			D>11–18	B	Reasonably free flow, effects of minor incidents still easily absorbed
			D>18–26	C	At or near free flow speed, queues may form behind any significant blockage
			D>26–35	D	Approaching unstable flow
			D>35–45	E	Unstable flow, operating at capacity
			D>45	F	Forced or breakdown flow, queues form behind breakdown points
	Multilane	Density, D (pcpmpl)	D≤11	A	Free flow
			D>11–18	B	Reasonably free flow, effects of minor incidents still easily absorbed
			D>18–26	C	At or near free flow speed, queues may form behind any significant blockage
			D>26–35	D	Approaching unstable flow
			D>35–40 to 45	E	Unstable flow, operating at capacity
	Two-lane, Class I	Percent-time-spent-following, P (%) Average travel speed, V (mph)	P≤35, V≥55	A	Low following, high speed
			P>35–50, V>50–55	B	Moderate following, relatively high speed
			P>50–65, V>45–50	C	Relatively high following, moderate speed
			P>65–80, V>40–45	D	High following, moderate–to-low speed
			P>80, V≤40	E	Extremely high following, low speed
	Two-lane, Class II	Percent-time-spent-following, P (%)	P≤40	A	Low following
			P>40–55	B	Moderate following
			P>55–70	C	Relatively high following
			P>70–85	D	High following
			P>85	E	Extremely high following
Intersection	Signalized	Control Delay, d (sec/veh)	d≤10	A	Free flow
			d>10–20	B	Stable flow
			d>20–35	C	Stable flow with acceptable delay
			d>35–55	D	Approaching unstable flow with tolerable delay
			d>55–80	E	Unstable flow, with intolerable delay
			d>80	F	Forced or breakdown flow
	Unsignalized	Control Delay, d (sec/veh)	d≤10	A	Free flow
			d>10–15	B	Stable flow
			d>15–25	C	Stable flow with acceptable delay
			d>25–35	D	Approaching unstable flow with tolerable delay
			d>35–50	E	Unstable flow, with intolerable delay
			d>50	F	Forced or breakdown flow

Table 8.2 Transit service availability assessment using LOS analysis

Element	*Performance measure*			*LOS*	*Description*
Stop	Frequency:	Headway (min)	Veh/h		
	Urban scheduled transit	<10	>10	A	Highly frequent service, riders do not need schedules
		10–14	5–6	B	Frequent service, passengers consult schedules
		15–20	3–4	C	Max desirable time to wait if a bus/train is missed
		21–30	2	D	Service unattractive to selective riders
		31–60	1	E	Service available during hour
		>10	<1	F	Service unattractive to all riders
	Frequency:	Access time (h)			
	Paratransit	0.0–0.5		A	Fairly prompt response
		0.6–1.0		B	Acceptable response
		1.1–2.0		C	Tolerable response
		2.1–4.0		D	Poor response, may need pre-trip planning
		4.1–24.0		E	Need pre-trip planning
		>24.0		F	Irregular service
	Frequency:	Trips/day			
	Intercity scheduled transit	>15		A	Frequent daily service
		12–15		B	Mid-day and frequent peak hour service
		8–11		C	Mid-day or frequent peak hour service
		4–7		D	Minimum service to provide choice of travel times
		2–3		E	Possible daily round-trip
		0–1		F	Impossible daily round-trip
Route segment	Service hours:	Hours/day			
		19–24		A	Night service provided
		17–18		B	Late evening service provided
		14–16		C	Early evening service provided
		12–13		D	Daily service provided
		4–11		E	Peak hour/limited mid-day service
		0–3		F	Very limited or no service
System	Service coverage:	Transit supportive area of coverage (%)			
		90–100		A	
		80–89.9		B	
		70–79.9		C	
		60–69.9		D	
		50–59.9		E	
		<50		F	

Source: Li (2018).

Table 8.3 Transit service quality assessment using LOS analysis

Element	*Performance measure*			*LOS*	*Description*
Stop	Passenger loads:	m²/p	p/seat		
	Bus	>1.20	0.00–0.50	A	No passenger needs to sit next to each other
		0.80–1.19	0.51–0.75	B	Passengers can choose where to sit
		0.60–0.79	0.76–1.00	C	All passengers can sit
		0.50–0.59	1.01–1.25	D	Comfortable standee load for design
		0.40–0.49	1.26–1.50	E	Maximum schedule load
		<0.40	>1.50	F	Crush loads
	Passenger loads:	m²/p	p/seat		
	Rail	>1.85	0.00–0.50	A	No passenger needs to sit next to each other
		1.30–1.85	0.51–0.75	B	Passengers can choose where to sit
		0.95–1.29	0.76–1.00	C	All passengers can sit
		0.50–0.94	1.01–2.00	D	Comfortable standee load for design
		0.30–0.49	2.01–3.00	E	Maximum schedule load
		<0.30	>3.00	F	Crush loads
Route segment	Reliability:	On-time operation (%), varying by 5–10 minutes for fixed route, 20 minutes for paratransit			
		97.5–100		A	One late transit vehicle per month
		95.0–97.4		B	Two late transit vehicles per month
		90.0–94.9		C	One late transit vehicle per week
		85.0–89.9		D	
		80.0–84.9		E	One late transit vehicle per direction per week
		<80		F	
System	Transit/auto time:	Travel time difference (min)			
		≤0		A	Faster by transit than by auto
		1–15		B	About as fast by transit as by auto
		16–30		C	Tolerable for selective transit riders
		31–45		D	Round-trip at least an hour longer by transit
		46–60		E	Tedious for all transit riders
		>60		F	Unacceptable to most riders

Source: Li (2018).

For pedestrian walking mode, the LOS analysis can also be utilized for mobility assessment on a walkway segment, crossing an intersection, and traveling on one or more arterial segments that cover multiple walkway segments and intersections (TRB, 2010). The rating of LOS C can be used as the threshold level for pedestrian walking improvements (Table 8.5).

8.5.2 Mobility improvement options

When dealing with a multimodal transportation system, mobility improvement options can be derived from travel demand management, capacity expansion, and efficient capacity utilization perspectives. Figures 8.2–8.4 depict typical mobility improvement options within each category.

Table 8.4 Shared use bicycle path mobility assessment using LOS analysis

Network component	*Performance measure*		*LOS*	*Description*
Bikeway	LOS^{Bike}_{score}	>4	A	Optimum condition, bikeway can accommodate more riders
		>3.5–4	B	Facility condition is good, bikeway can absorb more riders
		= 3–3.5	C	Fair condition with marginal ability to absorb more riders
		=2.5–3	D	Reduction in bicycle travel speed
		=2–2.5	E	Significant reduction in travel speed with crowded demand
		≤2	F	Significant conflicts

Source: Li (2018).

Table 8.5 Pedestrian walking mobility needs assessment using LOS analysis

Network component	*Performance measure*		*LOS*
Pedestrian walking on an arterial segment	LOS^{ped}_{link} score, average pedestrian space, S^{ped} (ft^2/p)	$LOS^{ped}_{link} \le 2.0$, $S^{ped} > 60$	A
		$LOS^{ped}_{link} = 2.0 - 2.75$, $S^{ped} = 40 - 60$	B
		$LOS^{ped}_{link} = 2.75 - 3.5$, $S^{ped} = 24 - 40$	C
		$LOS^{ped}_{link} = 3.5 - 4.25$, $S^{ped} = 15 - 24$	D
		$LOS^{ped}_{link} = 4.25 - 5.0$, $S^{ped} = 8 - 15$	E
		$LOS^{ped}_{link} > 5.0$, $S^{ped} \le 8$	F
Pedestrian walking on an arterial segment and an intersection	LOS^{ped}_{seg} score, S^{ped} (ft^2/p)	$LOS^{ped}_{seg} \le 2.0$, $S^{ped} > 60$	A
		$LOS^{ped}_{seg} = 2.0 - 2.75$, $S^{ped} = 40 - 60$	B
		$LOS^{ped}_{seg} = 2.75 - 3.5$, $S^{ped} = 24 - 40$	C
		$LOS^{ped}_{seg} = 3.5 - 4.25$, $S^{ped} = 15 - 24$	D
		$LOS^{ped}_{seg} = 4.25 - 5.0$, $S^{ped} = 8 - 15$	E
		$LOS^{ped}_{seg} > 5.0$, $S^{ped} \le 8$	F

Source: Li (2018).

8.6 MOBILITY IMPROVEMENT EVALUATION

Some mobility improvement benefits can be quantified, other benefits may only be qualitatively assessed. The quantifiable benefits of a mobility improvement option can be assessed by quantitative performance measures, such as travel time savings to highway users and transit riders. The quantities of travel time savings could be further converted to monetary values by applying the value of travel time (AASHTO, 1978, 2002, 2003, 2010; FHWA, 2000; Sinha and Labi, 2011; Li, 2018). However, the time cost of travel generally includes components of the resource cost reflecting the value to the traveler of an alternative use of time such as work and the disutility cost as the level of discomfort, boredom, or other negative aspects associated with time lost due to travel. The value of travel time can be largely estimated by mode choice, route choice, speed choice, dwelling choice, and wage rate-based methods, which could vary drastically. With the same level of travel time savings achieved by a mobility improvement option, the monetary benefits could be very different by adopting different values of travel time for analysis.

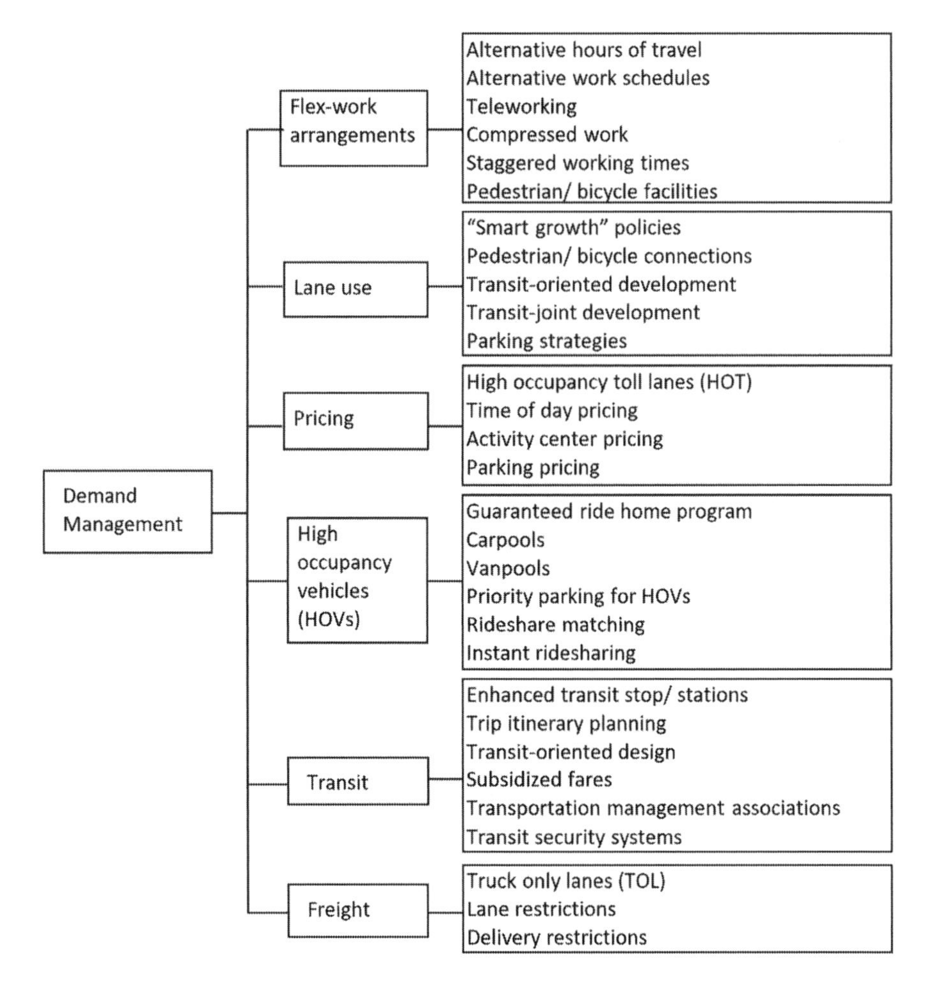

Figure 8.2 Travel demand management options. (Source: Li (2018).)

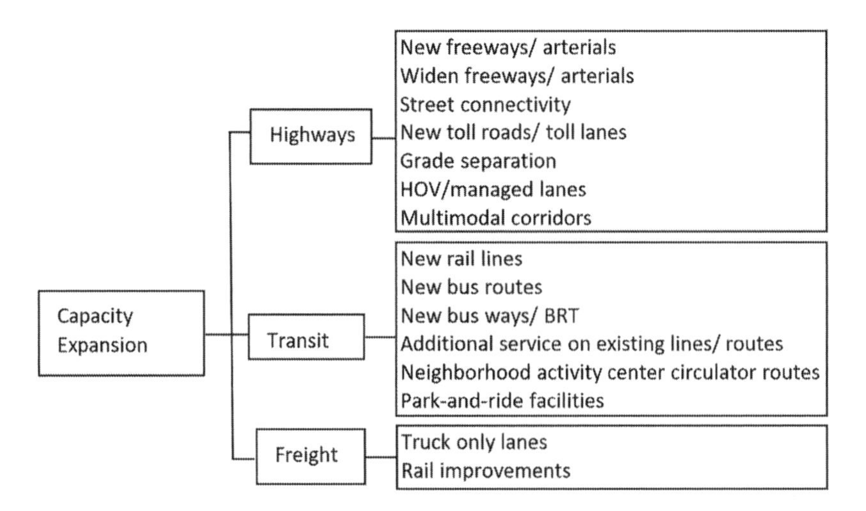

Figure 8.3 Transportation capacity expansion options. (Source: Li (2018).)

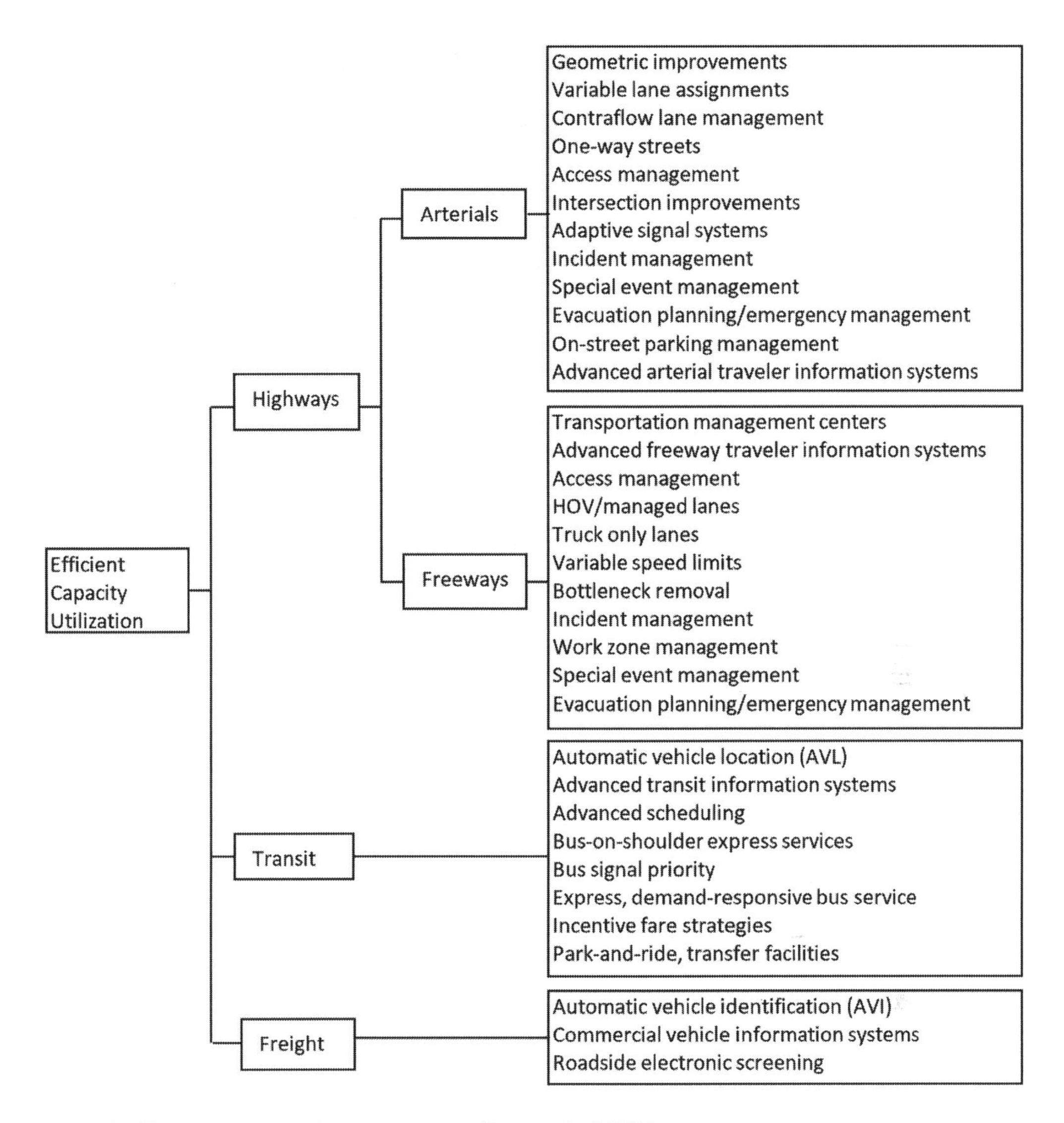

Figure 8.4 Efficient capacity utilization options. (Source: Li (2018).)

For this reason, two general approaches can be used for evaluating the benefits of mobility improvements. One approach is to convert the quantified mobility improvement benefits in the transportation facility service life cycle to dollar values by adopting generally accepted unit rates, such as values of travel time. To account for differences in service lifespans of different transportation facilities, the monetary benefits can be expressed in equivalent uniform annualized dollar benefits. The other approach is to convert the quantified mobility improvement benefits to utility values to reveal the level of preference from the transportation decision maker, user, or stakeholder associated with a mobility improvement option.

8.6.1 Mobility improvement benefits in monetary values

Mobility improvement options may be concerned with different travel modes, mode-specific facilities, or different traffic management strategies. Mobility benefits of a specific improvement option may be extended to a multiyear horizon in the transportation facility service lifespan (AASHTO, 2010). To ensure fully capturing the multiyear benefits of each mobility improvement option and equitably comparing the merits of different mobility improvement options with varying duration of impacts, the life-cycle analysis approach should be employed to quantify the benefits in the service life cycles of transportation facilities to which mobility improvements are made (AASHTO, 2003).

To establish the life-cycle benefits of a mobility improvement option, the annual benefits after the mobility improvement are estimated. Without loss of generality, a geometric growth rate of annual benefits is typically assumed for future years. The present worth of annual benefits in all years of the facility service life cycle can then be computed. This amount is further converted to equivalent uniform annualized benefits. For each mobility improvement option, the annualized benefits are compared with annualized costs to justify its economic feasibility or cost-efficiency. This helps identify the subset of economically feasible mobility improvement options. Among the economically feasible mobility improvement options, the corresponding annualized benefits could be used for cost-effectiveness comparisons.

In the analysis process, risks (multiple possible values with a known probability distribution) and uncertainties (multiple possible values with an unknown range or an unknown probability distribution) associated with input factors for computing mobility improvement benefits such as traffic forecasts, project costs, and discount rates may be incorporated to enhance the robustness of the benefit estimates.

8.6.2 Mobility improvement benefits in utility values

With benefits triggered by a mobility improvement option such as annual travel time savings quantified in the transportation facility service life cycle, instead of transforming the benefits to monetary values with which a general consensus might not be reached for the adopted unit value of time savings, the estimated benefits could be converted to utility values. In presence of multiple itemized benefits assessed by performance measures with non-commensurable units, they need to be separately converted to utility values using utility functions associated with individual performance measures. The respective utility functions are then synthesized to an overall utility function via a process of weighting, scaling, and amalgamation (Sinha et al., 2009; Li, 2018). The overall utility gain after implementing a mobility improvement option could be used as a surrogate measure of its benefits.

8.7 BUDGET ALLOCATION FOR MOBILITY-CENTERED PERFORMANCE IMPROVEMENTS

8.7.1 Issues

- *Risk tolerance of mobility improvement benefits.* In the budget allocation process, the decision maker aims to choose a portfolio of mobility improvement options that could minimize the total risk of the expected benefits of individual mobility improvement options.
- *Interdependent impacts of multiple mobility improvement options.* When multiple mobility improvement options are implemented in adjacent locations, along a major corridor, across multiple travel modes, or in an overlapping time sequence, they will trigger multiple rounds of traffic redistributions across the multimodal transportation network that would eventually change the mobility performance levels of all highway segments, urban streets, transit routes, and multimodal routes directly and indirectly affected by the mobility improvement options. Therefore, the overall benefits of

multiple mobility improvement options may be greater than, equal to, or smaller than the direct addition of individually estimated benefits of constituent mobility improvement options (Li et al., 2012, 2013).

- *Budget constraints.* Budget allocation for mobility improvements may be conducted for a multiyear period. The budget allocation decisions may be made by imposing yearly budget constraints or just considering the cumulative budget for all years combined.
- *Budget uncertainty.* Mobility improvement decisions are usually made based on an estimated budget many years ahead of the actual project implementation period. As time passes by, updated budget information would become available and budget allocation decisions need to be updated accordingly to remain realistic.
- *Budget allocation in the private-public partnership context.* Traditionally, budget available to the transportation agency is solely from the agency. In presence of the public-private partnership, a portion of the budget will come from the private partners. While mobility-centered, performance-based management of the transportation system treats mobility as the top priority, without sacrificing the overall performance of other transportation system management goals, the profit-driven nature of the private partners needs to be given full attention. Performance-based, mobility-centered budget allocation in the private-public partnership context should go one step beyond the conventional approach of maximizing the overall benefits that could be achieved from the available budget. Specifically, it should aim to identify a subset of economically feasible investment options to yield the maximized overall benefits of mobility gains, without leading to a net loss (by aggregating gains and losses) of performance levels of other transportation system management goals under two sets of constraints. The first set of constraint is the total budget jointly contributed by the transportation agency and the private entities. The second set of constraint is related to a prespecified narrow range of user charge rates (rendering revenue forecasts) to allow for a reasonable level of profit margins for private entities participating in transportation development and management and concurrently minimizing the costs of passenger travel and freight movements.

8.7.2 Budget allocation methods

Ranking, prioritization, and optimization are the typical methods that could be adopted for budget allocation with added complexity (FHWA, 1991; Haas et al., 1994).

- *Ranking.* It is the simplest form of priority setting for budget allocation for a single-year period, which is also called single-year prioritization. The ranking procedure mainly includes two steps. The first step is to determine the passenger or freight modes that should be considered for mobility improvements. For each set of mobility improvements, the best option is identified, and the corresponding cost is determined. The next step involves prioritization of mobility improvement options according to a given set of criteria. The ranking procedure may be implemented by using a single criterion, or composite criteria such as a ranking function combining travel mode, facility type, traffic, and mobility factors (Zimmerman, 1995). The ranking procedure produces a ranked list of mobility improvement options to be carried out, the cost associated with each option, and a cut-off line established based upon the expected budget level. As the timings of mobility improvement options are not considered in the ranking process, the long-term impacts of delaying or accelerating mobility improvement options from 1 year to another cannot be easily evaluated.
- *Multiyear prioritization.* It is a more sophisticated budget allocation approach for addressing transportation network mobility improvements. This method requires the use of mobility performance prediction models. It also requires the definition of trigger

points to identify needs and provisions that allow the acceleration or deferral of mobility improvements during the analysis period. Common approaches used to perform prioritization include marginal cost-effectiveness and incremental benefit–cost analyses. Multiyear prioritization differs from the ranking (i.e., single-year prioritization) procedure in several ways. First, different strategies that include mobility improvement options and timings are considered in multiyear prioritization. Another difference lies in the complexity of the analysis. In the ranking procedure, the most common criteria considered are current mobility performance and existing traffic levels. In a multiyear prioritization, future mobility performance levels need to be forecasted through the use of performance models with additional factors considered for analysis. With multiyear prioritization, the timing effect of mobility improvements can be analyzed. The impact of various budget levels can also be assessed (FHWA, 1991).

- *Optimization.* The optimization formulation for budget allocation to mobility improvements is known in the literature as the capital budgeting problem (Lorie and Savage, 1955; Weingartner, 1963). More generally, this problem falls into the category of the multi-choice multidimensional Knapsack problem (MCMDKP) (Sinha and Zoltners, 1979), where the budget comes from different sources and the analysis is conducted for a multiyear period. Notable deterministic optimization models developed in the last several decades include integer programming (Isa Al-Subhi et al., 1989; Weissmann et al., 1990; Zimmerman, 1995, 2009; Neumann, 1997), goal/compromise programming (Geoffroy and Shufon, 1992; Ravirala and Grivas, 1995), and multi-objective optimization (MOO) (Teng and Tzeng, 1996; Li and Sinha, 2009; Li, 2009; Li et al., 2010). In addition, some researchers investigated the stochasticity of factors affecting the results of budget allocation. For instance, Friesz and Fernandez (1979) introduced a stochastic model to handle the uncertainty of future demand.

8.7.3 Tradeoff analysis methods

Transportation decision-making involves multiple stakeholders that often possess conflicting preferences. Effective decision outcomes could only be reached by considering such conflicts in a holistic manner. Rarely, a truly optimal solution to all stakeholders exists, and finding the most compromised solution becomes essential. To this end, it often involves tradeoff analysis of candidate projects that assigns different levels of priorities for transportation system management goal of mobility versus other goals, different travel modes, or different types of facilities for a given budget level to determine which portion of the transportation system would benefit, which portion would be adversely affected, and what is the best combination of investments across modes, facilities, and mobility versus other system management goals that would produce the highest overall mobility improvement benefits without leading to a net loss of other system management goals, while achieving a the prespecified level of net profits for private entities in presence of the public-private partnership and concurrently without imposing excessive user charges to passengers and freight shippers (Li, 2018).

In the budget allocation process, different tradeoff methods may be employed to evaluate the impacts of such decision policies on the overall benefits, which will help identify the best combination of mobility improvement options. A wide range of tradeoff analysis scenarios may be considered within and between various travel modes, within mobility goal and between mobility and other system management goals, and within and between various isolated locations, corridors, and geographic areas, as well as considering deferred implementation of some large-scale investments. All of these tradeoff issues face the same basic question and involve the same basic elements. At the core, a generalized tradeoff in transportation budget allocation asks, “How much resource do I allocate to A versus B?”

The actual tradeoff issue itself is "What are the consequences of a particular allocation of resources to A and B?" and the choice becomes the allocation and set of consequences that the decision maker prefers. Because investment decisions are usually made based on an estimated budget year ahead of the investment implementation period, as time passes by updated budget information will be available, the resource allocation decisions must be updated accordingly.

8.7.4 Implementation of prioritized alternatives and feedback of effectiveness

In principle, the best combination of investment options that maximizes the overall mobility improvement benefits will be proposed for possible implementation. There might be organizational issues that could potentially impede the project deployments. Additional issues such as public support may also affect the sequence in which prioritized mobility improvement options are actually deployed. Also, the resource allocation decision-making is an iterative process that is expected to be responsive to the needs of the transportation agency, the user, private entities in presence of public-private partnership, and other stakeholders. It is important that the analysis be made flexible to keep abreast of the changing needs of the transportation system mobility, yet robust enough to be applicable in a wide variety of areas related to performance-based, mobility-centered transportation system management. Feedback evaluation involves the routine collection and analysis of field and predictive data, comparing the results with the previously established performance targets, and evaluating the performance of investment policies, strategies, and operational procedures. The feedback process allows the transportation decision maker to assess the effectiveness of mobility-centered system management efforts, identify areas for improvements, justify these improvements, and demonstrate the benefits of performance-based, mobility-centered resource allocation, and support requests for additional resources.

8.8 CASES IN ACTION

8.8.1 Budget allocation practices in U.S. state transportation agencies

In the current practice, analytical frameworks and methods used by the U.S. state transportation agencies for budget allocation vary greatly, which could be grouped into four categories: (i) legacy driven, (ii) fix-it-first, (iii) soft optimization, and (iv) performance based. For the legacy driven approach, budget allocation decisions are largely legacy driven, meaning that the amount of funding goes toward addressing certain performance deficiency issues is largely determined by the program structure in the past with very minimal changes. Some state transportation agencies utilize this approach are Kansas, Louisiana, and Mississippi (LADOTD, 2018; MDOT, 2019; KSDOT, 2019). The fix-it-first approach prioritizes available budget for system preservation, and the remaining portion of limited budget is allocated to needs for improving system usage performance. Colorado, Georgia, South Carolina, and Virginia are some of the states that use this approach for their budget allocation (GDOT, 2019; CODOT, 2019; SCDOT, 2019; VDOT, 2019). To improve from the fix-it-first and legacy-driven approach, which allocate the budget funding by historical frameworks and structures, some states such as Arizona, Ohio, and Michigan adopt the soft optimization approach that incorporates relative priorities of facility preservation and system operations are considered in the budget allocation process (ADOT, 2019; ODOT, 2019; MDOT, 2019). Moreover, very few states such as Indiana, Oregon, Utah, and Washington have moved toward performance-based budget allocation where budget allocation between facility preservation and system operational improvements across passenger and freight modes is generally based on data-driven analysis results (WSDOT, 2019; UDOT, 2019; ORDOT, 2019). This approach is technically advantageous over the previous three approaches that lack coherent, well-structured, flexible, and analytical rigor to holistically address multiple types of impacts of transportation investments to ensure achieving truly optimal returns.

The optimization of transportation budget allocation in the context of MCMDKP involving multiple system management goals and performance measures largely classifies the problem into (i) the multi-attribute decision-making (MADM) problem that involves a limited number of investment alternatives assessed by qualitative criteria; and (ii) the MOO problem that handles a large number of investment alternatives evaluated by quantitative criteria (Evans, 1984). Table 8.6 summarizes notable MCDM methods and models along with solution techniques developed over the years for transportation budget allocation.

8.8.2 Illinois tollways' investment decision-making in interdependent capital projects

The Illinois State Toll Highway Authority (ITA) manages over 300 centerline miles of open-toll freeways in the Chicago metropolitan area, Illinois. In 2010, the ITA Board of Directors identified six major tollway capital investments that accommodate a significant portion of regional travel. As depicted in Figure 8.5, these projects include Elgin-O'Hare/Western Bypass, IL-53/I-355 North Extension, I-294/I-57 Interchange, Illiana Expressway from I-65 in Indiana to I-55 in Illinois, Prairie Parkway connecting I-88 with I-80, and I-90 Corridor Expansion from O'Hare to Rockford.

As a practical matter, the six tollway projects could be implemented separately one at a time, or jointly from implementing two, three, four, and five projects at a time, to all six tollway projects together. As such, the number of possible project implementation scenarios becomes $2^6-1=63$. To estimate the potential benefits in decreases of life-cycle agency costs and tollway user costs, savings of travel times, reductions in vehicle crashes, and cutbacks of vehicle air emissions related to NMHC, CO, CO_2, NO_X, SO_2, $PM_{2.5}$, and PM_{10}, the Chicago regional multimodal travel demand forecasting model was executed to generate data on the traffic flow details including link-based traffic volumes, vehicle compositions, and speeds before and after project implementation. For each of the 63 project implementation scenarios, the geometric designs, capacities, and speed limits of highway segments within the project physical range were modified accordingly. In addition, a buffer zone extending significantly beyond the project physical range was created for each project implementation scenario to include all tollway and non-tollway segments expected to be indirectly affected by project implementation.

As documented in Li et al. (2013), for all joint project implementation scenarios, traffic volumes on project affected tollway segments were quite different between jointly implementing multiple projects and separately implementing the same set of projects one at a time. Also, the overall benefits of simultaneously implementing multiple projects were greater or smaller than the total of project benefits generated from separately implementing related projects one at a time, revealing either positive or negative interdependencies of project impacts. None of the scenarios exhibited an independent relationship.

The total cost of the six proposed projects was US$17.37 billion in 2010 dollars. The overall project benefits were estimated by increasing the budget level from 10% to 100% of US$17.37 billion. The overall project benefits generally followed an incremental trend along with budget increases. However, no additional network-wide benefits could be produced after the budget level reached 80% of US$17.37 billion (i.e., US$13.9 billion), suggesting

Table 8.6 Notable optimization methods and models for transportation budget allocation

Study	*Problem*		Method and model	Solution technique	Application
	MADM	*MOO*			
Tzeng et al. (2005)	✓		Analytic Hierarchy Process (AHP)	Technique for Order Preference by Similarity to Ideal Solution (TOPSIS)	Transit
Dabous and Alkass (2010)	✓	✓	Multiple attribute utility theory (MAUT)	Monte Carlo simulation	Bridges
Pirdavani et al. (2010)	✓		Delphi	TOPSIS	Safety
Zavadskas et al. (2014)	✓		AHP	Mathematical model applying log-normalization and multiplicative exponential weighting	Highways
Su and Hassan (2007)	✓		AHP	Optimization solver	Highways
Tamošaitien ✓ et al. (2013)	✓		Ranking method under fuzzy environment	Technique for Order Preference by Similarity to Ideal Solution with fuzzy criteria values (TOPSIS-F)	Highways
Kao et al. (2006)	✓		Resource-constrained project scheduling	High-level petri nets, activity-based costing, and TOPSIS	Highways
Bai et al. (2015)		✓	Pareto Frontiers	Genetic algorithm	Highways
Moussourakis and Haksever (2014)		✓	Mixed integer programming	Integer optimization solver	Highways
Pollack-Johnson and Liberatore (2006)	✓	✓	Goal programming	Goal programming model solver	Highways
Wey and Wu (2007)		✓	0/1 goal programming	Analytic Network Process (ANP) algorithm	Highways
Iniestra and Gutiérrez (2009)		✓	Knapsack model	Evolutionary algorithm	Highways
Hudson et al. (2014)		✓	Integer programming	Derivative free optimization algorithm	Pavements bridges
Li et al. (2010)		✓	Chance-constrained multidimensional Knapsack model	Heuristic algorithm	Highways
Maggiore and Ford (2015)		✓	Tool prototype	Generalized reduced gradient algorithm and heuristic algorithm	Highways
Porras-Alvarado et al. (2015)		✓	Fair division	Social welfare and collective utility functions	Pavements bridges
Zimmerman et al. (2016)		✓	Priority programming with cross-asset tradeoffs	Heuristic algorithm	Highways
Roshandeh et al. (2015)		✓	Surrogate worth tradeoff	Lagrange relaxation and ε-constraint	Tollways
Xiong et al. (2012)	✓	✓	Nadir compromise programming	Genetic algorithm	Highways
Truong et al. (2018)	✓	✓	e-STEP method	Mathematical model applying minimax algorithm	Tollways

Source: Truong et al. (2021).

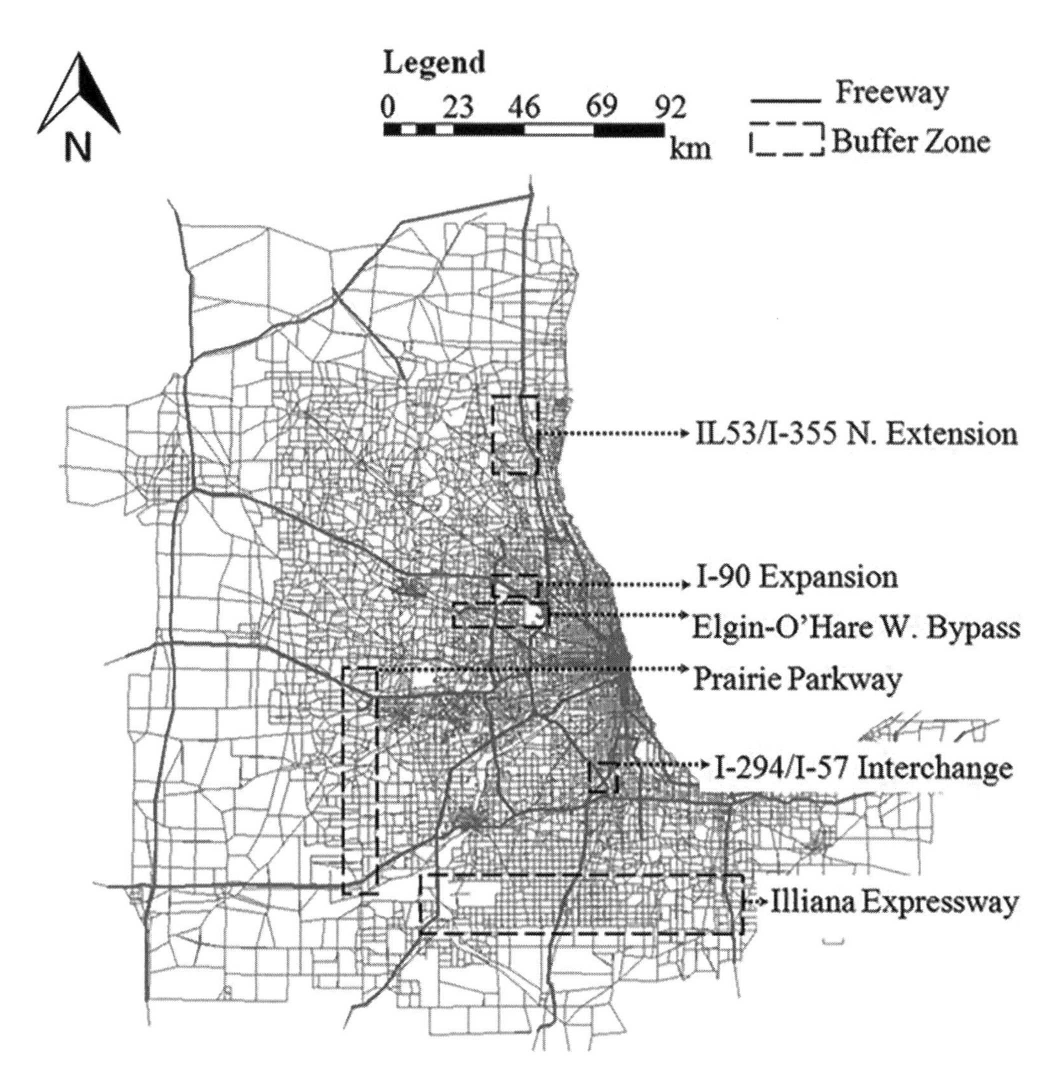

Figure 8.5 Illinois tollways capital investment projects proposed in the Chicago metropolitan area, Illinois.

that the total amount of tollway capital investments should be controlled within US$13.9 billion. This differs significantly from the conventional optimization modeling of project selection that neglects the effect of multi-project interdependencies—either positive or negative, leading to underestimating or overestimating the overall project benefits.

8.9 ISSUES AND CHALLENGES

Transportation systems are in a time of great changes in travel demand, multimodal transportation with flexible capacity, technology, transportation funding and financing, and public expectations. These changes have had a corresponding impact on how megacity mobility performance should be sustained. There are institutional, strategic, measurement, integration, and analytic challenges that the transportation agency must overcome to be successful.

8.9.1 Institutional issues

Because performance-based mobility management is holistic and proactive, it depends upon comprehensive coordination and communication of various offices and divisions of a transportation agency and the transportation agency with its partners in presence of the public-private partnership. Most transportation agencies are functionally segregated. It is imperative to promote understanding and appreciation in the benefits of practicing performance-based, mobility-centered resource allocation from the perspectives of the entire agency and potential private partners. Within the transportation agency, the commitment at both the executive and operations levels is critical to the success of implementing performance-based mobility management. As part of this commitment, qualified staffing and financial resources must be designated to facilitate performance-based mobility management.

8.9.1.1 Strategic challenges

The budget allocation for performance-based, mobility-centered management is guided by the setting of strategic system management goals. The transportation agency, the user, private entities in presence of the public-private partnership, and other stakeholders may have different or even conflict missions, agendas, and values. Such differences will lead to a range of differing goals, which are often in competition. Competing ideas have to be resolved through alignment with policy goals, an understanding of the tradeoffs, and consensus building.

8.9.1.2 Data management challenges

The data management challenges are concerned with measurement, generation, and integration challenges. Many of the performance measurement issues currently facing the transportation agency today lie in identifying appropriate and meaningful performance measures that will drive the collection, processing, and management of field data and generation of forecasted data that are ultimately used for analysis and proactive decision-making. Most transportation agencies find that they are data rich but information poor in that existing data are heterogeneous or in lack of information needed for performance-based analysis to make sound decisions. A reevaluation of data inventory and data items to be collected and generated using forecasting tools is often needed to resolve the measurement challenges.

All too often, field data collected over time are scattered across a variety of disconnected, incompatible data storage systems. This limits the ability of data accessing, processing, and analysis to identify existing mobility performance issues. Moreover, the lack of adequate forecasting tools to generate predictive data to help identify mobility performance issues to be expected in the future further limits the ability of data accessing and usage. A framework for integrating all data items using standard formats, models, tools, and protocols needs to be created to help perform the desired performance-based, mobility-centered management functions using transportation asset management principles. The framework should also address disparities in data sources and formats and respond flexibly to changing data requirements when new operational functions are introduced or when existing operational functions are modified. Although performance-based, mobility-centered analysis benefits greatly from continuously replenished fine-grained data, concerns may be raised regarding privacy of people travel and goods movement. This could potentially limit collecting certain types of data and restrict the geographic coverage, time frame, frequency, and time step interval of field data collection and predictive data generation, as well as removing some historical data from data repository in the future. The framework should be able to rigorously handle such situations.

8.9.1.3 Analytical challenges

The emphasis on performance modeling using statistical and econometric methods, traffic forecasts using simulation-based tools, evaluation of investment options using life-cycle analysis approach and decision theory, and selection of economically feasible investment options using optimization techniques in many transportation agencies is still under development. In this regard, a transportation agency must acknowledge the fact that the analytical capabilities can only be developed step by step. Continuing support and resources need to be given to develop the required analytical capabilities and update the analytical capabilities to keep abreast of changing analytical requirements over time. In presence of the potential risk of data shrinkage caused by private concerns in the future, the analytical methods, models, and tools need to be robust enough to continually generate predictive data with high accuracy and precision, and produce reliable results of data analysis that are essential to mobility management decision-making.

8.9.2 Strategies for implementing mobility-centered, performance-based budget allocation

While there is no standard approach or strategy to implementing performance-based, mobility-centered management for holistic, proactive, and cost-effective resource allocation, a variety of methods, models, tools, and techniques are available to assist in the efficient use of existing resources. The transportation agency must examine exactly where it stands, what information is available, and which approach to be chosen for implementing performance-based, mobility-centered management. The following strategies may be adopted to promote its implementation:

- Defining an appropriate sequence and priority of short-, mid-, and long-term tasks in the implementation strategy.
- Establishing support and confidence among agency's executive and operational management staff to effectively mitigate institutional barriers for implementing performance-based, mobility-centered management using transportation asset management principles. Open communication among agency personnel may be conducted to improve the basic understanding of the potential benefits of implementing performance-based management for mobility-centered resource allocation.
- Establishing support and confidence among agency's management staff on how and why decisions are made; reporting on successes and lessons learned toward established system performance goals; and linkage between the agency's vision and mission, goals, performance measures, performance targets, resources, and outcomes of mobility-centered resource allocation so that they have meaning to the transportation agency, the user, private entities in presence of the public-private partnership, and other stakeholders.
- Ensuring a long-term commitment by the agency's senior leadership and management to performance-based mobility management and providing resources necessary for implementing it.

REFERENCES

AASHTO. 1978. *A Manual on User Benefit Analysis of Highway and Bus-Transit Improvements*. American Association of State Highway and Transportation Officials, Washington, DC.

AASHTO. 1990. *Guidelines for Pavement Management Systems*. American Association of State Highway and Transportation Officials, Washington, DC.

AASHTO. 2002. *Transportation Asset Management Guide*. NCHRP Project 20-24(11). American Association of State Highway and Transportation Officials, Washington, DC.

AASHTO. 2003. *User Benefit Analysis for Highways*. American Association of State Highway and Transportation Officials, Washington, DC.

AASHTO. 2010. *User and Non-user Benefit Analysis for Highways*. American Association of State Highway and Transportation Officials, Washington, DC.

ADOT. 2019. *Arizona Transportation Asset Management Plan*. Arizona Department of Transportation, Phoenix, AZ.

Akcelik, R. 1991. Travel time functions for transport planning purposes: Davison's function, its time-dependent form and an alternative travel time function. *Australian Road Research* 21, 49–59.

Ardekani, S., and Herman, R. 1987. Urban network-wide traffic variables and their relations. *Transportation Science* 21, 1–16.

Bai, Q., Ahmed, A., Li, Z., and Labi, S. 2015. A hybrid Pareto frontier generation method for analyzing trade-offs among transportation performance measures. *Wiley Journal of Computer–Aided Civil and Infrastructure Engineering* 30(3), 163–180.

Bauer, J., Evans, J., Jeannotte, K., Vandervalk, A., and Cambridge Systematics. 2016. *The Use of Data in Planning for Operations: State-of-the-Practice Review*. Report No. FHWA-HRT-15-071. Federal Highway Administration, U.S. Department of Transportation, Washington, DC.

Ben-Akiva, M., and Lerman, S. 1985. *Discrete Choice Analysis: Theory and Application to Travel Demand*. The MIT Press, Cambridge, MA.

Ben-Akiva, M., Humplick, F., Madanat, S.M., and Ramaswamy, R. 1991. Latent performance approach to infrastructure management. *TRB Journal of Transportation Research Record* 1311, 188–195.

Boardman, A.E., and Lave, L.B. 1977. Highway congestion and congestion tolls. *Journal of Urban Economics* 4, 340–359.

BPR. 1964. *Traffic Assignment Manual*. Bureau of Public Roads, U.S. Department of Commerce, Washington, DC.

Carter, M., Rakha, H., and Aerde, M.V. 1999. Variability of traffic-flow measures across freeway lanes. *Canadian Journal of Civil Engineering* 26(3), 270–281.

CODOT. 2019. *Colorado Risk-Based Asset Management Plan*. Colorado Department of Transportation, Denver, CO.

Dabous, S.A., and Alkass, S. 2010. A multi-attribute ranking method for bridge management. *Engineering, Construction and Architectural Management* 17(3), 282–291.

Drake, J., Schofer, J., and May, A. 1967. A statistical analysis of speed-density hypotheses in vehicular traffic science. In *Proceedings of the 3rdInternational Symposium on the Theory of Traffic Flow*. Highway Research Board, New York, 112–117.

Ehrgott, M., and Ruzika, S. 2008. Improved ε–constraint method for multiobjective programming. *Journal of Optimization Theory and Applications* 138(3), 375.

Evans, G.W. 1984. An overview of techniques for solving multiobjective mathematical programs. *Management Science* 30(11), 1268–1282.

Feighan, K.J., Shahin, M.Y., Sinha, K.C., and White, T.D. 1989. An application of dynamic programming and other mathematical techniques to pavement management systems. *TRB Journal of Transportation Research Record* 1215, 101–114.

Fekpe, E.S., Windholz, T., Beard, K., and Novak, K. 2003. *Quality and Accuracy of Positional Data in Transportation*. NCHRP Report 506. Transportation Research Board, National Academies Press, Washington, DC.

FHWA. 1985. *Traffic Control Systems Handbook*. Report No. FHWA-IP-85-11. Federal Highway Administration, U.S. Department of Transportation, Washington, DC.

FHWA. 1987. *Bridge Management Systems*. Demonstration Project No. 71. Federal Highway Administration, U.S. Department of Transportation, Washington, DC.

FHWA. 1991. *Pavement Management Systems*. Federal Highway Administration, U.S. Department of Transportation, Washington, DC.

FHWA. 2000. *Highway Economic Requirements System*. Federal Highway Administration, U.S. Department of Transportation, Washington, DC.
FHWA. 2001. *FHWA SMART Van*. Federal Highway Administration, U.S. Department of Transportation, Washington, DC.
FHWA. 2003. *Freeway Management and Operations Handbook*. Report No. FHWA-OP-04-003. Federal Highway Administration, U.S. Department of Transportation, Washington, DC.
FHWA. 2004a. *Mitigating Traffic Congestion: The Role of Demand-Side Strategies*. Federal Highway Administration, U.S. Department of Transportation, Washington, DC.
FHWA. 2004b. *Traffic Congestion and Reliability: Linking Solutions to Problems*. Federal Highway Administration, U.S. Department of Transportation, Washington, DC.
FHWA. 2005a. *Elements of a Comprehensive Signals Asset Management System*. Federal Highway Administration, U.S. Department of Transportation, Washington, DC.
FHWA. 2005b. *Roadway Safety Hardware Asset Management Systems Case Studies*. Report No. FHWA-HRT-05-073. Federal Highway Administration, U.S. Department of Transportation, Washington, DC.
FHWA. 2005c. *Traffic Congestion and Reliability: Trends and Advanced Strategies for Congestion Mitigation*. Federal Highway Administration, U.S. Department of Transportation, Washington, DC.
FHWA. 2007. *Maintaining Traffic Sign Retroreflectivity*. Report FHWA-SA-07-020. Federal Highway Administration, U.S. Department of Transportation, Washington, DC.
FHWA. 2009. *Manual on Uniform Traffic Control Devices*. Federal Highway Administration, U.S. Department of Transportation, Washington, DC.
FHWA. 2013. *Practical Guide for Quality Management of Pavement Condition Data Collection*. Federal Highway Administration, U.S. Department of Transportation, Washington, DC.
Fielding, G.J. 2013. *Managing Public Transit Strategically: A Comprehensive Approach to Strengthening Service and Monitoring Performance*. Jossey-Bass, San Francisco, CA.
Friesz, T., and Fernandez, E.J. 1979. A model of optimal transport maintenance with demand responsiveness. *Transportation Research Part B: Methodological* 13(4), 317–339.
Fritzsche, H.T. 1994. A model for traffic simulation. *Traffic Engineering and Control* 35(5), 317–321.
Gazis, D., Herman, R., and Rothery, R. 1961. Nonlinear follow-the-lead models of traffic flow. *Operations Research* 9(4), 545–567.
GDOT. 2019. *Georgia Transportation Asset Management Plan*. Georgia Department of Transportation, Atlanta, GA.
Geoffroy, D.N., and Shufon, J.J. 1992. Network level pavement management in New York State: A goal–oriented approach. *TRB Journal of Transportation Research Record* 1344, 57–65.
Gipps, P.G. 1981. A behavioral car-following model for computer simulation. *Transportation Research* 15B, 105–111.
Greenberg, H. 1959. An analysis of traffic flow. *Operations Research* 7(1), 79–85.
Greenshields, B.D. 1935. A study of highway capacity. *Proceedings of Highway Research Record* 14, 448–477.
Guzzella, L., and Sciarretta, A. 2007. *Vehicle Propulsion System: Introduction to Modeling and Optimization*. Springer, New York.
Haas, R., Hudson, W.R., and Zaniewski, J.P. 1994. *Modern Pavement Management*. Krieger Publishing Company, Melbourne, FL.
Harper, W.V., Lam, J., Al-Salloum, A., Al-Sayyari, S., Al-Theneyan, S., Ilves, G., and Majidzadeh, K. 1990. Stochastic optimization subsystem of a network-level bridge management system. *TRB Journal of Transportation Research Record* 1268, 68–74.
Hudson, S., Galenko, A., and Scheinberg, T. 2014. *Trade-off Analysis for Transportation Agencies: Two Approaches to the Cross-Asset Problem*. Kos Island, Greece.
Iniestra, J.G., and Gutiérrez, J.G. 2009. Multicriteria decisions on interdependent infrastructure transportation projects using an evolutionary–based framework. *Applied Soft Computing Journal* 9(2), 512–526.
Inman, R.P. 1978. A generalized congestion function for highway travel. *Journal of Urban Economics* 5, 21–34.

Isa Al-Subhi, K.M., Johnston, D.W., and Farid, F. 1989. *Optimizing System-Level Bridge Maintenance, Rehabilitation, and Replacement Decisions*. North Carolina State University, Raleigh, NC.

Jiao, S., Zhang, S., Li, Z., Zhou, B., and Zhao, D. 2020. An improved optimal velocity function considering the preceding vehicle's speed, space headway, and driver's reaction time. *Wiley Journal of Advanced Transportation* 2020. doi: 10.1155/2020/2797420.

Kao, H.P., Wang, B., Dong, J., and Ku, K.C. 2006. An event-driven approach with makespan/cost tradeoff analysis for project portfolio scheduling. *Computers in Industry* 57(5), 379–397.

Keeler, T.E., and Small, K.A. 1977. Optimal peak load pricing, investment, and service levels on urban expressways. *Journal of Political Economy* 85, 1–25.

Keeney, R.L., and Raiffa, H. 1993. *Decision with Multiple Objectives: Preferences and Value Tradeoffs*. Wiley, New York.

KSDOT. 2019. *Kansas Transportation Asset Management Plan*. Kansas Department of Transportation, Topeka, KS.

LADOTD. 2018. *Louisiana Federal NHS Transportation Asset Management Plan*. Louisiana Department of Transportation and Development, Baton Rouge, LA.

Laval, J.A., and Daganzo, C.F. 2006. Lan-changing in traffic streams. *Transportation Research Part B: Methodological* 40(3), 251–264.

Li, Z. 2009. Stochastic model and O(N2) algorithm for highway investment decision-making under budget uncertainty. *ASCE Journal of Transportation Engineering* 135(6), 371–379.

Li, Z. 2018. Transportation Asset Management: Methodology and Applications. CRC Press, Boca Raton, FL, USA.

Li, Z., and Sinha, K.C. 2004. Methodology for Multicriteria decision making in highway asset management. *TRB Journal of Transportation Research Record* 1885, 79–87.

Li, Z., Sinha, K.C., and McCarthy, P.S. 2001. Methodology to Determine Load and Non-Load Shares of Highway Pavement Rehabilitation Expenditures. *TRB Journal of Transportation Research Record* 1747, 79–88.

Li, Z., Sinha, K.C., and McCarthy, P.S. 2002. A Determination of Load and Non-Load Shares of Highway Pavement Routine Maintenance Expenditures. *Journal of Road and Transport Research* 11(2), 3–13. ARRB Transport Research Ltd., Melbourne, Australia.

Li, Z., and Sinha, K.C. 2009. Methodology for the determination of relative weights of highway asset management system goals and of performance measures. *ASCE Journal of Infrastructure Systems* 15(2), 95–105.

Li, Z., Madanu, S., Zhou, B., Wang, Y., and Abbas, M. 2010. A heuristic approach for selecting highway investment alternatives. *Wiley Journal of Computer-Aided Civil and Infrastructure Engineering* 25(6), 427–439.

Li, Z., Kaul, H., Kapoor, S., Veliou, E., and Zhou, B. 2012. A new model for transportation investment decisions considering project interdependencies. *TRB Journal of Transportation Research Record* 2285, 36–46.

Li, Z., Roshandeh, A.M., Zhou, B., and Lee, S.H. 2013. Optimal decision-making of interdependent tollway capital investments incorporating risk and uncertainty. *ASCE Journal of Transportation Engineering* 139(7), 686–696.

Lighthill, M.J., and Whitham, G.B. 1955. On kinematic waves II: A theory of traffic flow on long crowded roads. *Proceedings of the Royal Society of London A: Mathematical, Physical and Engineering Sciences* 229(1178), 317–345.

Lorie, J.H., and Savage, L.J. 1955. Three problems in rationing capital. *Journal of Business* 28(4), 229–239.

Maggiore, M., and Ford, K.M. 2015. *Guide to Cross–Asset Resource Allocation and the Impact on Transportation System Performance*. No. Project 08-91. The National Academies Press, Washington, DC.

May, A. 1990. *Traffic Flow Fundamentals*. Prentice–Hall, Englewood Cliff, NJ.

MDOT. 2019. *Michigan Transportation Asset Management Plan*. Michigan Department of Transportation, Lansing, MI.

Meyer, M.D., and Miller, E.J. 2000. *Urban Transportation Planning*, 2ndEdition. McGraw–Hill, New York.
Moussourakis, J., and Haksever, C. 2014. Flexible model for time/cost tradeoff problem. *ASCE Journal of Construction Engineering and Management* 130(3), 307–314.
MSDOT. 2019. *Mississippi Transportation Asset Management Plan*. Mississippi Department of Transportation, Jackson, MS.
Neumann, L.A. 1997. *Methods for Capital Programming and Project Selection*. NCHRP Synthesis of Highway Practice 243. National Cooperative Highway Research Program, Transportation Research Board, National Academies Press, Washington, DC.
Newell, G.F. 1971. *Applications of Queueing Theory*. Chapman and Hall, London, UK.
Newell, G.F. 1998. A moving bottleneck. *Transportation Research Part B: Methodological* 32B(8), 531–537.
Ni, D., and Henclewood, D. 2008. Simple engine models for VII-enabled in-vehicle applications. *IEEE Transactions on Vehicular Technology* 57(5), 2695–2702.
ODOT. 2019. *Ohio Transportation Asset Management Plan*. Ohio Department of Transportation, Columbus, OH.
Oppenheim, N. 1995. *Urban Travel Demand Modeling: From Individual Choices to General Equilibrium*. Wiley-Interscience, Hoboken, NJ.
ORDOT. 2019. *Oregon Transportation Asset Management Plan*. Oregon Department of Transportation, Salem, OR.
Patten, R.S., Schneider, R.J., Toole, J.L., Hummer, J.E., and Rouphail, N.M. 2006. *Shared-Use Path Level of Service Calculator: A User's Guide*. Report No. FHWA-HRT-05-138. Federal Highway Administration, U.S. Department of Transportation, Washington, DC.
Pipes, L.A. 1953. An operational analysis of traffic dynamics. *Applied Physics* 24, 274–287.
Pipes, L.A. 1967. Car-following models and the fundamental diagram of road traffic. *Transportation Research* 1, 21–29.
Pirdavani, A., Brijs, T., and West, G. 2010. A multiple criteria decision-making approach for prioritizing accident hotspots in the absence of crash data. *Transport Reviews* 30(1), 97–113.
Pollack-Johnson, B., and Liberatore, M.J. 2006. Incorporating quality considerations into project time/cost tradeoff analysis and decision making. *IEEE Transactions on Engineering Management* 53(4), 534–542.
Porras-Alvarado, Juan, D., Han, Z., and Zhanmin, Z. 2015. A fair division approach to performance–based cross–asset resource allocation. *In 9th International Conference on Managing Pavement Assets*, Virginia Polytechnic Institute and State University, Washington, DC.
Rakha, H. 2016. Traffic flow theory. In *The Routledge Handbook of Transportation*. Editor: Teodorovic, D. Taylor & Francis, New York, 5–24.
Rakha, H., Lucic, I., Demarchi, S.H., Setti, J.R., and Aerde, M.V. 2001. Vehicle dynamics model for predicting maximum truck acceleration levels. *ASCE Journal of Transportation Engineering* 127(5), 418–425.
Rakha, H., Pasumarthy, P., and Adjerid, S. 2004. Modeling longitudinal vehicle motion: Issues and proposed solutions. *Transport Science and Technology Congress*, Athens, Greece.
Ravirala, V., and Grivas, D.A. 1995. Goal-programming methodology for integrating pavement and bridge programs. *ASCE Journal of Transportation Engineering* 121(4), 345–351.
Richards, P.I. 1956. Shock waves on the highway. *Operations Research* 4(1), 42–51.
Romero, C., Tamiz, M., and Jones, D.F. 1998. Goal programming, compromise programming and reference point method formulations: Linkages and utility interpretations. *Journal of the Operational Research Society* 49(9), 986–991.
Roshandeh, A.M., Li, Z., Neishapouri, M., Patel, H., and Liu, Y. 2015. A tradeoff analysis approach for multiobjective transportation investment decision-making. *ASCE Journal of Transportation Engineering* 141(3), 04014085.
Salem, O., AbouRizk, S., and Ariaratnam, S. 2003. Risk-based life-cycle costing of infrastructure rehabilitation and construction alternatives. *ASCE Journal of Infrastructure Systems* 9(1), 6–15.

SCDOT. 2019. *South Carolina Transportation Asset Management Plan*. South Carolina Department of Transportation, Columbia, SC.

Sinha, K.C., and Labi, S. 2011. *Transportation Decision Making: Principles of Project Evaluation and Programming*. John Wiley & Sons, New York.

Sinha, P., and Zoltners, A.A. 1979. The multiple-choice knapsack problem. *Operations Research* 27(3), 503–515.

Sinha, K.C., Patidar, V., Li, Z., Labi, S., and Thompson, P.D. 2009. Establishing the weights of performance criteria: Case studies in transportation facility management. *ASCE Journal of Transportation Engineering* 135(9), 619–631.

Small, K.A. 1982. The scheduling of consumer activities: Work trips. *American Economic Review* 72, 467–479.

Small, K.A., and Verhoef, E.T. 2007. *The Economics of Urban Transportation*, 2ndEdition. Routledge, New York.

Smeed, R.J. 1968. Traffic studies and urban congestion. *Journal of Transportation Economics and Policy* 2, 33–70.

Su, M., and Hassan, R. 2007. *Application of the Analytic Hierarchy Process in Road Asset Management: User Manual*. Austroads, Sydney, NSW, Australia.

Tamošaitienė, J., Zavadskas, E.K., and Turskis, Z. 2013. Multi-criteria risk assessment of a construction project. *Procedia Computer Science* 17, 129–133.

Teng, J.Y., and Tzeng, G.H. 1996. A multiobjective programming approach for selecting non-independent transportation investment alternatives. *Transportation Research Part B* 30(4), 291–307.

TRB. 2010. *Highway Capacity Manual*, 5thEdition. Transportation Research Board, National Academies Press, Washington, DC.

Trieber, M., and Kesting, A. 2013. *Traffic Flow Dynamics: Data, Models and Simulation*. Springer, Berlin, Germany.

Truong, T., Li, Z., and Kepaptsoglou, K. 2018. Entropy–STEP multiobjective trade–off analysis method for optimal transportation investment decisions. *ASCE Journal of Transportation Engineering Part A: Systems* 144(1), 04017065.

Truong, T., Zhang, J., Li, Z., and Wang, L. 2021. Incorporating Herfindahl-Hirschman index to compromise programming for optimal transportation investment decisions. *TRB Journal of Transportation Research Record* 2021. doi: 10.1177/03611981211011648.

Turner, S.M., Eisele, W.L., Benz, R.J., and Holdener, D.J. 1998. *Travel Time Data Collection Handbook*. Report No. FHWA-PL-98-035. Federal Highway Administration, U.S. Department of Transportation, Washington, DC.

Tzeng, G.H., Lin, C.W., and Opricovic, S. 2005. Multi-criteria analysis of alternative-fuel buses for public transportation. *Energy Policy* 33(11), 1373–1383.

UDOT. 2019. *Utah Transportation Asset Management Plan*. Utah Department of Transportation, Salt Lake City, UT.

Underwood, R.T. 1961. Speed, volume and density relationships. In *Quality and Theory of Traffic Flow*. Bureau of Highway Traffic, Yale University, New Haven, CT. 141–187.

Van Aerde, M. 1995. Single regime speed-flow-density relationship for congested and uncongested highways. *Presented at the 74th Annual Transportation Research Board (TRB) Meeting* (Paper No. 95080). Washington, DC.

Van Aerde, M., and Rakha, H. 1995. Multivariate calibration of single regime speed-flow density relationships. In *Proceedings of the 1995 Vehicle Navigation and Information Systems (VNIS) Conference: 6th International VNIS. A Ride into the Future*, Seattle, WC. doi: 10.1109/VNIS.1995.518858.

VDOT. 2019. *Virginia Transportation Asset Management Plan*. Virginia Department of Transportation, Richmond, VA.

Vonderohe, A.P., Travis, L., Smith, R.L., and Tsai, V. 1993. *Adaptation of Geographic Information Systems for Transportation*. NCHRP Report 359. National Cooperative Highway Research Program, Transportation Research Board, National Academies Press, Washington, DC.

Vuchic, V.R. 2005. *Urban Transit: Operations, Planning, and Economics.* John Wiley and Sons, Hoboken, NJ, USA.

Weingartner, H.M. 1963. *Mathematical Programming and the Analysis of Capital Budgeting Problems.* Prentice–Hall, Englewood Cliffs, NJ.

Weissmann, J., Harrison, R., Burns, N.H., and Hudson, W.R. 1990. *Selecting Rehabilitation and Replacement Bridge Projects, Extending the Life of Bridges.* ASTM STP, West Conshohocken, PA.

Wey, W.M., and Wu, K.Y. 2007. Using ANP priorities with goal programming in resource allocation in transportation. *Mathematical and Computer Modelling* 46(7–8), 985–1000.

Wiedemann, R. 1974. Simulation des Strassenverkehrsflusses. Technical Report. Institut fur Verkehrswesen, Universitat Karlsruhe, Karlsruhe, Baden-Württemberg, Germany, In German.

Wiedemann, R., and Reiter, D.J. 1992. Microscopic traffic simulation: The simulation system MISSION, background and actual state. CEC Project ICARUS (V1052) Final Report, Volume 2, Appendix A. Commission of the European Communities (CEC), Brussels, Belgium, EU.

Wilson, A.G. 1974. *Urban and Regional Models in Geography and Planning.* John Wiley and Sons, Hoboken, NJ.

Wong, J.Y. 2001. *Theory of Ground Vehicles.* John Wiley & Sons, New York.

WSDOT. 2019. *Washington State Transportation Asset Management Plan.* Washington State Department of Transportation, Olympia, WA.

Xiong, H., Shi, Q., Tao, X., and Wang, W. 2012. A compromise programming model for highway maintenance resources allocation problem. *Mathematical Problems in Engineering* 2012. doi: 10.1155/2012/178651.

Zavadskas, E.K., Vilutienė, T., Turskis, Z., and Šaparauskas, J. 2014. Multi–criteria analysis of Projects' performance in construction. *Archives of Civil and Mechanical Engineering* 14, 114–121.

Zimmerman, K.A. 1995. *Pavement Management Methodologies to Select Projects and Recommend Preservation Treatments.* NCHRP Synthesis of Highway Practice 222. National Cooperative Highway Research Program, Transportation Research Board, National Academies Press, Washington, DC.

Zimmerman, K.A., Allen, B.W., Ram, P.V., Duncan, G.M., Smadi, O., Smith, K.L., Manda, K.R., and Bektas, B.A. 2009. Identification of effective next generation pavement performance measures and asset management methodologies to support MAP-21 performance management requirements. DTFH6115C00045. Federal Highway Administration, U.S. Department of Transportation, Washington, DC.

Zimmerman, K.A., Allen, B.W., Ram, P.V., Duncan, G.M., Smadi, O., Smith, K.L., Manda, K.R., and Bektaş, B.A. 2016. Identification of effective next generation pavement performance measures and asset management methodologies to support MAP-21 performance management requirements, DTFH6115C00045, Federal Highway Administration, U.S. Department of Transportation, Washington, DC, USA.

Chapter 9

The path to sustainable megacity mobility

Cities typically develop in organic and incremental ways over time. In cities experiencing rapid growth, policymakers and planners struggle to keep up with the pace of changes and the seemingly incessant and rapidly increasing demand on public services. However, as time and development progress, cities reach a point where the entire transportation system and network needs to be re-examined. Villages don't work, economically or politically, the same way as large, sprawling urbanized areas do. In large cities, isolated nodes become integrated components of a complex, continually evolving urban ecosystem. Not surprisingly, transportation planning and policymaking need to be considered in a different context more suitable to their level of evolution. Megacities need to be considered in this context, as cities of such scale and scope their transportation planning systems and networks need to be re-thought and re-calibrated.

While how megacities make this transition will differ based on economic, political, and social contexts, this book has attempted to present a more integrated vision and framework for urban transportation in these massive urban areas. These cities, many an integral part of a global economic network, face unique challenges in scale, size, and scope. Each of these challenges also provides opportunities for policymakers, elected officials, planners, and engineers willing to tackle them in a holistic and comprehensive way. Even if their transportation systems are complex and sophisticated, these regions often have unprecedented access to resources to improve the functioning of their networks and adopt the latest technology to improve efficiency and productivity. Indeed, the scale of the investments and networks provides unique opportunities to tap into innovative mechanisms for funding and managing transportation facilities and networks using innovations such as real-time transportation technologies, state-of-the-art construction techniques, variable rate pricing for infrastructure of all sorts, and access to private capital and expertise through tools such as public-private partnerships (PPPs).

An important theme in this book is the role mobility should play in anchoring planning and network development in megacities. Mobility, broadly speaking, is the ability and speed of travel for people and goods from one point to another point in the network. The ability to move people and goods safely, efficiently, and effectively is crucial to the economic sustainability of megacities. These large cities, however, are no longer attempting to optimize one mode over another. Rather, they are juggling ways to deploy and coordinate multiple modes to meet the unique and often dynamic needs and preferences of users in a safe, efficient, and effective manner. Thus, previous chapters have attempted to examine and evaluate a wide range of potential strategies, frameworks, solutions, and applications that might allow megacities to achieve the highest possible level of efficiency, effectiveness, and equity, using the overall framework of optimizing mobility to anchor the discussion of the role for the transportation system.

DOI: 10.1201/9780429345432-9

Transportation systems and networks, of course, are expected to consider and often optimize goals other than mobility. Within a transportation system, policymakers and managers also consider facility preservation, implementation and other costs associated with operating the system, burden on users and nonusers, and environmental impacts, as well as supporting economic prosperity. Optimizing goals that are in conflict requires evaluating the tradeoffs involved with emphasizing one or more of these goals—including mobility—over another or others. However, degrading mobility carries significant downstream consequences that may fundamentally compromise a megacity's ability to grow, accommodate new growth, or generate the wealth necessary to address other goals. By monetizing the benefits of the transportation system, additional resources can be invested in achieving these other goals. Thus, a holistic approach to transportation development and management can make these tradeoffs explicit and the consequences more transparent, aiding the ability of policymakers to address (and balance) competing goals within the transportation system and network.

While putting mobility as the goal with top priority, we do not dismiss or trivialize other goals, such as environmental sustainability, economic development, or maximizing the built environment. Rather, we emphasize that the primary goal of a transportation network is to ensure goods and people dependent on transportation infrastructure move efficiently and safely within a region. The transportation system can, and often should, be calibrated to work within larger goals and policy frameworks, but the primary goals of the transportation network should not be ignored or subverted if megacities are expected to prosper.

With the sustained growth and the passage of time, the character and nature of travel change. The needs, goals, and objectives of travelers, whether passengers or commercial in nature, will change as priorities change and technology makes benefits, costs, and consequences of travel more transparent to the system users, even nonusers affected by the transportation system. The characteristics of mobility, system performance, congestion (as a part of the mobility issues), and their causal factors are likely to change and differ among modes, facilities, locations, and operations. The transportation and mobility needs of a town of 100,000 will be different in fundamental ways from a city of 1 million, which will be different from a city of 10 million or a city of 30 million or more.

The differences in travel and mobility may also be influenced by the development stage of the city as well. Cities in lower-income countries that are growing rapidly will often not have the access capital to build the facilities they need to ensure timely and efficient build out of their transportation system. The experience of other countries with low levels of development may be instructive for these cities (and countries). China, for example, started from an extremely low industrial and income base as an agrarian, nearly subsistence economy prior to 1978. As the nation loosened its controls over the economy, encouraged entrepreneurship, and allowed exports to higher income nations, China's urban economies and ports exploded. To facilitate this growth, China courted and aggressively encouraged private capital to invest in the nation's infrastructure, mostly highways, bridges, and ports, during the 1980s and 1990s. The nation created independent transportation authorities, many of which traded on international stock exchanges, in order to raise capital and build facilities to accommodate growing passenger and commercial vehicle traffic. They implemented policies such as tolling to generate revenue to pay for the infrastructure.

This model was a success. User fees funded an unprecedented expansion of the nation's infrastructure using private capital. While the nation also learned valuable, sometimes difficult, lessons on how to structure these agreements, the overall success of the approach is difficult to dismiss. Starting from a very low manufacturing and services base, national and provincial policies unleashed the economic potential for economic growth and development. A key part of this strategy was recognizing the role that infrastructure played in ensuring the success of the broader local, regional, and national economy. The key for lower income nations and cities is to learn the lessons from China (and other nations) experiences to create and implement a more robust and resilient framework for financing the transportation infrastructure necessary to support this growth.

The discussions in the previous chapters suggest several critical keys to efficiently and productively managing megacity transportations systems: planning in a three-dimensional (3D), integrated framework; engineering facilities and networks to optimize multimodal network capacity; adopting new technologies to manage demand and fund facilities and systems; reforming budgeting and financing systems to create sustainable investments in critical transportation infrastructure; and conducting performance-based budget allocation using transportation asset management principles. When combined into an overarching strategy or policymaking framework, the transportation system contributes to these building blocks and provides a powerful platform for sustaining growth and furthering human (and economic) development.

9.1 KEYS TO BUILDING A MOBILITY-SUSTAINED MEGACITY IN THE 21ST CENTURY

As the world emerges from the COVID-19 pandemic, transportation planners will be challenged in unprecedented ways. However, most research suggests changes in work arrangements and supply chains are likely to reinforce existing trends rather than fundamentally change trajectories. Greater use of technology will facilitate faster adoption of work-at-home arrangements but will not fully displace offices. Overall, pre-pandemic trends toward more flexible work arrangements in the service sector, just-in-time manufacturing, and nimble logistical systems will accelerate, challenging transportation networks built on less flexible systems to adapt.

But these challenges, while creating real-time uncertainty and concern about the future of cities, do not fundamentally alter the transportation planning/development and management for megacities. On the contrary, the vast scale and scope of megacity economies and communities implies they are constantly in flux as a matter of practice even without the unprecedented disruption of a pandemic. The transportation policy question before megacity policymakers is fundamentally about how they manage the growth and development of their networks and how they will accommodate the changing desires, needs, and preferences of their residents and business interests.

This book has argued for an integrated approach to megacity transportation planning. Megacities do not have the luxury of becoming an "all transit," "all auto/private vehicle," "all rail," or all anything transportation city. Their economies are so diverse and broad that policymakers and planners need to avoid falling into a one mode way of thinking. On the contrary, they must think about building a robust, resilient economic region that is flexible and efficient enough to meet the mobility needs of all its constituencies and stakeholders. This lens implies a radically multimodal-integrated transportation system that uses the latest technology to accommodate the organic and increasingly complex needs of the 21st century megacity. The previous chapters have scoped out some of these overarching principles, strategies, and tactics that can be used to manage these dynamic needs and their aspirations.

9.1.1 3D spiderweb multimodal transportation network planning

One of the more important elements of transportation planning will be embracing a 3D, fully integrated multimodal network, what we call the *3D Spiderweb*. While key features of transportation systems remain (e.g., corridors, urban centers), transportation planning in a contemporary megacity recognizes that most urban areas will be organized around multiple urban "cores" that spontaneously and dynamically organize subregional and neighborhood economies and travel. Megacities of today and the future are a network of town centers and villages, rather than cities with a single dominant downtown core that imposes a monocentric organizational structure on the entire region. While traditional downtowns may be able to retain critical functions, economic activities will continue to decentralize.

The critical tasks for transportation planners will be to facilitate connections between these centers along corridors and improve travel across neighborhoods and regions. In addition, these connections will need to be designed to facilitate and encourage economic and social activity consistent with different levels and layers of urban life. For example, some corridors will need to stitch together dominant residential, commercial, recreational, or industrial hubs. Other corridors will have to facilitate the coordination of multiple layers of services at densities suitable for providing variety and economically diverse options at the neighborhood or even block levels. All these layers will have to be knit together within the tapestry of a transportation network that, in the ideal, seamlessly folds a wide range of travel modes together in a complementary but coherent system.

Not surprisingly, regional connections and travel in megacities might be more likely suited to underground or above ground facilities that minimize the number of interruptions in travel for certain modes with specific uses and purposes (e.g., connecting employment, commercial, or residential hubs). These limited access routes for cars, trains, and buses of various travel modes would provide access to surface and neighborhood streets (and avenues) at critical intersections and junctions. Shanghai's existing and Beijing's planned network of underground and above ground highways are examples. Chongqing's underground traffic circle in its Yuzhong district provides an example of a more limited application with significant regional impacts on circulation and mobility beyond the district itself.

Technologies and innovations in engineering have significantly improved the efficiency of underground and elevated transportation facilities—tunnels, queue jumpers (and queue duckers), elevated highways, and subterranean intersections as clearly demonstrated in places such as Paris and Sydney as well as regional plans for cities like Chicago. While large-scale underground expressway networks such as those planned in Beijing are often thought of as solutions for automobiles, in truth these facilities can serve a variety of modes, including bus, rail, and freight truckways. Indeed, these facilities are expanding and filling in gaps for many cities with a century or more of experience with underground transit networks. London's underground opened in 1863, New York City opened its first subway lines in 1904, Buenos Aires opened its underground transit network in 1913, and Tokyo opened its first stations, which currently serve 8.7 million passengers a day, in 1927. Now, the task before megacity planners and engineers is to transform these systems into networked facilities that effectively integrate a wide range of modes based on origins, destinations, distance, and access.

9.1.2 Engineering innovations for multimodal network capacity

Building new facilities, however, meet only one element, and perhaps not even the most important, of a megacity's transportation and mobility needs. Traditional transportation planning tends to be overly siloed when it comes to integrated transportation systems. Critical to the success of these megacities will be harnessing the engineering innovations that enable an unprecedented degree of connectivity using technology and tactical investments in system improvements. Real-time data collection and processes can help manage existing facilities through various types of "demand leveling" which strive for minimizing the single-occupancy-vehicle use and distribute the total traffic volumes more evenly throughout the day and night. Reversible lanes, traffic signal optimization, and various types of market pricing can influence and, in some cases, direct traveler behavior. Various Intelligent Transportation Systems (ITS) technologies can encourage travelers to shift trips to less-intensive times of the day as well as encourage the use of alternative travel modes by providing real-time information about route options parking availability, travel demand, bottlenecks, and incidents. A critical component of these strategies will be to respect the individual traveler's choice about mode and destination rather than trying to manipulate specific outcomes. Travelers will find ways to get to their destination via their preferred mode, although the mode preference itself often changes based on situation, context, and travel goals. Technology can help improve the system efficiency by encouraging access use of the network at periods where capacity can be optimized and travel times to destinations minimized.

Optimizing system capacity, however, implies collecting, analyzing, and using data consistently and reliably for specific transportation goals and objectives. Fundamentally, this framework deemphasizes a default emphasis for a particular mode (e.g., automobile, bus, or train) and emphasizes mode choice based on the goal-oriented decisions of travelers. Thus, research on traveler behavior is remarkably robust in recognizing the travelers' preferences determine mode choice. For example, bus travelers typically prefer relatively less expensive travel options, accepting a tradeoff with longer commute times. Fixed guideway travelers put a higher premium on-time savings while traveling to specific destinations with high densities for retail, residential development, or office commuting. Automobile travelers prefer the speed and flexibility of the automobile, even if the implied cost of transportation is higher than other mode choices. Personal mobility devices, microtransit, shared mobility and autonomous vehicles are likely to reinforce the options available to travelers as well as present new alternatives in situations that cannot be predicted reliably today. A transportation system that relies on accurate predictions or static assumptions about mode choice is likely to be less robust and resilient, reducing transportation network efficiency and, by extension, mobility.

Similarly, planners will need to give more attention to the integration of freight and passenger travel. Creative ways to separate short- and long-haul freight movement, usually, but not always via truck, will improve delivery times and access to goods and services. Dedicated routes for freight, separate from passenger travel modes, will improve the efficiency of the network in large cities attempting to link commercial hubs. Passenger travel, in contrast, is easier to disperse over a wide range of diverse modes, whether automobile, minibus, scooter, walking, ridesharing, or fixed guideway transit alternative. The emergence of autonomous vehicle technologies will challenge the resilience of the transportation network but also likely improve the public safety, efficiency, reliability of travel times, and for both freight and passenger travel modes. These dimensions to planning and interconnectedness, triangulating across the needs of residents, commuters, and freight haulers, will provide a multilayered framework for connecting blocks, to neighborhoods, to town centers, and to major commercial and residential hubs.

9.1.3 New technologies for demand management and efficient capacity utilization

Achieving these ambitious levels of system performance is possible largely because of recent advances in technology that allow people, goods, and vehicles to be tracked and balanced in a coordinated transportation network. Underground parking facilities can create capacity for greater movements among people while capacity limits are monitored and conveyed to motorists, transit systems, and trucking companies. Traffic signals can be coordinated to improve flows on the margins of corridors, creating downstream improvements in traffic flow based on traffic events or unexpectedly heavy flows at intersections upstream. Transport for London coordinates more than one thousand traffic signals at key intersections to manage flow through the byzantine networks of local roads. Real-time variable tolling can use prices to convey signals to travelers about time delays and the relative costs of using alternative modes through reports on mobile apps or variable message signing along routes. GPS technology allows individuals to track their progress, map out alternative routes, or direct autonomous vehicles (public transit or private) to improve overall system performance.

One can easily envision a 3D transportation network in a megacity that uses a reliable revenue base generated through user fees to manage traffic flows and speeds as deteriorating assets are replaced or entirely new facilities are put in place. While still far in the future, 3D printing may allow on-site construction of major facilities components. This technology is already being used to build housing and in some commercial spaces.

9.1.4 Innovative financing and asset management-based budget allocation

Current and emerging technology can also transform the way megacities plan for and implement system-informed transportation facility management. Rather than focus on specific segments or components of the transportation network—a bridge or an intersection for example—in an isolated setting, planners and engineers can consider the wear, use, and replacement of facilities within a dynamic, long-term setting. The ability to monetize some facilities through tolling, or the entire network through distance-based user fees, provides historically unique opportunities to consider wide ranges of tools for managing these facilities. The classic "public good" properties of transportation networks and components are quickly eroding. Already, entire networks of highways such as those in Chile and France are now managed by private companies because the specific facilities can be monetized for tolling. Now, even local facilities can be tolled efficiently using license plate recognition and transponders. For example, a queue jumper can monetize the value of the benefits it provides to travelers at a local intersection, creating a revenue stream dedicated to maintaining and operating the specific facility, much in the way ski lifts were traditionally operated and maintained by families in the early days of skiing in the Alps.

While economies of scale and scope likely limit the ability of one entity to manage or operate an intersection, megacities will likely have scores if not hundreds of these facilities which can be organized into competitive bundles suitable for competitive bidding among public and private transportation organizations. Thus, PPP arrangements do not have to be limited to major facilities or megaprojects to maximize efficiency, create accountability in the system, and finance missing components or build entirely new layers. Megacities, due to their size and scale, have the ability to fundamentally rearrange and restructure PPPs in a way that can use private capital to build facilities and the expertise of the private sector to operate and manage them once they have been created. Therefore, the same technologies that allow for the optimizing management of an existing travel network provide opportunities to more effectively manage and finance transportation facilities.

9.2 EMERGING ISSUES AND CHALLENGES

As the preceding chapters indicate, the framework for planning and conceptualizing transportation system performance and management for megacities is only one, broad brush step. Scores of different tactics and strategies will need to be deployed to achieve the sustained levels of mobility that ensure economic growth continues to increase. Coordinating these components of the network and investing in the technologies to ensure its smooth operation is a complex and difficult task. Even if they adopt current technology and state-of-the-art planning frameworks, planners and public officials will still face daunting challenges and will have to grapple with emerging problems in novel ways. Perhaps one of the most important emerging issues to address is the design and application of price-based strategies for revenue generation and system management.

9.2.1 Challenges to pricing-based funding

Pricing-based funding aims to apply user charges to various types of vehicles (or users) to recover the total costs of building and preserving transportation facilities. These costs are estimated based on expected wear and tear on the facilities from traffic loading and deterioration caused by repetitive use over their service life cycles. In addition, pricing-based funding imposes user charges to individual system user cost components, including costs of vehicle operation, travel time, crashes, air emissions, and noise pollution with user charge rates set as the marginal external user costs to eliminate inefficient system usage. The pricing-based funding methodology will help secure funding needed to sustain transportation facility conditions and system usage performance. To fully benefit from the proposed pricing-based methodology for transportation funding, the following issues need to be explicitly addressed.

9.2.1.1 Consideration of traffic dynamics

The fairness of user charges determined by the pricing-based methodology is governed by the accuracy (the extent to which data measurements reflect the truth) and precision (the consistency of repeated measurements) of data on transportation facilities and system usage. Accurately computing travel behavior and impacts depends crucially on the availability of quality data with sufficient measurement detail to monitor and evaluate physical facilities and system usage based on vehicle operation, travel time, safety, air emissions, and noise pollution. Systems and engineers will have to manage massive amounts of data on traffic detail such as vehicle volumes and composition, speed, and speed changes. These data are currently collected by a wide range of equipment, predicted by different models, and processed from integration, interpolation, and extrapolation. These data include inconsistencies and errors, an inherent problem endemic to collecting data from diverse sources. These data need to be cleaned, checked, tested, and analyzed in real time in order to be effective in meeting system management and development goals.

As mobile devices become more ubiquitous, fine-grained cross-sectional data on traffic counts from counting stations and data on traveler's trajectories are increasingly available. For example, during a fuel-shortage panic in North Florida, gasoline availability and traffic volumes were calculated in real time for thousands of different locations across hundreds of miles of highway and reported by mobile apps such as Waze, Gas Buddy, Good Maps, and Yahoo maps. Thus, the traffic impacts on transportation facilities and system usage can be calculated in a much more refined manner. For instance, the time intervals of traffic and travel data measurements could be reduced from daily, hourly, 15 minutes, 5 minutes, 1 minute, 1 second, to a fraction of 1 second. As individual vehicles and travelers are traced on a fraction of 1-second basis, shifting the analysis from macroscopic level to microscopic level is practical and feasible. Consequently, the microscopic estimation of impacts will help determine more effective user charges to recover the losses of facility preservation and mitigate inefficiency of system usage. As such, the time dependency of traffic dynamics needs to be incorporated into the pricing-based funding framework.

Further, user charges will influence travel demand as travelers adjust. As prices go up, for example, the system should experience reductions in total demand. As prices change across the system, travelers should modify certain types of trip frequencies, shift departure times, and alter destination location and arrival times, travel modes, and trip paths. Tradeoffs not only exist in travel schedule changes aimed to minimize the weighted total travel time cost but also costs of facility construction and preservation versus system usage in relation to travel time, vehicle crashes, air emissions, and noise pollution that are to be minimized holistically. Consequently, tradeoffs need to be considered in establishing user charges to preserve transportation facilities and to ensure system usage efficiency. Traffic dynamics in (current and future) time and space domains need to be included into the multilayer tradeoffs involved with impacts analysis with added complexity to ensure that the user charges are highly efficient in impacts mitigation that will sustain the overall transportation system performance.

9.2.1.2 Advanced vehicle technologies

The market penetration of advanced technology vehicles like battery-operated vehicles, biofuel vehicles, and connected and automated/autonomous vehicles (CAVs) will only increase as more cities, states, and nations experiment with their application. This will lead to significant changes in traffic flow patterns and traffic stream characteristics. Traffic impacts on physical facilities and system usage will change accordingly. As a result, user charges and how they are applied to technological innovations need to be kept abreast of changes in advanced vehicle technologies to ensure efficiency in managing transportation systems. For example, fuel taxes long served as a reliable means for funding transportation infrastructure. With the phasing out of fossil fuels, new fees will need to provide the bulk of funding, assisted by innovative financing mechanisms, to sustain the existing and future transportation infrastructure.

9.2.1.3 Integration of physical facility, vehicle, and user/nonuser components

Perhaps the most profound challenge facing planners and transportation policymakers over the next several decades will be how to more effectively and seamlessly integrate life cycle analysis across different transportation system components. The facility, vehicle, and user components of a transportation system are becoming increasingly integrated as data collection, processing, analysis incorporate the exchange of pertinent information extracted from data throughout the system and different points and nodes. The facility-based life-cycle costing approach for impact analysis will be extended to cover life-cycle cost analysis of vehicles during their service lives and system users in their life spans. That is, life-cycle cost analysis will collectively consider physical facility, vehicle, and user components of an entire transportation system.

The system integration offers opportunities to create a bottom-up iterative feedback process for top-level asset management decisions with the determination of efficient user charges being a key analytical component. This will help achieve adaptive management of the transportation system while also providing a more holistic perspective and approach to transportation network management. The user charges based on the pricing-based methodology should act as a catalyst to capture impacts of temporal and spatial dynamics of vehicles on transportation facility and system usage performance in support of system component integration, adaptive transportation system management, and sustainable transportation development.

9.2.2 Institutional issues

In addition to these merging technical issues and concerns, transportation policymakers will need to consider significant reforms or modernizations of institutions, agencies, and transportation authorities. Top-down, compliance-driven agencies will need to adopt more bottom-up, iterative approaches to transportation system dynamics, management, and development. While we have talked a lot in this book about the need for good planning, planners need to also be humble, acknowledging the information they don't, and can't have, and seeking to integrate knowledge and ideas up, rather than force them down from the top. This is particularly relevant for megacities, where transportation systems are expected to manage the travel for millions of peoples and vehicles on a daily basis involve hundreds of millions of decisions at the micro level. Top-down approaches to managing this system are inevitably going to be less nimble and adaptive, compromising the ability of the network to remain focused on overarching goals just as improving and sustaining mobility.

9.2.2.1 Performance-based agency reorganization

The transportation system and the institutional issues within this complex dynamic system are three-fold and involve long-term issues, mid-term issues, and short-term issues based on the temporal effects of the issue (Rietveld and Stough, 2005). Local transportation agencies have traditionally had a compartmentalized institutional framework (often modal), where each department dealt with a specific activity. The current institutional framework and hierarchy in the transportation sector specifically is not conducive for inter-agency data sharing, dynamic decision-making, and system optimization. This limits the scope of achieving a sustainable transportation system.

For example, the governance of the transportation system in the United States is stratified in terms of jurisdiction as being under federal, state, and local governance. The demarcation of the highway network between the federal and other departments is straightforward in terms of the federal government operating and maintaining the freeway system. The state and local transportation agencies have tasks which might overlap, creating inherent redundancy and duplication owing to an institutional lack of collaborative transparency. The agencies were formed and evolved primarily to meet the growing demand with objectives of mobility and accessibility when the automobile industry was in its early stages of growth in the 20th century.

As the system grew larger owing to rising demand and use by non-commercial travelers, safety emerged as a system goal and priority for decision-making. By the turn of the 20th century, growing concerns for the environment and social equity resulted in their inclusion as the system goals. Eventually, with the advancement in technology, automotive engineering, computers, electrical engineering, and mechanical and construction engineering, the various fields of sciences were applied to the transportation system as well and the agencies had to expand and fit to serve the changing user demand. Today, information technology and computers have created an inseparable and important component to the transportation system which is the ITS sector. ITS has forced many agencies to collaborate, but still there are areas where there is duplication of efforts owing to lack of institutional framework which encourages data sharing.

The demand of transportation users in today's times is not only growing exponentially but also changing everyday with the technological advancements in the tangential fields. To match up and adapt to these changes is a very difficult task for the agencies in the United States as they are formed based on policy and legislature. Legislative changes require time to be altered and refined. This lag, and the politics involved with securing legislative changes, is one of the most significant institutional barriers to optimizing system performance.

The current U.S. transportation legislation—Moving Ahead for Progress in the 21st century (MAP-21)—mandates the use of performance-based management which requires collaboration between all the agencies and departments involved for the transportation system management. The performance-based system management mandates the use of network-level analysis versus the traditionally used project-level analysis, which requires the collaboration between local, state, and federal agencies, as in an urban land use these three are interacting to form the network. The network-level analysis will ensure that the impact of a project translated in terms of induced demand in the network is analyzed rather than only using estimated traffic volumes at the local project zone. This approach will ensure an optimal system performance, even though the performance indicators might not be encouraging at project-level impacts.

All megacities must grapple with policy formulation and alignment across the entire vertical hierarchy, and standardized performance measures to ensure uniformity in performance evaluation of the system performance. Laws and legislatures mandate the use of performance measures and standards to achieve the system goals which are outlined in the policy formulated for the future vision of the transportation system of a nation. This not only support the future vision for the entire nation's system, but it also ensures that the decision-making process is also refined, to make the system standardized, more effective, and speedier. The future vision, however, is based on forecasted demand as well as availability of the resources and funding.

One of the important resources of a transportation system, particularly highways, is land, which is difficult to acquire especially in a built environment where the demand would be higher. Acquiring the necessary land and building infrastructure are the two most costly components in a transportation project. In an urban setting, the perception that capacity can be augmented should be marginal, and the outlook should be focusing on operating the existing system efficiently before capacity augmentation as a mitigation strategy. This outlook has to be integrated along with policy goals across all agencies involved. When the subject of policy formulation is broached, local and national policy interests are not always consistent. The two interests must be balanced, or at least aligned, to ensure local agencies' priorities are integrated within the policy and collaborate with the broader jurisdiction. At times, the national interest will supersede the local interest. This national interest should not be a hindrance for the future collaborative efforts between the national and local agencies.

When the inter-modal transportation system is examined, something critically important to megacities, the collaboration and institutional transparency is even more important and challenging. Adding to these complex dynamics is the threat of natural disasters, emergency situations, and terrorism which demand the collaboration of not only transportation management agencies but the medical, law enforcement, and other related services as well.

The agencies are still trying to adapt to the ever-changing scenario of user demand where most sciences are merging, in the quest to achieve a sustainable livable community with transportation at its core. In order to optimize any system, each dollar spent must yield the maximum benefits. But this level of optimization cannot be achieved until the vertical and horizontal hierarchies of decision-making are revamped to focus on performance-based collaboration in terms of intra-agency as well as inter-agency goals. The barriers that have to be addressed for achieving a sustainable transportation system are (i) policy barriers, (ii) legal barriers, (iii) institutional barriers, (iv) social and cultural barriers, (v) resources barriers, and (vi) physical barriers.

9.2.2.2 Intra- and inter-agency coordination

Coordination and its effect on the decision-making process is important. Within transportation, siloed decision-making and planning have often led to unbalanced and poorly matched transportation systems. Certain modes or constituencies may be favored over others without a practical understanding of how resources might be better managed to serve larger regional goals or objectives such as improving system-wide efficiency or mobility. Decision-making is almost always working within a zero-sum game as advocates for specific modes or constituencies attempt to capture as many resources as possible regardless of the regional impacts.

Megacities, however, need to break through these silos in order to build and optimize their transportation system and networks. In theory, these coordination problems could be addressed through an integrated, unitary transportation agency. However, if the information and knowledge collected and used at the local level are not adequately communicated or understood by those in agency leadership, the system's performance will likely suffer or fail to meet its goals, see for example, Fuller et al. (2004) and Richards et al. (1999). Stoker and Mossberger (1994) suggest that the alignment of several common priorities is required to achieve effective change along with a not very complex institutional framework.

The coordination within an agency and across agencies can be complex. Each subunit has a different set of policy issues and objectives related to specific goals for their transportation area; funding and resource allocation; needs for data collection, storage, analysis, and distribution; operations and maintenance; work-zone management; inventory management, etc. Full integration requires policies to be reconciled within an agency or institution (vertical integration) as well as interactions between similar levels at other agencies or organizations (horizontal integration). In addition, transportation agencies need to consider the spatial dimensions of their programs and services: which organizations or subunits provide which services and where. Consistency between national and local goals is essential in order for collaboration and integration to be successful. However, a system in which a superregional entity (e.g., federal or provincial government) coordinates multiple public or private entities providing services requires a much more complex environment for managing policy and implementation (Marsden and May, 2006).

- *Intra-agency coordination.* Goals, objectives, and implementation strategies must be aligned within an agency, or among agencies providing service at equivalent levels and to similar beneficiaries in order for the transportation network to work optimally. The primary concern is the internal resistance by existing stakeholders within the unit to accept change. In order to overcome this hurdle, policymakers must reconcile the goals and objectives of each unit, so they are focused on achieving the same outcomes. Once the objectives are created, a horizontal alignment is achieved and a general framework and set of policy objectives can be created. Agencies and units can then be given flexibility as to how to use resources, adapt their own policies, and set performance goals within their own organizational norms and structures. This process of policy reconciliation can be greatly facilitated by establishing personnel who can coordinate these efforts, ensuring that the objectives are realized in a positive manner.

  For example, when transportation legislation mandated the use of performance-based management by the local agencies in the United States to be eligible for federal funding, the law required collaboration among all state transportation agencies. Switching over from traditional methods long in use to a new approach and methodology takes persistence and commitment. Management and staff need to be trained on the use and application of the new methodology. They also need to ensure that the past projects and data are converted to allowed standardized comparisons. These requirements will often be viewed as burdens by the affected agency in the beginning. Resistance to a paradigm change is natural and should be expected. Change and the adoption of new methods can be facilitated by allowing for flexibility and providing training in the early phase of change.

- *Inter-agency coordination.* When different agencies are interacting, issues of territorialism, institutional friction, the informal setup of the agencies, aligning the agency goals, interest groups, and power distribution among the stakeholders are a few of the issues which emerge. An example of such a dynamic situation is when a mega project involving an urban area is being dealt with, wherein the highway system, transit agencies, sub-contracting consultants, and state and local transport agencies are involved. Such a myriad composition will lead to power struggle for responsibilities and power of decision-making. The collaboration will require sharing of data, duties, and responsibilities for success. The subject of data sharing in inter-agency collaboration is especially sensitive, as each agency is skeptical and wants to protect their agency's interest primarily. Here the formulation of the project and its intended benefits to each involved agency is of prime importance for successful data collaboration. An important point to remember in such collaborations is that each involved agency has to be treated as a separate entity and the notion of collaboration should not have expectation of seamless functioning of the agencies. Each agency will help achieve the collaborative task as individual agencies by sharing only as demanded by the task.

9.2.2.3 Agency workforce development

A very eminent problem facing transportation agencies is the development of their workforce to keep abreast of technological advancements. Skills will need to be added and their technical knowledge updated to provide the necessary support for optimal system performance. As megacities more fully integrate their regional role as policymakers and implementers, their transportation policy, planning, and implementation staff will have to manage many components, many of which will involve public, private, contract, and subcontracted players. Private consulting firms already provide much of the required technological and technical expertise for transportation projects and provide critical support to agencies. Cash-strapped agencies have limited resources, and the cost is eventually borne by the agency for the manpower hired by the consulting firms. The consulting firms being a private entity are profit driven may not prioritize cost containment unless incentivized to do so. Well-defined contracts with clear goals, expectations, and objectives are essential to ensure resources are directed to the highest priority areas (as set by the transportation planning agency). So is ensuring competition among consultants and vendors and ensuring you have the right talent in-house in balance with contracted capabilities. Similarly, agencies need to ensure their in-house staff are well trained and up to speed on technology, best practices, and project management tools and techniques.

Transportation agencies and the industry have always provided their feedback to the academic institutions to aid in the formulation of the course curriculum to ensure the industry's future workforce requirements are incorporated into curricula and class work. But technological advancements and the consolidation of various fields into emerging industry clusters with the transportation systems also means experts in each relevant field must be hired and retained as in-house staff for the transportation agencies.

Data collection and analysis efforts are a major part of the transportation agency tasks for maintaining efficient operation of the system. With the big data sets being collected, stored, and analyzed, a large number of consultants are hired ranging from electrical, computers, hardware, data analysts, and various other experts. Agencies surely need to develop their workforce gearing toward the needs for data processing and analysis. Agencies should retain control of their data and ensure it is available to the public.

This process of development cannot be done in a simple manner by hiring alone, but now along with academic institutes, the agencies need to fashion a learning database from their varied experiences with the consultants and subcontractors. This will help the agency to learn about the job expectations, and their manpower to deal with the tasks, along with the technology being used and their dependence on manpower and personnel versus technology to achieve the task. Too often an agencies best staff are attracted to the consulting firm and hired away. Agencies need to be competitive in order to retain crucial staff.

9.2.2.4 Communication with stakeholders

The first step of conceptual development of a public project requires dealing with various issues ranging from land acquisitions, alignments, environmental impacts, social acceptance, financing options and alternatives, public benefits, social equity, job creations, etc. Once a project is conceptualized and choices pertaining to these factors are made, stakeholders must be brought into the process to reach the next-phase feasibility study of all possible alternatives.

Stakeholders are all parties involved in the initiated project and have an interest directly or indirectly either as users or nonusers and can decide the acceptance and success of the project. The essence of this process is transparency and information regarding each involved parameter and its impact on the stakeholder. Starting with examining information on benefits and dis-benefits related to the stakeholders and the methods of quantification of these impacts.

The goal of this communication is for all the involved stakeholders to be as well informed as possible about the agency's outlook, goals and objectives, the parameters of the project, performance measures that will be used to estimate the measure of goals to be achieved, the tangible and intangible impacts of the project, and their quantification process. The stakeholders should also be aware of the funding sources and alternatives to the projects. Once the stakeholders are all well informed, they offer feedback based on facts and not heuristics. The focus of this effort should be only relaying the information, rather than convincing them. The information will eventually help them decide on their approval or disapproval of the project.

Once the stakeholders provide their input, the next step examines their conclusions, suggestions, advice, or decisions. This process can help in bridging the expectation of the stakeholders with the agency's goals and objectives, allowing the project to move to the next phase, such as project design, or will help the agency on the next project or an alternate venture for the same goal achievement.

This process is not simple or easy. Successful stakeholder engagement typically involves months and years of ongoing commitment and discussion. Often interaction with the public is influenced by local or national politics, which may or may not be based on facts. Often, there is lobbying by one faction of stakeholders who prioritize their personal interests over the public interest.

Many methods are used by agencies, departments, and units to interact with stakeholders. Tools include charrettes, focus groups, surveys, interviews, and public forums. Many of these tools are deployed by trained facilitators and mediators to help arrive at consensus or mutual agreement on various aspects and parameters of the projects. Transportation agencies should hold meetings on a regular basis to gather information, insight, and feedback from various interest groups and stakeholders to align their goals and objectives to their expectations and demand. Some meetings may include the local political representatives at times when required. These meetings also serve to help survey stakeholder interests and opinions of future projects.

9.2.2.5 *Acceptance of users and nonusers*

User acceptance of any transportation initiative is the core that forms its success. Acceptance by various factions of stakeholders is related to the benefits they perceive from the venture. For any individual or group of stakeholders, perceptions about who will benefit will influence their decision on whether to accept the project. Granovetter (1985) found that the social relations of the interacting entities often determined their decisions, rather than an independent social context and preconceptions of expected behavior scripted into the interaction by expectations. He also observed that interpersonal ties and not institutional structure decide the order and honesty in the relations. But trust in interpersonal ties can also create opportunities for dishonesty or malfeasance. This might not protect the public interest as intended at the start of the task. McGuire et al. (1993) recommended the use of historical case studies to observe how those dynamics have played out in a given community in the past.

Keeping these hurdles and the objective of public interest in mind, the agency tries to communicate and clarify for stakeholders the outcomes of the project. None of this is to say that getting stakeholder acceptance is easy or always achievable, given competing interests. But whatever acceptance is achieved can lead to feedback that can make a project more successful.

9.3 TOWARD THE TRANSFORMATION OF TRANSPORTATION DEVELOPMENT AND MANAGEMENT

Megacities are poised to lead a new wave of transportation planning, policy development, and system management. Their size, scale, and scope provide unprecedented opportunities to use the latest technology and thinking about transportation planning, policy, and implementation to re-envision regional transportation planning. Given the complexity of the transportation system megacities need to maintain, policymakers will need to think creatively and boldly about ways to seamlessly integrate an entire range of transportation solutions and design in order to accommodate the diverse, dynamic needs of metropolitan areas with access to vast resources and global connectivity.

Moreover, the importance of addressing mobility issues for residents and businesses in megacities carries significant implications for social equity. The burdens of lower mobility are felt most keenly by the urban poor and communities on the margins of society such as immigrants and ethnic minorities. As economies become more prosperous and build wealth, these communities will benefit from improvements in transportation networks most in the long run. Those working on the margins of society will have broader access to job opportunities as their labor market "opportunity circle" or "travel shed" broadens (Balaker, 2006; Balaker and Staley, 2006; Staley and Moore, 2009). Access to more job opportunities most often leads to higher wages, the ability to build personal wealth over time, escape poverty, and reduce exposure to economic insecurity. These social equity impacts are non-trivial. Empirical research covering national urban hierarchies finds that urban life becomes more diverse, robust, and vigorous as cities scale upward (Bettencourt et al., 2007). Given these effects, megacities are uniquely positioned to help their national economies grow while providing more opportunities for those on the margins of their economies and societies.

Moreover, megacities are uniquely positioned to accelerate this growth and wealth creation. In order for cities to reach 10 million in population, their economies must already have achieved a level of complexity, diversity, and scope that sustains value creation. They have, in essence, provided "proof of concept" for a sustainable urban area or region. Attracting capital into these areas is much less risky to outside investors as long as they continue to grow. Contrary to much conventional wisdom, financing for transportation infrastructure does not occur as part of a zero-sum game among investors. Quite the contrary. Private capital is looking for opportunities to grow their financial assets. They expect to earn a positive rate of return on their investments. Thus, megacities, unlike their smaller urban counterparts, are able to tap into this vast global asset base to build infrastructure of all types, modes, and scopes as long as their transportation planning and policy commitments are tied to reasonable goals and objectives to increase economic productivity. Indeed, McKinsey Global Institute estimates that US$4 trillion in annual investment is required to keep pace with economic growth through 2035 (MGPMR, 2019). Infrastructure fundraising, the institute estimates, grew 17% annually in recent years with more than 8,000 private equity firms fielding hundreds of billions of dollars in investment funds. Transportation policies that recognize the economic potential of improving mobility will be well positioned to tap into this capital much as China did this in the 1980s, 1990s, and 2000s. These investments created the logistical platforms necessary for a very poor country to become an economic powerhouse and global leaders just 50 years later.

Thus, while this book has argued that the primary objective of a regional transportation system should focus on improving mobility—the increase in the capacity for a network to move people and goods quickly and efficiently—we also recognize that other policy goals are also in play, including ensuring that facility construction and preservation continue to be cost-efficient and system remains safe and environmentally sustainable. We do not believe these goals are in conflict. In fact, they are complimentary. Market-based pricing for infrastructure and access to transportation facilities, for example, help balance the demand for different modes and restructure traveler decisions based on the cost of providing these travel options within a given network. Similarly, life-cycle asset management tools help fully load the costs of facilities at the appropriate times in design, construction, and replacement. Using private capital to build (and manage) transportation facilities also ensures the full costs of these facilities are transparent and incorporated into the cost–benefit calculations of policymakers. Megacities are able to integrate these planning, operational, and system development decisions comprehensively and holistically because of their regional scale, priority within national transportation systems, and connectedness to global markets and networks.

Megacity transportation planners will need to consider the many tradeoffs involved in prioritizing facilities, the vehicles that depend on their transportation network in all its forms, the demands of users, and the impacts on nonusers. While the complexity and challenges faced by policymakers in these vast urban transportation systems are daunting, contemporary tools and strategy provide the tools they need to strike the proper balance. To achieve this balance, of course, priorities must be well established and serve as the benchmark against which decisions by other actors within the transportation system can evaluate their own progress toward the general goal.

We hope this book at least provides a useful framework for thinking through ways policymakers and students of urban transportation systems can begin to tackle these challenges and develop planning frameworks that allow for effective deployment of tactics, strategies, and facilities that will sustain and grow these megacities through the 21st century.

REFERENCES

Balaker, T. 2006. *Why Mobility Matters*. Policy Brief No. 43. Reason Foundation, Washington, DC.

Balaker, T., and Staley, S.R. 2006. *The Road More Travelled: Why the Congestion Crisis Matters More Than You Think, and What We Can Do About It*. Rowman & Littlefield, New Kingstown, PA, USA.

Fuller, C., Bennett, R.J., and Ramsden, M. 2004. Local government and the changing institutional landscape of economic development in England and Wales. *Environment and Planning C: Politics and Space* 22(3), 317–347.

Granovetter. M. 1985. Economic Action and Social Structure: The Problem of Embeddedness. *American Journal of Sociology* 91, 481-510.

McGuire, P., Granovetter, M., and Schwartz, M. 1993. Thomas Edison and the Social Construction of the Early Electricity Industry in America. *Explorations in Economic Sociology*, Edited by Swedberg, R. Russell Sage Foundation, New York, NY.

MGPMR. 2019. *Private Markets Come of Age*. McKinsey Global Private Markets Review. McKinsey & Company, Chicago, IL.

Richards, S., Barnes, M., Sullivan, H., Gaster, L., Leach, B., and Coulson, A. 1999. *Cross Cutting Issues in Public Policy and Public Service*, Report to the Department of Environment Transport and the Regions, London, United Kingdom.

Rietveld, P., and Stough, R. 2005. *Barriers to Sustainable Transport: Institutions, Regulations and Sustainability*. Spon Press, New York, NY.

Staley, S.R., and Moore, A.T. 2009. *Mobility First: A New Vision for Transportation in a Globally Competitive Twenty-First Century*. Rowman & Littlefield, New York, NY, USA.

Stoker, G., and Mossberger, K. 1994. Urban Regime Theory in Comparative Perspective, *Environment and Planning C: Government and Policy* 12(3) 195-212.

Index